Sources and Studies in the History of Mathematics and Physical Sciences

Sources and Studies in the History of Mathematics and Physical Sciences was inaugurated as two series in 1975 with the publication in Studies of Otto Neugebauer's seminal three-volume History of Ancient Mathematical Astronomy, which remains the central history of the subject. This publication was followed the next year in Sources by Gerald Toomer's transcription, translation (from the Arabic), and commentary of Diocles on Burning Mirrors. The two series were eventually amalgamated under a single editorial board led originally by Martin Klein (d. 2009) and Gerald Toomer, respectively two of the foremost historians of modern and ancient physical science. The goal of the joint series, as of its two predecessors, is to publish probing histories and thorough editions of technical developments in mathematics and physics, broadly construed. Its scope covers all relevant work from pre-classical antiquity through the last century, ranging from Babylonian mathematics to the scientific correspondence of H. A. Lorentz. Books in this series will interest scholars in the history of mathematics and physics, mathematicians, physicists, engineers, and anyone who seeks to understand the historical underpinnings of the modern physical sciences.

More information about this series at http://www.springer.com/series/4142

Jacques Sesiano

Magic Squares

Their History and Construction
from Ancient Times to AD 1600

 Springer

Jacques Sesiano
Geneva, Switzerland

ISSN 2196-8810   ISSN 2196-8829   (electronic)
Sources and Studies in the History of Mathematics and Physical Sciences
ISBN 978-3-030-17995-3   ISBN 978-3-030-17993-9   (eBook)
https://doi.org/10.1007/978-3-030-17993-9

This Springer imprint is published by the registered company Springer Nature Switzerland AG
The registered company address is: Gewerbestrasse 11, 6330 Cham, Switzerland

Preface

The original purpose of this book was to provide an English translation of my study on the science of magic squares in Islamic countries, published first in French (*Les carrés magiques dans les pays islamiques*, Lausanne 2004) and then in Russian (*Magicheskie kvadraty na srednevekovom vostoke*, Saint Petersburg 2014). It was to follow my *Magic squares in the tenth century* published quite recently (2017), which edited the two earliest texts then known, by Anṭākī and Būzjānī.

These two texts were of a rather different nature. Whereas the second author had produced an easily readable text, providing a didactic introduction to the science of magic squares, the first gave, between a set of propositions mostly taken from Euclid and a collection of problems, a very concise study on the construction of certain types of magic square. First of all, it was surprising to consider the difference in level between the three parts of Anṭākī's text: while the first and the third gave the impression of having been compiled without much insight by a second-rate mathematician, the middle one, on magic squares, was of an amazingly high level and obviously the result of serious mathematical studies. At that time I already expressed my conviction that Anṭākī could not be the author of it, and that he was just copying a text ultimately going back to Greek antiquity. Next, the difference in level between Anṭākī's and Būzjānī's texts was also surprising: although their authors were contemporaries, it seemed as if the second was providing the basic elements of a nascent science, whereas the first treated specialized topics certainly not accessible to general readers. Not having any explanation for this discrepancy, I could only write that "the tenth century, for the history of magic squares, gives the impression of being both a beginning and an end".

The discovery in a London manuscript of several fragments of the earliest Arabic writing on magic squares, that by Thābit ibn Qurra (836-901), explained these incongruities. First, it appeared that Anṭākī had just copied verbatim Thābit's text (without making any mention of him) and that Anṭākī's work could thus certainly not be considered as a tenth-century achievement. Furthermore, from Thābit's introduction, it appeared that his source was indeed one Greek text —obviously, then, a particular study devoted to just specific aspects of magic squares, perhaps indeed a culmination in antique times although we lack any elements of comparison. Since this text does not explain the basic elements of the science of magic squares, Būzjānī's treatise is an attempt to do it.

The origin of Arabic studies on magic squares is thus T̲h̲ābit's translation. Incidentally, we may note that the transmission of this Greek study was itself some kind of miracle: as T̲h̲ābit tells us at the beginning of his translation, it was preserved by two manuscripts, both in rather poor condition, of which he could make sense only because what was damaged in one copy was legible in the other.

This discovery had of course an impact on the present translation since it changed drastically the earliest history of magic squares. The immediate effect was to change the title, originally to be 'Magic squares, their history and construction from early times to AD 1600'; for I could now safely replace 'from early times' by 'from ancient times'. As to my choice of the upper limit, that is because the seventeenth century saw the beginning of a new epoch in the history of magic squares, as Europe began, with Fermat, to study magic squares on new bases. In the two centuries before, Europe had known only two sets of seven magic squares of different sizes, all received from late mediaeval Latin translations of Arabic texts, where each square, being associated with one of the seven planets then known, could have, used as a talisman, the same influence as its planet. There was no single indication on the construction of these few squares, but lengthy descriptions of their use for good or evil purposes. This first European encounter with these figures and their magical use thus gave rise to their modern designation (which also accounts for the little consideration they often subsequently received). That had not always been the case: the Arabic denomination *wafq al-aʿdād*, 'harmony of numbers' —corresponding perhaps to something like ἁρμονία τῶν ἀριθμῶν— did far more justice to the science of these squares.

Table of contents

Chapter I. Introduction

Chapter II. Ordinary magic squares

Chapter III. Composite magic squares

Chapter IV. Bordered magic squares

Introduction

§1. Definitions

A *magic square* is a square divided into a square number of cells in which natural numbers, all different, are arranged in such a way that the same sum is found in each horizontal row, each vertical row, and each of the two main diagonals.

A square with n cells on each side, thus n^2 cells altogether, is said to have the *order n*. The constant sum to be found in each row is called the *magic sum* of this square. Thus the squares of the figures 1 and 2 have the order 6. But their constant sum is different, with 111 for the first and 591 for the other. The first being filled with the first 36 natural numbers, its sum is the smallest possible for a 6×6 magic square. In the second, the numbers are not consecutive. Though both are magic squares, what is usually considered is the first kind, thus a square of order n filled with the first n^2 natural numbers. In this case the constant sum is easily determined: since the sum of all these numbers equals

$$\frac{n^2\left(n^2+1\right)}{2},$$

the sum in each row (line, column, main diagonal), thus the magic sum for such a square, will be

$$M_n = \frac{n(n^2+1)}{2}.$$

1	32	34	3	35	6
30	8	27	28	11	7
19	23	15	16	14	24
18	17	21	22	20	13
12	26	10	9	29	25
31	5	4	33	2	36

131	101	92	107	36	124
191	94	100	115	85	6
3	137	50	135	72	194
192	123	134	75	62	5
1	40	110	69	175	196
73	96	105	90	161	66

Fig. 1   Fig. 2

A square displaying the same sum in the $2n+2$ aforesaid rows is an *ordinary* or *common magic square*. It meets the minimal number of required conditions, and such a square can be constructed for any given order $n \geq 3$. (There cannot be a magic square of order 2 with different

© Springer Nature Switzerland AG 2019
J. Sesiano, *Magic Squares*, Sources and Studies in the History of Mathematics and Physical Sciences, https://doi.org/10.1007/978-3-030-17993-9_1

numbers, as we shall see below.) But there are magic squares which display further properties.

A *bordered magic square* is one where removal of the successive borders leaves each time a magic square (Fig. 3). With an odd-order square, after removing each (odd-order) border in turn, we shall finally reach the smallest possible square, that of order 3. With an even-order square, that will be one of order 4 (the border cannot be removed since there is no magic square of order 2). For order $n \geq 5$, bordered squares are always possible.

92	17	4	95	8	91	12	87	16	83
99	76	31	22	77	26	73	30	69	2
1	20	64	41	36	63	40	59	81	100
3	19	67	58	47	51	46	34	82	98
96	80	33	52	45	57	48	68	21	5
7	78	35	49	56	44	53	66	23	94
90	27	62	43	54	50	55	39	74	11
13	72	42	60	65	38	61	37	29	88
86	32	70	79	24	75	28	71	25	15
18	84	97	6	93	10	89	14	85	9

Fig. 3

As seen above, the magic sum for a square of order n filled with the n^2 first natural numbers is

$$M_n = \frac{n(n^2 + 1)}{2}.$$

Clearly, the *average* sum in each cell is $\frac{n^2+1}{2}$. Accordingly, for m cells, this average sum should be m times that quantity; this will be called the *sum due* for m cells. Thus, the inner square of mth order ($m \geq 5$) within a bordered square of order n, being a separate entity, will contain in each row its sum due, namely

$$M_n^{(m)} = \frac{m(n^2 + 1)}{2}$$

if the main square is filled with the n^2 first natural numbers. The magic sum in any inner square will therefore differ from the sum in the next one, surrounding it, by $n^2 + 1$. With this in mind, we clearly see the structure of bordered squares: the elements of the same border which are opposite (horizontally, vertically, or diagonally for the corner cells), which we shall call *complements*, add up to $n^2 + 1$ if the main square is of order n. For example, in each border of Fig. 3, the sum of opposite elements is 101, and the magic sums of the successive squares are 505, 404, 303, and 202 for the smallest. Again, since for an odd-order bordered square there is just one central cell, it must necessarily contain $\frac{n^2+1}{2}$.

Particular cases of ordinary magic squares

It is a condition for magic squares that the main diagonals will contain the magic sum. But there are ordinary magic squares in which the magic sum is also found in the broken diagonals —thus pairs of diagonal rows, on either side of and parallel to a main diagonal, comprising n cells altogether. Such squares are called *pandiagonal* (Fig. 4 where, for example, the broken diagonals $47, 31, 20, \ldots, 21, 26, 42$ and $47, 59, 18, 8, \ldots, 14$ make the magic sum). They are possible for any odd order from 5 on, and for any even order *divisible by 4*, from 4 on.

11	22	47	50	9	24	45	52
38	59	2	31	40	57	4	29
18	15	54	43	20	13	56	41
63	34	27	6	61	36	25	8
3	30	39	58	1	32	37	60
46	51	10	23	48	49	12	21
26	7	62	35	28	5	64	33
55	42	19	14	53	44	17	16

Fig. 4

Such squares remain magic if we move any lateral row to the other side: the main diagonals of the new square will be magic since they were already magic as broken diagonals. By repeating such vertical and/or horizontal moves we are able to place any chosen element in any given cell of the square; for example, the cell containing 1 can be made to occupy any place in the square. This feature seems to have been highly prized in the early days of magic squares.

This property may be simply represented in the case of the smallest possible pandiagonal square, that of order 4, one form of which is seen in Fig. 5. Supposing it repeated in the plane, as in Fig. 6, any square of order 4 taken in it will be magic (and pandiagonal).

13	2	7	12	13	2	7	12	13	2	7	12
3	16	9	6	3	16	9	6	3	16	9	6
10	5	4	15	10	5	4	15	10	5	4	15
8	11	14	1	8	11	14	1	8	11	14	1
13	2	7	12	13	2	7	12	13	2	7	12
3	16	9	6	3	16	9	6	3	16	9	6
10	5	4	15	10	5	4	15	10	5	4	15
8	11	14	1	8	11	14	1	8	11	14	1
13	2	7	12	13	2	7	12	13	2	7	12
3	16	9	6	3	16	9	6	3	16	9	6
10	5	4	15	10	5	4	15	10	5	4	15
8	11	14	1	8	11	14	1	8	11	14	1

1	8	11	14
12	13	2	7
6	3	16	9
15	10	5	4

Fig. 5   Fig. 6

Another class of ordinary squares is that of *composite* squares (Fig. 7): the subsquares arranged within the main magic square are themselves magic with each having its own magic sum. Clearly, the possibility of such an arrangement depends on the divisibility of the order of the main square. Moreover, it will appear that the smallest possible composite square is that of order 9. Composite magic squares of lesser order are not possible or particular: that of order 8 may well be divided into four subsquares, but, because of the impossibility of a 2×2 magic square, they cannot have different magic sums; as to that of order 6, its four subsquares filled with different numbers could not even make equal sums.

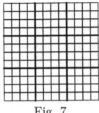

Fig. 7

§2. Categories of order

As mentioned above, no magic square of order 2 is possible. Indeed (Fig. 8) the conditions would be

$$a+d=b+c=b+d=c+d=c+a=a+b$$

from which we infer that $a = b = c = d$. A magic square of order 2 is therefore not possible with numbers which are to be different.

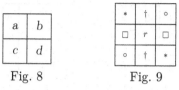

Fig. 8   Fig. 9

Consider now a square of order 3 (Fig. 9). Surrounding the central cell is a border comprising four pairs of opposite cells, horizontally ($\square$), vertically ($\dagger$) or diagonally ($*, \circ$). Clearly, with M designating the magic sum and r the central element, the sum in each pair of opposite cells must equal $M - r$. The sum in the whole square must therefore equal $4(M - r) + r$; it also equals $3M$. Equating these two quantities gives $M = 3r$.

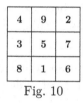

Fig. 10

Let us now fill the 3×3 square with the nine first natural numbers. Their sum being 45, the magic sum M_3 will be 15; then $r = 5$ and each pair of opposite cells must add up to $M_3 - 5 = 10$. Let us begin by considering the place of 1. If it is in a corner, 9 will be in the opposite corner and we shall need, to complete the two lateral rows meeting in 1, two pairs of numbers adding up to 14. Now there is just one such pair available, 6 and 8. Therefore 1 must necessarily be in the middle of a lateral row, with 6 and 8 as its neighbours. With these four numbers in place the remainder of the square is determined (Fig. 10).

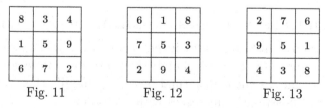

Fig. 11   Fig. 12   Fig. 13

We could exchange the places of 6 and 8, but this would just invert the figure. As a rule, we shall not consider as different the seven squares obtained from a given one by rotation or by inversion, as here, from Fig. 10, Fig. 11-13 and Fig. 14-17. We infer from that and the above stepwise construction that there must be just one form of the magic square of order

3 filled with the nine first natural numbers. This is an exception: for all other squares the possibilities are numerous.

2	9	4
7	5	3
6	1	8

Fig. 14

4	3	8
9	5	1
2	7	6

Fig. 15

8	1	6
3	5	7
4	9	2

Fig. 16

6	7	2
1	5	9
8	3	4

Fig. 17

We have thus far considered individually the squares of order 2 and order 3. Now it is not necessary to do that for higher orders. For there exist *general construction methods*, that is, ways of directly constructing a square of given order, either ordinary or bordered, once the empty square is drawn: a few more or less easily remembered instructions will enable us to place the sequence of consecutive numbers without any computation or recourse to trial and error. These general methods, though, are not applicable to all orders but, as a rule, to just one among three main categories, which are:

— The squares of *odd* orders, also called *odd squares*, thus with orders 3, 5, 7, ..., generally $n = 2k + 1$, of which the smallest is the square of order 3.

— The squares of *evenly-even* orders, or *evenly-even squares*, thus with orders 4, 8, 12, ..., generally $n = 4k$, of which the smallest is the square of order 4.

— The squares of *evenly-odd* orders, or *evenly-odd squares*, thus with orders 6, 10, 14, ..., generally $n = 4k + 2$, of which the smallest is the square of order 6.

Note, though, that these general methods may require some adapting for squares of lower orders, such as 3 or 4, sometimes also 6 and 8.

§3. Banal transformations of ordinary magic squares

As said above, rotations and inversions of the same square are not seen as different. On the other hand, there are for *ordinary* squares of orders larger than 3 some transformations which modify their aspect, and the number of possibilities increases with the order.[1]

First, each number i may be replaced by $n^2 + 1 - i$. The result corresponds to placing the numbers in reverse order. For a *symmetrical magic square*, in which the sum of two diagonally opposite elements is $n^2 +$

---

[1] We omit here the particular feature of pandiagonal squares seen above (p. 3).

1, the square will just be rotated by 180°. See Fig. 18a & b. Otherwise, the square will be different, with some of the numbers having new neighbours. See Fig. 19a & b.

23	2	19	6	15
4	8	25	12	16
17	21	13	5	9
10	14	1	18	22
11	20	7	24	3

Fig. 18a

3	24	7	20	11
22	18	1	14	10
9	5	13	21	17
16	12	25	8	4
15	6	19	2	23

Fig. 18b

22	18	1	14	10
9	5	13	21	17
15	6	19	2	23
16	12	25	8	4
3	24	7	20	11

Fig. 19a

4	8	25	12	16
17	21	13	5	9
11	20	7	24	3
10	14	1	18	22
23	2	19	6	15

Fig. 19b

Second, we may, for an even order, exchange the quadrants diagonally (Fig. 20a & b). If the order is odd, part of the median rows must be exchanged as well (Fig. 21a & b).

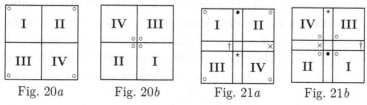

Fig. 20a Fig. 20b Fig. 21a Fig. 21b

Third, considering two pairs of rows, with both the horizontal and vertical ones at the same distance from their median axis, we may exchange them in turn (Fig. 22a & b).

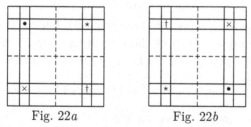

Fig. 22a Fig. 22b

The magic property will hold since lines and columns keep their elements, and diagonals as well. For a symmetrical square, we may perform just *one* exchange of two symmetrical rows: the two pairs of diagonally opposite elements will move to the other diagonal, but this will not change the sum in the diagonals since each such pair adds up to $n^2 + 1$.

Applying the second transformation to bordered squares will not preserve their character. As to the first, it will modify the places of elements within the border rows (Fig. 23a & b). Now this is a banal change: we may arbitrarily modify the place of elements *within* the border rows of such a square provided each opposite undergoes the same change.

4	24	23	8	6
19	10	17	12	7
21	15	13	11	5
1	14	9	16	25
20	2	3	18	22

Fig. 23a

22	2	3	18	20
7	16	9	14	19
5	11	13	15	21
25	12	17	10	1
6	24	23	8	4

Fig. 23b

§4. Historical outline

When the science of magic squares began is not known. No ancient Greek text dealing with magic squares is still extant, no allusion whatsoever is made to them in antiquity, and the only Greek text preserved is a Byzantine work by Manuel Moschopoulos (ca. 1300), who appears to draw his knowledge from some Arabic or Persian source.[2] Furthermore, allusions to Greek studies by later Arabic authors seem unreliable, such as the attribution of the discovery of magic squares to Archimedes by the 13th-century geographer Qaswīnī[3] despite the fact that the list of Archimedes' works is well known since Greek times.

Still, it now appears certain that there existed Greek books on the subject, and at a remarkable stage of advancement. This we know from a text attributed to the prolific translator of Greek works Thābit ibn Qurra (836-901). He tells us at the beginning how he became acquainted with the science of magic squares, obviously hardly known at that time, and the difficulty he had to interpret the text of this (apparently anonymous) work.[4] *My first acquaintance with the subject of magic square was the*

---

[2] Moschopoulos' text has been edited and translated by Tannery; on its origin, see our study of it.

[3] See Wiedemann's *Beitrag V*, pp. 451-452.

[4] This and other parts of Thābit's translation are found in the anonymous MS.

3 × 3 figure mentioned by Nicomachos in the Arithmetic.[5] *Then Abū al-Qāsim al-Ḥijāzī came across a 4 × 4 square, (filled with the numbers) from 1 to 16 with constant difference, which caused his admiration.*[6] *Next I found a 6 × 6 figure on the back of Euclid's book (of the Elements) in the handwriting of Isḥāq.*[7] *Afterwards I came across a book containing three or four figures less than (the order) 10. Then I found in the Royal library (in Baghdad) two books, for the greater part damaged by insects so that one could understand just a small part of them; the notice about them was in al-Māhānī's handwriting and the first page for the most part in Ḥusayn ibn Mūsā al-Nawbakhtī's handwriting.*[8] *Then I examined them, found elucidating them very laborious, and it seemed to me that it should be possible to make sense of those parts which had been damaged in one by what had been preserved in the other.*

With Thābit's translation general methods for constructing bordered and composite squares may thus be considered as available in the late ninth century. For ordinary squares, however, we do not know of any ancient source, and the documents we have are all, as it seems, originally

---

London BL Delhi Arabic 110, a 12th-century compilation of works by various authors from the 9th to 12th centuries. For the Arabic text of this passage, see fol. 60ᵛ in this MS or below (p. 281), Appendix 1. (Note our use of slanted writing, not italics, throughout for quotations.)

[5] Nicomachos' *Introduction to arithmetic* (Ἀριθμητικὴ εἰσαγωγή), written towards the end of the first century, is —as indicated by its title— an elementary introduction to number theory (such is indeed the Greek meaning of 'arithmetic'). It does, though, depart from the sound mathematical works of the classical Greek period in that its author tends at times to take his conjectures for theorems. Still, the *Introduction* was one of the main channels of transmission of Greek mathematics, and for early mediaeval Europe the only one, namely through the Latin adaptation by Boëtius, while in the East it was translated twice: into Syriac and, by Thābit ibn Qurra also, into Arabic (ed. Kutsch). There is no mention of a magic square in it (or in the Arabic translation), but a 3 × 3 square may have appeared as a marginal addition. In a recent study, N. Vinel has quoted (pp. 551-552) a fragment by Iamblichos, a 4th-century commentator on Nicomachos, which seems to allude to the 3 × 3 magic square.

[6] Abū'l-Qāsim al-Ḥijāzī was a theologian of the first half of the 9th century. The description being rather naive, Thābit must be quoting him. His name and those of the other scholars appear in the *Fihrist* of Ibn al-Nadīm, written towards the end of the 10th century, which contains in particular a list of Arabic books or translations.

[7] Isḥāq ibn Ḥunayn, a contemporary of Thābit ibn Qurra, translated the *Elements*, a translation then revised by Thābit himself.

[8] Muḥammad ibn 'Īsā ibn Aḥmad al-Māhānī was a Persian mathematician and astronomer working in Baghdad around the middle of the 9th century; Ḥasan (as commonly) ibn Mūsā al-Nawbakhtī was a theologian and philosopher of the second half of the 9th century, personally acquainted with Thābit (*Fihrist*, ed. Flügel *et al.*, I, 177, 11). Incidentally, Ibn al-Nadīm also mentions the damage caused by 'insects' (or termites: *araḍa*) to manuscripts (I, 243, 27).

Arabic. General methods are seen to appear in the late tenth or early
eleventh century for odd and evenly-even orders, and then, at the end of
the eleventh, for evenly-odd ones. (With the evenly-odd order presenting
most difficulty, such a stepwise development was only to be expected.)

This then leaves open the question of the existence of ordinary magic
squares in Greece. Given the superior level of the Greek methods reported
in Thābit's translation, it would be surprising, to say the least, to find the
Greeks not acquainted with their construction. My suggestion of twenty
years ago may still be considered: that this science remained more or less
confined to (neo-Pythagorean?) circles, just as, more than a thousand
years before, the teaching of the Pythagorean school had been restricted
to its members, and only later began to spread.[9] This, incidentally, might
explain why so little of this Greek science has been transmitted.

$\bar{\alpha}$	$\bar{\beta}$	$\bar{\gamma}$	$\bar{\delta}$	$\bar{\epsilon}$	$\bar{\digamma}$	$\bar{\zeta}$	$\bar{\eta}$	$\bar{\vartheta}$
١	ب	ج	د	ه	و	ز	ح	ط
1	2	3	4	5	6	7	8	9
$\bar{\iota}$	$\bar{\varkappa}$	$\bar{\lambda}$	$\bar{\mu}$	$\bar{\nu}$	$\bar{\xi}$	$\bar{o}$	$\bar{\pi}$	$\bar{\mathrm{q}}$
ى	ك	ل	م	ن	س	ع	ف	ص
10	20	30	40	50	60	70	80	90
$\bar{\rho}$	$\bar{\sigma}$	$\bar{\tau}$	$\bar{\upsilon}$	$\bar{\varphi}$	$\bar{\chi}$	$\bar{\psi}$	$\bar{\omega}$	$\bar{\lambda}$
ق	ر	ش	ت	ث	خ	ذ	ض	ظ
100	200	300	400	500	600	700	800	900
$,\bar{\alpha}$								
غ								
1000								

Fig. 24

Returning to Arabic times, we do see continuous development in the
11th and early 12th centuries, with the discovery of various other constructions for ordinary magic squares, sometimes pandiagonal as well.
Squares filled with non-consecutive numbers also begin to appear. The
origin of that has to do with the association of Arabic letters with numerical values —an adaptation of the Greek numerical system (Fig. 24); this
adaptation appeared in early Islamic times, before the adoption of the Indian numerals, but remained in use later. Since to each letter of a word,
or to each word of a sentence, was attributed a numerical quantity, thus
to the word or the sentence a certain sum, this word or sentence could be

---

[9] *Les carrés magiques*, p. 270; *Magicheskie kvadraty*, pp. 277-278.

placed as such in the cells of a certain row of a square. The task was then to complete the square numerically so that it would display in each row the sum in question —a mathematically interesting problem since this is not always possible.

Meanwhile, however, popular use of magic squares as talismans grew apace. Authors then followed this trend: many late, shorter texts are just not interested in teaching the construction of magic squares —to say nothing of their mathematical fundament. They give the figures of a few magic squares, most commonly squares of the orders 3 to 9 associated with the seven then known planets (including Moon and Sun) of which they embodied the respective, good or evil, qualities —abundantly described and commented in these texts. The reader is taught on what material and when he is to draw each of such squares; for both the nature of the material and the astrologically predetermined time of drawing are presumed to increase the square's efficacy. It merely remains to put this object in the vicinity of the chosen person, beneficiary or victim.

Of such kind were the first Arabic texts translated into Latin in 14th-century Spain. A characteristic example of theory and application is the following, which uses the properties of Jupiter, mostly favourable, and Mars, mainly not.[10]

The figure of Jupiter is square, four by four, with 34 on each side. If you wish to operate with it: make a silver plate in the day and the hour of Jupiter, provided Jupiter is propitious,[11] and engrave upon it the figure; you will fumigate it with aloes wood and amber. When you carry it with you, people who see you will love you and you will obtain from them whatever you request. If you place it in the store-house of a merchant, his trade will increase. If you place it in a dovecote or in a hive, a flock of birds or a swarm of bees will gather there. If someone unlucky carries it, he will prosper and be always more successful. If you place it in the seat of a prelate, he will enjoy a long prelature and will not fear his enemies, but be successful among them.

The figure of Mars when unfavourable means war and exactions. It is a square figure, five by five, with 65 on each side. If you wish to operate with it, take a copper plate in the day and hour of Mars when Mars is decreasing in number and brightness, or malefic and retrograding, or in

---

[10] This and other examples in our *Magic squares for daily life*. For the original Latin text, see below, Appendix 2. The two magic squares represented are those of our figures 27 and 29 below.

[11] Thus on Thursday, 1st and 8th hours of the day and 3rd and 10th of the night, and with Jupiter on a direct course and increasing in brightness.

any way unfavourable, and engrave the plate with this figure; and you will fumigate it with the excrement of mice or cats. If you place it in a new building, it will never be completed. If you place it in the seat of a prelate, he will suffer daily harm and misfortune. If you place it in the shop of a merchant, it will be wholly destroyed. If you make this plate with the names of two merchants and bury it in the house of one of them, hatred and hostility will come between them. If you happen to fear the king or some powerful person, or enemies, or have to appear before a judge or a court of justice, engrave this figure as said above when Mars is favourable, in direct motion, increasing in number and brightness; fumigate it with one drachma ($= \frac{1}{8}$ ounce) of carnelian stone. If you put this plate in a piece of red silk and carry it with you, you will win in court and against your enemies in war, for they will flee at the sight of you, fear you and treat you with deference. If you place it upon the leg of a woman,[12] she will suffer from a continuous blood flow. If you write it on parchment on the day and the hour of Mars and fumigate it with birthwort and place it in a hive, the bees will all fly away.

It was thus the arrival of such texts in late mediaeval Europe which first aroused interest in, and led to the study of, such squares there (the figures of these squares are given below, §6). This incidentally explains the use of the term 'magic' —formerly also 'planetary', which we find still employed by Fermat.[13] The original Arabic (perhaps originally Greek) denomination, 'Harmonious arrangement of numbers' (*wafq al-a'dād*), which had a more mathematical connotation, remained unknown, as well as the various constructions abundantly described in Arabic and Persian manuscripts.

We may infer from this that 'magic' applications do not seem to have played a significant rôle at the outset. As the original name indicates, such squares were then being studied as a branch of number theory, just as were the perfect and amicable numbers so highly esteemed by Pythagoreans and their followers in late antiquity. In short, such research was at the time considered as being perfectly serious and had not yet been tainted with the sad reputation later acquired through popular use of these squares.

Finally, we may note that magic squares are also found elsewhere. The magic square of order 3 appears in China at the beginning of our era; to our knowledge, higher-order squares do not occur there before the

---

[12] Reverting to evil uses.

[13] *Varia opera mathematica*, p. 176; or *Œuvres complètes*, II, p. 194.

13th century, and are clearly of Arabic or Persian origin. The same holds for Indian magic squares.

§ 5. Main sources considered

In the tenth century, 'Alī ibn Aḥmad al-Anṭākī (d. 987) wrote a commentary on Nicomachos' *Introduction to arithmetic*, of which only the third (and last) part is extant.[14] It contains first a list of definitions and propositions chiefly taken from Euclid's *Elements*, then a section on magic squares, finally a set of recreational problems, namely how to determine through indirect questions a number being thought of by someone. None of all this has any relation to Nicomachos' work, and we have here a compilation of *various* sources from late antiquity. The part on magic squares is taken verbatim from Thābit's translation, but without any mention of it.[15] The construction of bordered magic squares of any order was thus known in the Islamic world in the middle of the tenth century.

The second 10th-century text, entirely devoted to magic squares, is by the Persian Abū'l-Wafā' Būzjānī (940-997/8), who was one of the most famous mathematicians of Islamic civilization. His treatise is clear and mostly didactic, for every step is explained and justified. We are told how to construct ordinary magic squares, but only for small orders and with particular methods, then bordered squares, but with only one general method, namely for odd squares: for the two even orders, the reader is just told how to *find* the elements in each border. At the end of the treatise, the reader is taught how to form composite squares, and then odd-order bordered squares displaying separation by parity, this mainly relying on trial and error. All this is clearly explained, and gives the impression of reporting the early steps in the science of magic squares for the benefit of the reader (above, p. vi). It is thus a nice introduction, but, as inferred from the information provided by the previous treatise, it does not give an overview of the general methods known at that time.[16]

Another famous mathematician was Abū 'Alī al-Ḥasan ibn al-Hay-

---

[14] Edited in our *Magic squares in the tenth century*.

[15] Thus in our edition we could only surmise that the whole of it was taken from ancient Greek sources (above, p. v). Our *A*.II.7-*A*.II.35 are found on fol. 80ᵛ-84ᵛ of the London manuscript Delhi Arabic 110, while *A*.II.44-*A*.II.54 occur on fol. 99ʳ-100ᵛ, followed by larger squares not reproduced by Anṭākī (fol. 102ᵛ-107ʳ). Other sources are also used in this manuscript, including three of the authors mentioned below (Būzjānī, Ibn al-Haytham, Asfizārī); this may account for sections omitted from Thābit's translation, with the author selecting his excerpts.

[16] The most significant parts are found in our *Magic squares in the tenth century*; for the complete edition, see our *Traité d'Abū'l-Wafā'*.

tham (b. ca. 965 in Basra, d. 1041 in Egypt). We do not know his treatise on magic squares, but significant excerpts from it are found in an anonymous 12th-century source which considered him to be one of the best authors on the subject.[17] Ibn al-Haytham is the first to have discovered, or reported, or justified, some general methods for constructing ordinary squares, obviously unknown to the two previous authors. After him, mathematical justification of methods seems to have been neglected.

The anonymous *Harmonious arrangement of the numbers*, from the early 11th century as it seems, is one of our main sources.[18] Its (unknown) author tells us that he turned to the science of magic squares as a source of relief from certain disappointments. He further tells us that the main part of his work is his own findings, while anything taken over from his 'predecessors' was ameliorated by him. In short, *oratio pro domo*. His treatise teaches various ways of constructing bordered squares for all three orders, ordinary squares for odd and evenly-even orders, composite squares, and finally squares with a set of given numbers occupying one of the rows. To this last part, which is very elaborate and complete, we shall give particular attention in our sixth chapter.

Contemporary with it, also anonymous but much shorter, is the *Brief treatise teaching the harmonious arrangement of numbers*.[19] The extent of its author's knowledge is much the same, but he restricts himself to one method, sometimes different, for each kind.

With the Persian Abū Ḥātim Muẓaffar Asfizārī (d. before 1121/2), we reach the end of the 11th century. We know two of his methods from the anonymous 12th-century source already mentioned and from the next author to be mentioned. The first is a method for constructing ordinary evenly-even squares, said to be *easier than that by Anṭākī and that by Ibn al-Haytham*, and the second teaches how to obtain an odd-order ordinary magic square displaying separation by parity.[20]

His compatriot Jamāl al-Zamān 'Abd al-Jabbār Kharaqī (d. 1138/9) does not claim any originality, but his treatise is the first to teach general methods for all three types of the two main categories of magic squares, thus now including the previously omitted case of ordinary squares of evenly-odd orders. His text thus shows that the problem of constructing

---

[17] Publication of this source in our *Une compilation arabe*. This source has other excerpts than the London manuscript (above, n. 4).

[18] See our *Un traité médiéval*.

[19] Our *L'Abrégé enseignant la disposition harmonieuse des nombres*.

[20] Obtaining this separation is easy with ordinary squares —unlike in the case of bordered ones. All these methods will be discussed below.

ordinary magic squares was completely solved around 1100.[21]

Another Persian author writing about magic squares in the next century is ʿAbd al-Wahhāb ibn Ibrāhīm Zanjānī. His short treatise is limited to the construction of bordered squares as well as that of order 4 containing in its upper row four given numbers or, if this arrangement is not possible, their sum. Its importance is not due to any originality but to its success, as witnessed by the number of manuscript copies extant. Zanjānī himself tells us in the introduction that he wrote it at the request of some friends, who just wanted to know the rules for constructing squares of any order.[22] This indeed illustrates the later characteristics of treatises on magic squares: written in response to public demand, they are to teach methods aiming at results, without bothering the reader with questions of fundaments or feasibility.

The first trace of magic squares in the West is the treatise of the Byzantine Manuel Moschopoulos, written at the very beginning of the 14th century (above, p. 8). He constructs ordinary squares of odd orders, of evenly-even orders (taken by him, as it seems, to mean $n = 2^k$), but none of evenly-odd orders. It would seem that he attempted to reconstruct the methods from examples of squares he saw in some manuscript, either Persian or Arabic.

The same happened later, from the 14th century on, in Western Europe, when the first examples of squares arrived through Latin translations of magic texts (see above, pp. 11-12, and below, § 6). They contained two sets of seven squares attributed to the seven planets then known (in the geocentric system, thus including Sun and Moon), and the only explanations were, as we have seen, about their magic use. No Arabic text describing methods was then translated, or studied, and the first attempts to explain or generalize them were based solely on these examples. With these first studies are associated the names of Cornelius Agrippa, Girolamo Cardano and Claude-Gaspar Bachet de Méziriac in the 16th and early 17th centuries. A curious exception is the case of Michael Stifel (1487-1567), who gives instructions for constructing bordered squares of all orders;[23] there is no mention of any source, nor does he give himself any credit, and his methods are not exactly like the ones we know from Arabic sources.

Meanwhile, treatises expounding the construction of magic squares

---

[21] See our *Herstellungsverfahren III*.

[22] See *Herstellungsverfahren II, II'*.

[23] In his *Arithmetica integra*, fol. 24^v-30^r.

continued to be written in Arabic, which makes it all the more strange
that these methods remained unknown in Europe (though some were re-
discovered there in the following centuries). A very complete and mostly
clear treatise is the *Rising of the illumination for arranging magic squares*,
written around 1600 by the Egyptian Muḥammad Shabrāmallisī.[24] This
does not make, though, the author highly competent: sometimes he describes at length something very simple, sometimes he wrongly attempts
to generalize a particular method (see below, p. 95). He carefully trans-
mits known constructions, but cannot go beyond that. Through his work
we may nevertheless learn the complete treatment of other magic figures,
such as literal squares and magic circles. He thus helped to keep alive the
knowledge of magic squares, as witnessed by the references to him more
than one century later by the Sudanese Muḥammad al-Fulānī al-Kishnāwī
(d. 1741) in his own work.[25]

§ 6. Squares transmitted to the Latin West

The following magic squares are those which were transmitted to Europe in the 14th century and frequently copied or printed during the 15th
to 17th centuries.[26]

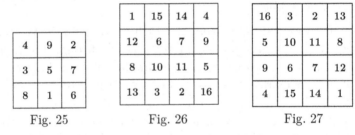

Fig. 25   Fig. 26   Fig. 27

The 3 × 3 square of Fig. 25 appears both as such and in inverted (right-
to-left) form. As to the 4 × 4 square, that of Fig. 26 appears mostly in
the inverted form while the second (which is the previous one rotated and
with median columns switched) will be used by Dürer in his *Melencolia*
in order to date it (1514).

The two 5 × 5 squares of Fig. 28 and 29 result from well-known constructions, as we shall see later (II, §§ 2 and 4). Here they are, however,
quite different in aspect, for the second square separates the odd numbers
from the even ones, all placed in the corners.

---

[24] *Ṭawāliʿ al-ishrāq fī waḍʿ al-awfāq*, MS. Paris BNF Arabe 2698.

[25] MS. 65496 of the London School of Oriental and African studies.

[26] Their occurrences in manuscripts is given in Folkerts' *Zur Frühgeschichte*.

11	24	7	20	3
4	12	25	8	16
17	5	13	21	9
10	18	1	14	22
23	6	19	2	15

Fig. 28

14	10	1	22	18
20	11	7	3	24
21	17	13	9	5
2	23	19	15	6
8	4	25	16	12

Fig. 29

The first 6 × 6 square (Fig. 30), which results from a well-known method of construction (p. 91), occurs in inverted form. The second arrangement is more particular, and seems to have been obtained stepwise.[27]

1	35	34	3	32	6
30	8	28	27	11	7
24	23	15	16	14	19
13	17	21	22	20	18
12	26	9	10	29	25
31	2	4	33	5	36

Fig. 30

1	32	34	3	35	6
30	8	27	28	11	7
20	24	15	16	13	23
19	17	21	22	18	14
10	26	12	9	29	25
31	4	2	33	5	36

Fig. 31

22	47	16	41	10	35	4
5	23	48	17	42	11	29
30	6	24	49	18	36	12
13	31	7	25	43	19	37
38	14	32	1	26	44	20
21	39	8	33	2	27	45
46	15	40	9	34	3	28

Fig. 32

The configuration of this 7 × 7 square (Fig. 32), also found in inverted form, is the most commonly seen in mediaeval Arabic texts (II, § 2).

---

[27] In the way described in our *Les carrés magiques dans les pays islamiques*, p. 258 (Russian edition, pp. 266-267).

Both 8×8 squares of Fig. 33-34 also occur in inverted form. Whereas the construction of the first is well known (below, p. 46), that of the second is particular and seems to have been obtained by trial and error.[28]

1	63	62	4	5	59	58	8
56	10	11	53	52	14	15	49
48	18	19	45	44	22	23	41
25	39	38	28	29	35	34	32
33	31	30	36	37	27	26	40
24	42	43	21	20	46	47	17
16	50	51	13	12	54	55	9
57	7	6	60	61	3	2	64

Fig. 33

1	2	62	61	60	59	7	8
16	10	51	53	12	54	15	49
48	47	19	20	21	22	42	41
25	39	38	28	29	35	34	32
33	31	30	36	37	27	26	40
24	23	43	44	45	46	18	17
56	50	11	13	52	14	55	9
57	58	6	5	4	3	63	64

Fig. 34

Finally, the 9×9 square of Fig. 35 is also very common, and easy to construct (II, § 2).

37	78	29	70	21	62	13	54	5
6	38	79	30	71	22	63	14	46
47	7	39	80	31	72	23	55	15
16	48	8	40	81	32	64	24	56
57	17	49	9	41	73	33	65	25
26	58	18	50	1	42	74	34	66
67	27	59	10	51	2	43	75	35
36	68	19	60	11	52	3	44	76
77	28	69	20	61	12	53	4	45

Fig. 35

Remark. The fact that the Arabic sources of these 14th-century Latin texts contain squares constructed by trial and error may seem surprising, for by that time general methods were well known and widespread. But these squares might be much older in origin. Thus, we find in the 10th-century *Epistles of the Brothers of Purity (Rasā'il ikhwān al-*

---

[28] See our reconstruction of it *ibid.* p. 261 (Russian edition, p. 268).

ṣafā'), and thus in its later copies, a set of seven squares attributed to the seven planets, the last three of which are constructed —as expected for ordinary squares at that early time— by trial and error.[29]

---

[29] Their correct forms (very corrupt in the printed editions) have been established by Hermelink; on the way these squares were obtained, see *Les carrés magiques dans les pays islamiques*, pp. 263-266 (Russian edition, pp. 269-272).

Chapter II

Ordinary magic squares

A. SQUARES OF ODD ORDERS

§ 1. First attempts

We shall see in this chapter that some methods for constructing ordinary magic squares, in particular of odd and evenly-even orders, are quite simple and easy to memorize; whereas the construction of bordered magic squares, to be studied in Ch. IV, requires more attention, whatever the method. But we know that general constructions of bordered squares were found first, and only later those of ordinary magic squares. There is an explanation: it is easier to discover a construction for bordered squares, for we start from a known square and add successively borders around it; whereas constructing an ordinary magic square requires a global examination of the arrangement of n^2 numbers.

Indeed, the construction of an ordinary magic square is based on the *natural square*, that is, a square of the same order as the one to be constructed but containing the first numbers in their natural order. Such a square has two notable properties: first, its main diagonals already contain the magic sum; second, pairs of symmetrically placed lines or columns display differences with the magic sum having the same amount but with a different sign. To obtain a magic square, one would then normally, without modifying the diagonals, proceed to exchange numbers between symmetrically placed lines and columns until reaching the magic sum.

As the study of the two extant 10th-century texts shows, such exchanges turned out to be successful only for small orders, that is, they did not give rise to general methods. That is why one of these two authors, Anṭākī, writes:[30] *Some people begin by placing these numbers according to the succession of the natural order, from 1 to the number of squares in the figure where they wish to construct the magic square. Next they move its numbers —(which are) always in excess in some rows and in deficit in the rows opposite— and then arrange all the rows in a certain manner. This is a method presenting difficulty for the beginner.*

---

[30] *Magic squares in the tenth century*, A.II.2; or below, Appendix 3. In the translation, brackets enclose our additions.

© Springer Nature Switzerland AG 2019
J. Sesiano, *Magic Squares*, Sources and Studies in the History of Mathematics
and Physical Sciences, https://doi.org/10.1007/978-3-030-17993-9_2

Other people proceed with that in a different, and easier, way. He then proceeds with the construction of bordered squares.

This is the only allusion of Anṭākī to ordinary magic squares. It is clear that he did not know of any general method for putting directly the sequence of consecutive numbers in the square to be constructed; for that does not in fact present any difficulty once the general method is known.

The second 10th-century text, that of Abū'l-Wafā' Būzjānī, confirms this absence of a known general method at that time. He indeed considers exchanging numbers in the natural square but only, besides the order 3, for the order 5. And the two ways he then suggests are particular to that order and cannot be extended to higher orders. That is indeed in keeping with Anṭākī's remark: the construction of magic squares from the natural square relies on general principles (for equalizing the individual rows) which are not easy to apply to individual orders: for what works for one will not for the other.

1	2	3	4	5
6	7	8	9	10
11	12	13	14	15
16	17	18	19	20
21	22	23	24	25

Fig.36

Here are, with left-to-right orientation, the constructions of the two examples of Abū'l-Wafā' Būzjānī. Consider the natural square of order 5 in Fig. 36.

1	2	3	4	5
6	7	8	9	10
11	12	13	14	15
16	17	18	19	20
21	22	23	24	25

Fig. 37

1	2	3	12	5
18	7	20	9	10
11	4	13	22	15
16	17	6	19	8
21	14	23	24	25

Fig. 38

1	23	24	12	5
18	7	20	9	11
10	4	13	22	16
15	17	6	19	8
21	14	2	3	25

Fig. 39

1. We leave the diagonals unchanged (Fig. 37). We exchange each of the other numbers in the inner square of 3 with that in its bishop's cell by turning in the same direction (Fig. 38). Exchanging finally the remaining numbers in the border with those of the opposite rows, keeping their succession, we obtain a magic square (Fig. 39).

1	14	3	4	5
18	7	20	9	10
11	24	13	2	15
16	17	6	19	8
21	22	23	12	25

Fig. 40

1	14	22	23	5
18	7	20	9	11
10	24	13	2	16
15	17	6	19	8
21	3	4	12	25

Fig. 41

2. Leaving once again the diagonals unchanged, we invert the pairs of numbers adjacent to (say) the descending diagonal (Fig. 36 and 40). Then, as before, we exchange the remaining numbers of the border with those of the facing row (Fig. 41). We thus obtain a slightly different magic square.

We shall expound a reasoned way of proceeding after teaching some mediaeval methods generally applicable to odd orders. For the time being, let us examine once again the 10th-century situation for ordinary magic squares. According to what Anṭākī says, a magic square of a given order was obtained from the natural square of that order by exchanges; from this we gather that no general method, valid for any odd order and without recourse to the natural square, was known. From Abū'l-Wafā''s example we see that these exchanges left the two main diagonals unchanged, for they already contain the magic sum. This was in fact, as we shall see, the main obstacle to finding a general method.

Remark. The game of chess was invented in India and soon after appeared in Persia, probably in the seventh century. Now many Arabic texts describe the construction of magic squares by means of the chess moves in use at that time. That of the queen (*firzān*, Persian *farzīn*) was to any next cell diagonally, that of the knight (*faras*, horse), two cells in one axial direction and one in the other, that of the bishop (*fīl*, Persian *pīl*, elephant), two cells diagonally (Fig. 42).

	B	K		K	B	
	K	Q		Q	K	
			⋆			
	K	Q		Q	K	
	B	K		K	B	

Fig. 42

§2. Method of diagonal placing

1. Description

The most common and, as it seems, oldest general method for constructing ordinary odd-order squares has received several names, changing as historical research progressed. It was first called 'Bachet's method', after Bachet de Méziriac, who expounded it (but without taking credit for it) in the second edition (1624) of his classical treatise on recreational problems. It was then called 'Cardano's method' for the same reason, the treatise being then the *Practica arithmeticæ* of 1539 (chapters 42 & 66).[31] When the work on magic squares by the Byzantine Manuel Moschopoulos (above, pp. 8, 15) became known (de la Hire, 1705), the construction was attributed to him. It must now be considered to date at least from the early 11th century since Ibn al-Haytham (above, p. 14) gives a justification of it. From then on it is frequently found in mediaeval treatises.

This construction already shows the two characteristics we shall see in the other methods of ordinary magic for odd-order squares: the sequence of natural numbers is placed *continuously*, thus term by term, in the square to be constructed by means of *two moves* which divide the n^2 numbers to be placed into n sequences of n consecutive numbers; and in all these methods the starting cell must be set.

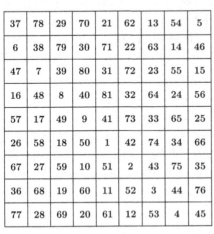

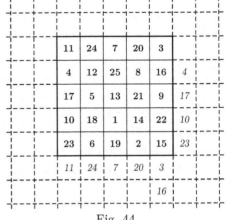

Fig. 43   Fig. 44

Thus, in order to obtain a magic square of odd order without having recourse to the natural square, we proceed as follows.

---

[31] By the way, in a letter to Mersenne, Fermat notes, about the method expounded by Bachet: *pour la regle des quarrez impairs, je dis premierement qu'elle n'est pas de son invention, car elle est dans l'Arithmetique de Cardan* (see his *Varia opera mathematica*, p. 174).

(1) We write 1 in one of the four cells adjacent to the central cell, say that below (Fig. 43, with $n = 9$). We move diagonally away from it, cell by cell, writing the sequence of numbers; when we reach a side of the square, we move to the opposite side as if we were to continue the diagonal move (in order to determine the cell and the subsequent diagonal moves, we may draw on each side auxiliary rows, as in Fig. 44, with $n = 5$).

(2) When a sequence of n numbers has been thus placed, the next cell is occupied; we then move two cells down in the same column, and resume the diagonal placing.

The magic square obtained by this method displays a further property: it is *symmetrical*; that is, two numbers placed symmetrically relative to the centre of the square make the sum $n^2 + 1$, with $\frac{n^2+1}{2}$ in the central cell. We may also observe that this method is applicable to the smallest odd-order square, that of 3 —though as a very particular case.

This is the most widespread method, and the first to be known in Europe since three of the squares then transmitted ($n = 5, 7, 9$, dismissing the particular case of order 3) were constructed in this way (above, pp. 16-18).

2. Discovery of this method

The extant documents suggest that this method appeared between the end of the 10th and the beginning of the 11th century. For, as already said, it is established and justified by Ibn al-Haytham starting from two properties of the natural square of odd order —which indeed proves that this method was not yet in common use. A source of the 12th century (already mentioned, see p. 14), gives the following account of this.[32] *Here is what Abū 'Alī ibn al-Haytham explained: (...) If the order of the square is odd, then, after writing in it the numbers in their natural sequence from 1 to the last number to be comprised in the larger square, which is the number of cells it contains, one will find that the contents of the two (main) diagonals are equal, and that the content of each pair of diagonals on either side of a main diagonal having together the same number of cells as the main diagonal is equal to the content of the main diagonal. This is one of the natural arithmetical properties of these numbers when written in the cells in natural order.*

This is the illustration of what he has explained, in this figure (Fig. 45) in which the side has been divided into five parts, so that (the square) is divided into twenty-five cells, where the numbers from 1 to 25 have been

---

[32] *Une compilation arabe*, pp. 163-164 in the translation, Arabic text lines 242-262; or below, Appendix 4. But here our figures read from left to right.

written in their natural order. Summing the content of the main diagonal composed of 1, 7, 13, 19, 25, the sum will be 65. Summing the content of the other diagonal, thus 5, 9, 13, 17, 21, one also obtains 65. Considering now each pair of broken diagonals on either side of the main diagonals having together the same number of cells as the (main) ones, we shall find that they contain the same quantity. Indeed, summing the content of the diagonal formed by 2, 8, 14, 20, thus four cells, and adding to it the content of the corner cell, thus 21, which makes altogether five cells, as many as the number of cells of the (main) diagonal, we shall again obtain 65. Likewise, summing the content of the diagonal formed by 3, 9, 15, thus with three cells, and adding to it, on the other side, the content of the two-cell diagonal formed by 16 and 22, we shall again obtain 65. Likewise, taking the content of the diagonal formed by 4 and 10, and adding to it, on the other side, the content of the three-cell diagonal formed by 11, 17, 23, the sum of it will be 65. If, in the same manner, one considers cells on either side of the second diagonal, one will find the same.

1	2	3	4	5
6	7	8	9	10
11	12	13	14	15
16	17	18	19	20
21	22	23	24	25

Fig. 45

As we see, Ibn al-Haytham first considered the following two properties of the natural square (our source omits the second, but it must have been mentioned).

I. The sum in the diagonals, either main or broken, of any natural square equals the magic sum for the order considered.

This is clear, for in each of them one finds each of the units from 1 to n and each of the multiples of the order from $0 \cdot n$ to $(n-1)n$. This may be verified in our Fig. 46, where the numbers of the natural square of Fig. 45 are represented in basis 5, thus with the five first units and the multiples of 5. Therefore, the sum in each diagonal, main or broken, will be

$$1 + 2 + \ldots + n + n(0 + 1 + \ldots + (n-1))$$
$$= \frac{n(n+1)}{2} + n \cdot \frac{(n-1)n}{2} = \frac{n(n^2+1)}{2} = M_n.$$

II. The sum in the median rows of any odd-order natural square equals the magic sum for the order considered.

Indeed, it appears from Fig. 46 that the median column contains on the one hand n times the quantity $\frac{n+1}{2}$ and, on the other, each of the multiples of the order, thus $0 \cdot n$, $1 \cdot n$, ..., $(n-1)n$; their sums, equal to $\frac{n(n+1)}{2}$ and $\frac{n^2(n-1)}{2}$, respectively, indeed equal the magic sum M_n. The same holds for the median horizontal row since it contains each of the units from 1 to n and n times the quantity $\frac{n(n-1)}{2}$.

0,1	0,2	0,3	0,4	0,5
1,1	1,2	1,3	1,4	1,5
2,1	2,2	2,3	2,4	2,5
3,1	3,2	3,3	3,4	3,5
4,1	4,2	4,3	4,4	4,5

Fig. 46

Our source's text continues with the account of Ibn al-Haytham's investigations: *Having thus found that this property is inherent in any odd-order table, he prescribed to draw two squares, to write in one the numbers according to their natural sequence, to transfer the content of the two median rows, vertical and horizontal, to the diagonals of the other square. He then carries out the displacement of the content of the remaining diagonals towards their opposite under lengthy conditions, reporting which would take time and the realization of which presents for the beginner difficulties.*

Here ends the account. The mention of 'taking time' and 'difficulties for the beginner' is quite in line with Antākī's reluctance to deal with the construction of ordinary magic squares (above, p. 21). Still, this mention here is odd; for the essential step was to put the natural median row and column in the diagonals of the square to be constructed: the rest is straightforward, as we shall see. Perhaps Ibn al-Haytham justified at length why the resulting square would be magic.

Indeed, let us consider the case of the 5×5 square. Fill first the diagonals as said above (Fig. 47-48); according to the second property, they will satisfy the magic condition. We now know, for each row and column, except the median ones, two elements; and this enables us to fill, cell by cell, the remainder of the square using Property I. Thus, the row containing 11 and 3 must, to make the magic sum, contain the elements of the corresponding broken diagonal of the natural square, thus 3, 7, 11, 20, 24, while the column containing 8 and 14 must contain 2, 8, 14, 20, 21; their intersection must thus contain the common element, 20, the place of which in the magic square is thus determined. In this manner we can fill all remaining cells, the elements left being for the median rows and columns. The resulting square has then the known form (Fig. 49).

1	2	3	4	5
6	7	8	9	10
11	12	13	14	15
16	17	18	19	20
21	22	23	24	25

Fig. 47

11			*20*	3
	12		8	
		13		
	18		14	
23				15

Fig. 48

11	24	7	20	3
4	12	25	8	16
17	5	13	21	9
10	18	1	14	22
23	6	19	2	15

Fig. 49

As we see, the basis for this method is the transfer of the median natural rows to the diagonals of the square to be constructed. The weakness and the limitation of earlier methods such as those we have seen was leaving the natural diagonals unchanged.

3. Another way

Ibn al-Haytham's construction required two squares, one filled with the natural sequence and the other to receive the magic square. This prompted our anonymous source to invent another way to obtain the magic square, namely by constructing, in the auxiliary square itself, an oblique square with as many cells (Fig. 50).[33] *From this method I have deduced a general method, for any odd-order square whatsoever. It is the following.*[34] *If, in any (empty) odd-order square, the side is halved, the point of section will fall in the middle (of the side) of the median (lateral) cell. We join then each point on the middle of a side with that of the perpendicular side. There thus appears an oblique square. Now we find that the sides of this oblique square cut the sides of the (smaller) squares in the main square. Joining each point with the corresponding one on the opposite side, the oblique square will also contain twenty-five cells. Having written in the main square the numbers in their natural sequence, we shall find that thirteen of them fall inside cells of the oblique square while the twelve remaining cells are (occupied) with diagonals. We shall (also) find that in the main square four triangles have been separated, each of which is comprised by one side of the oblique square and two half-sides of the main square. If (now) we imagine that we detach each of*

---

[33] *Une compilation arabe*, p. 165 of the translation, Arabic text lines 267-281; or below, Appendix 5.

[34] Another source attributes this discovery to the 12th-century mathematician Khāzinī; see MS. London BL Delhi Arabic 110, fol. 53ᵛ - 54ʳ.

these triangles and put that one of its sides which is the side of the lesser square on its opposite side, we shall find that the three symbols which are in it fall in the three cells with diagonals of the lesser square. Writing then the content of each cell in the cell it covers, the magic property will be obtained in the lesser square. This is a general rule, (valid) for all odd-order squares.

In other words, we are to construct within the square of the considered order, divided into cells, an oblique square the corners of which will bisect each side. Drawing the parallels through the points of intersection, there appears a square of the same order as the original one (Fig. 50). Write now the natural numbers in the larger square (Fig. 51). Some of the cells of the oblique square will then contain numbers of the larger (natural) square, the others being empty. These will be filled by moving, grouped, the numbers from each corner triangle to the opposite side of the oblique square (Fig. 52). The oblique square will then be magic, with an arrangement the same as was obtained by the previous method.

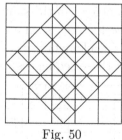

Fig. 50

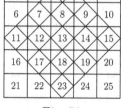

Fig. 51

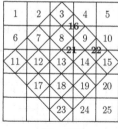

Fig. 52

From a practical point of view this construction of an auxiliary, oblique square is unnecessary since, as we have seen, the sequence of numbers may be placed directly. For us, though, it has the advantage of clearly showing that the magic property obtained depends only on the two previously seen properties of the natural square. Indeed, the two diagonals of the constructed square are the median rows of the natural square, and moving the corner triangles into the oblique square connects the two parts of the natural square's broken diagonals.

Our 12th-century author may have been, or may have thought himself to be, the inventor of this construction. In any event, his inspiration is to be found in a treatise, which he himself mentions, by the Persian Asfizārī, who indeed draws such an oblique square, but so as to find another arrangement (see below, p. 35). Whatever its origin, this drawing of an auxiliary square remained in use until the 18th century, as seen in Fig. 53, taken from a treatise by Kishnāwī (who reproduced it from a copy of Shabrāmallisī's work he often uses, and mentions).

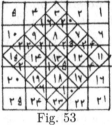

Fig. 53

4. A related method

The advantage of putting the initial cell next to the central cell is that this makes filling easy and continuous. But we might just as well place each beginning of a sequence of n numbers, except for the median one, in any of the cells immediately below the ascending diagonal. Such constructions are seen in the figures 54 and 55. The horizontal rows and the columns of a square thus formed will be magic since, as before, they contain each unit and each multiple of the order; the two diagonals will be magic as well since they contain the same elements as before, in the same order for the descending diagonal and a different one for the ascending one (it contains all middle terms of the sequences of n elements, in an order which depends on that of the starting points of these sequences).

37	33	20	16	3	80	67	63	50
51	38	34	21	17	4	81	68	55
56	52	39	35	22	18	5	73	69
70	57	53	40	36	23	10	6	74
75	71	58	54	41	28	24	11	7
8	76	72	59	46	42	29	25	12
13	9	77	64	60	47	43	30	26
27	14	1	78	65	61	48	44	31
32	19	15	2	79	66	62	49	45

Fig. 54

37	33	74	52	3	62	13	72	23
24	38	34	75	53	4	63	14	64
65	25	39	35	76	54	5	55	15
16	66	26	40	36	77	46	6	56
57	17	67	27	41	28	78	47	7
8	58	18	68	19	42	29	79	48
49	9	59	10	69	20	43	30	80
81	50	1	60	11	70	21	44	31
32	73	51	2	61	12	71	22	45

Fig. 55

Now this less simple but more general method seems to be as old as the common one. Indeed, examples of it are found among the squares of the 11th-century author Zarqālī, who lived in Spain;[35] furthermore, the

---

[35] In his treatise on the seven 'planetary squares', Arabic MSS. Vienna NB 1421, fol. 1^v - 8^r (order 5, 7, 9), and London BL 977, fol. 133^r - 145^r (order 7).

Harmonious arrangement, of the early 11th century, seems to consider the common method as a particular case of it.[36] Appearances of the general case are then seldom seen to occur.

Remarks

(1) The square of Fig. 54 is symmetrical, unlike the other one. The reason is that in the first case the starting points of conjugate sequences (1, 73; 10, 64; 19, 55; 28, 46) are in cells which are placed symmetrically below the ascending diagonal. Thus the terms of one sequence and those of its conjugate (taken in inverse order) will be in diagonally symmetrical cells. The descending diagonal will contain the only unplaced sequence, namely the median one beginning with 37.

(2) We may more generally observe that it is easy to obtain the magic sum in lines and columns using sequences of n numbers by shifting their starting points in a regular manner, provided this distributes equitably in each horizontal and vertical row units and multiples of the order. The real difficulty will then be the filling of the two main diagonals.

5. Modifying the square's aspect

We may modify the aspect of the square obtained by the common method since the property of symmetry therein enables us to exchange rows (above, p. 8). As we see (Fig. 56-58), its transformation is such that the two characteristic steps are no longer recognizable. The elements in each horizontal or vertical row remain, however, the same and there is still symmetry (in the main diagonals, elements have been exchanged).

22	47	16	41	10	35	4
5	23	48	17	42	11	29
30	6	24	49	18	36	12
13	31	7	25	43	19	37
38	14	32	1	26	44	20
21	39	8	33	2	27	45
46	15	40	9	34	3	28

Fig. 56

22	35	16	41	10	47	4
5	11	48	17	42	23	29
30	36	24	49	18	6	12
13	19	7	25	43	31	37
38	44	32	1	26	14	20
21	27	8	33	2	39	45
46	3	40	9	34	15	28

Fig. 57

The more general construction seen in the previous section as well as the transformations mentioned in this one show that a single method may engender a large variety of squares, the number of which will increase with the order.

---

[36] *Un traité médiéval*, pp. 32-34, lines 139-168 in the Arabic text.

22	35	16	41	10	47	4
5	11	48	17	42	23	29
38	44	32	1	26	14	20
13	19	7	25	43	31	37
30	36	24	49	18	6	12
21	27	8	33	2	39	45
46	3	40	9	34	15	28

Fig. 58

§3. A method brought from India

The common method of §2 consisted in placing sequences of n consecutive numbers by means of two moves, starting from a set cell. The next method relies on the same principles. We put (Fig. 59) 1 in the median cell of the upper row, then we write diagonally, but this time ascending, the numbers of the first sequence. In order to proceed with the following sequence, we move, from the last number written, one cell down, and place the numbers of the second sequence as before. The resulting square is, like most of those already seen, symmetrical. Moreover, the elements found in its columns and in one diagonal are the same as with the previous method (Fig. 56).

30	39	48	1	10	19	28
38	47	7	9	18	27	29
46	6	8	17	26	35	37
5	14	16	25	34	36	45
13	15	24	33	42	44	4
21	23	32	41	43	3	12
22	31	40	49	2	11	20

Fig. 59

This other method became known in Europe towards the end of the 17th century. Simon de la Loubère, sent by Louis XIV to Thailand for the years 1687-1688, tells us that he learned it from one of his fellow travellers while returning to France.[37] *Monsieur Vincent, of whom I have often spoken in my account, seeing me once on the boat during our return*

---

[37] See his *Du royaume de Siam*, II, pp. 237-239; or below, Appendix 6.

constructing for amusement magic squares in the manner of Bachet,[38] told me that the Indians of Surat[39] construct them much more easily, and he taught me their method for odd-order squares, having, he said, forgotten that for even ones.

The first square, that of nine cells, turned out to be that of Agrippa, only rotated; but the other odd squares were quite different from those of Agrippa.[40] He arranged the numbers in the cells directly, without hesitation, and I take the liberty of giving the rules and the illustration of this surprisingly simple method, with which one very easily reaches a result though deemed difficult by all our mathematicians.

(1) After dividing the whole square into its cells, we put in it the numbers in their natural sequence, I mean beginning with the unit and proceeding with 2, 3, 4 and all other consecutive numbers. And we (first) put the unit, or the first number of the given arithmetical progression, in the median cell of the upper row.

(2) (Generally,) after putting a number in the top cell of a column, we put the next in the bottom cell of the next column on the right, that is, we descend from the upper line directly to the lower one.

(3) (Generally,) after placing a number in the last cell of a line we put the following one in the first cell of the line immediately above, that is, we turn back from the last right-hand column directly to the first on the left.

(4) In all other cases, after placing a number we put the next ones in the cells following diagonally, or sideways, from bottom to top and from left to right, until we reach one of the cells of the top line or the last column on the right.[41]

(5) Whenever we find that the way is blocked by some cell already filled by some number, we take the cell immediately below that we have just filled, and we continue as before diagonally from bottom to top and from left to right.

These few rules, easy to remember, are sufficient to construct generally all odd squares.

This construction, the essence of which is given in the middle of the

---

[38] Which leads to the magic squares of § 2.

[39] North of Bombay.

[40] Agrippa's odd-order squares are those obtained by the common method (with that of order 3 of course unchanged). Cornelius Agrippa popularized some of the squares received from Spain by reproducing them in his 1533 edition of the *De occulta philosophia* (thus before Cardano, see above, note 31); these are the squares of our figures 25, 26, 28, 30, 32, 33, 35.

[41] Then we are to apply the two previous rules.

14th century by the Indian mathematician Nārāyaṇa, is clearly explained two centuries later, in India again.[42]

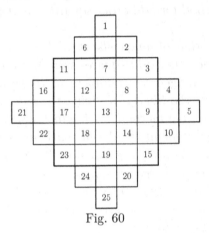

Fig. 60

Remark. La Loubère's assertion that the Indian method is easier than the common one seen above (§ 2) may seem odd. As a matter of fact, that is because Bachet did not teach it as we did: in order to construct his odd-order magic squares, he draws each time an auxiliary figure (Fig. 60 —recalling the oblique square we have seen) of which, after writing in the natural sequence, he moves the outer parts towards the central square. This is also why la Loubère makes a distinction between Agrippa's squares and their construction by Bachet, although the result is the same.

§ 4. Separation by parity

Let us revert to the construction of an oblique auxiliary square within an empty odd-order square (Fig. 50-52). After filling the empty square and moving the numbers out of its corner triangles, we obtained in the oblique square the magic property. Fill now the *oblique* square with the natural numbers (Fig. 61). We observe that some numbers fall in cells of the surrounding square while others are on the crossings. Let us move the latter ones, divided into four groups, into the empty corner triangles opposite (Fig. 62). We thus obtain a magic square (Fig. 63). But this resulting figure has a remarkable additional property: the even numbers are separated from the odd ones, the first being in the corners whereas the others remain within the oblique square. (As seen, the odd numbers have stayed in place while the others have moved out of the oblique figure.)

---

[42] See pp. 156 & 196-197 in Hayashi's study.

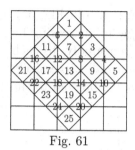

14	10	1	22	18
20	11	7	3	24
21	17	13	9	5
2	23	19	15	6
8	4	25	16	12

Fig. 61   Fig. 62   Fig. 63

This method and that of § 2 are related not only as to realization (use of an auxiliary oblique square) but also fundamentally: lines and columns, as well as diagonals, contain the same elements, only their order is modified (see Fig. 49 & 63). Thus the origin of their magic property is the same, namely putting together elements belonging to the median rows and broken diagonals of the natural square.

This procedure is described in Kharaqī's (d. 1138/9) treatise.[43] He attributes its discovery to his younger contemporary Asfizārī (p. 14) and calls it, probably as did the latter, 'method of the knots and leaps' (each number to be moved in Fig. 62 is located on what we call a crossing, and will find its new place by jumping towards the opposite corner over as many cells of the oblique square —extended, if need be— as is its order). The discovery must in fact be earlier, for this placing is explained, though less simply, in the early 11th century.[44]

It often happened that a method for obtaining a certain configuration becomes over time simpler, or more elegant. Thus another way to obtain this square was to extend diagonally one of the corners of the empty square to form a further figure (Fig. 64). We then fill an imaginary oblique square in the main square with the sequence of odd numbers and the other one with the sequence of even numbers. The three outer parts correspond to what is needed in the empty corner triangles and are moved accordingly. This will give the square of Fig. 65. Fig. 66 shows this construction in a manuscript copy of Shabrāmallisī's treatise.[45]

As a matter of fact, we may even do without any auxiliary construction, as explained by the later Kishnāwī (ca. 1700) just before expounding the above construction. After putting the odd numbers in the oblique square, drawn or imaginary, we write the first even numbers, in quantity k if the order is $n = 2k + 1$, along the hypotenuse of the triangle oppo-

---

[43] *Herstellungsverfahren III*, pp. 200-201, lines 249-271 of the Arabic text.

[44] *Un traité médiéval*, pp. 35-39, lines 172-220 of the Arabic text. The attribution to Asfizārī is confirmed elsewhere (MS. London BL Delhi Arabic 110, fol. 41^r - 41^v).

[45] MS. Paris BNF Arabe 2698, fol. 15^r.

site to the first sequence of odd numbers (placing of 2, 4, 6 in Fig. 65). We next fill the hypotenuse of the other lower triangle by increasing by 1 the odd numbers of the cells just above. Adding repeatedly, to each of these even numbers, the quantity $n + 1$ enables us to fill successively the columns, first going down and then jumping from the bottom to the top.[46]

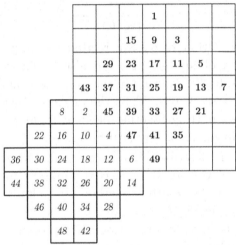

						1			
					15	9	3		
				29	23	17	11	5	
			43	37	31	25	19	13	7
		8	2	45	39	33	27	21	
	22	16	10	4	47	41	35		
36	30	24	18	12	6	49			
44	38	32	26	20	14				
	46	40	34	28					
	48	42							

Fig. 64

26	20	14	1	44	38	32
34	28	15	9	3	46	40
42	29	23	17	11	5	48
43	37	31	25	19	13	7
2	45	39	33	27	21	8
10	4	47	41	35	22	16
18	12	6	49	36	30	24

Fig. 65

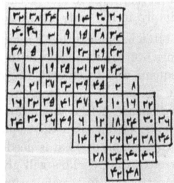

Fig. 66

§5. Use of the knight's move

In two of the foregoing methods a sequence of n numbers was placed by moving diagonally from one cell to the next, thus by a queen's move to the next column. The two methods which follow differ in that they use the knight's move to pass from one column to the next.

---

[46] This alternative way is, however, older since it occurs directly after the description of Asfizārī's method in MS. London BL Delhi Arabic 110 (thus perhaps by Asfizārī himself).

(1) The first is described in the early 14th century by the Byzantine Manuel Moschopoulos as follows (Fig. 67-70).[47]

8	1	6
3	5	7
4	9	2

Fig. 67

10	18	1	14	22
4	12	25	8	16
23	6	19	2	15
17	5	13	21	9
11	24	7	20	3

Fig. 68

38	14	32	1	26	44	20
5	23	48	17	42	11	29
21	39	8	33	2	27	45
30	6	24	49	18	36	12
46	15	40	9	34	3	28
13	31	7	25	43	19	37
22	47	16	41	10	35	4

Fig. 69

This is another method, by three and five.[48] We draw the square and trace in it the cells (in a number equal to the quantity) of the square number. Then we put the unit always in the middle of the upper cells. We count three cells, one being that which has the unit and two subsequent downward of it, and we put horizontally, on the right of the third, 2.[49] Then again we count from there likewise three cells, and put, on the right, 3 (and we continue in the same way). If we do not have a cell on the right, we go the other way horizontally to the left, as in the previous method (...). We do the same until we reach the side (= the root) of the square in question. Attaining this, we count five cells, one being that containing the side and four downward of it, and put then in the fifth cell, without turning sideways, the number following that of the side. Then we count again by (steps of) three as far as the (number of the) side, (and so on) going round through the cells again, as in the previous method, until we get to their end. This method is throughout like the previous one, except that there the unit was put each time in a different cell whereas here it is always in the middle of the upper cells;[50] furthermore, there we proceeded by two and three, and here by three and five.

Thus, as with the Indian method, we start in the middle of the upper row with 1, then descend sideways with the knight's move. After a sequence of n numbers has been placed, we go each time down, thus in

---

[47] Tannery, *Le traité de* (...) *Moschopoulos*; or below, Appendix 7.

[48] 'Another method': he has just explained the common method (our §2). 'Three and five': this is to characterize the two moves.

[49] As usual in Greek and Arabic texts, the length of each step includes the point of departure.

[50] With the previous method the starting cell varies in distance from the side of the square (but is always below the central cell).

the same column, by four cells. We obtain squares which, excepting the
3 × 3 square, are not symmetrical.

26	58	18	50	1	42	74	34	66
6	38	79	30	71	22	63	14	46
67	27	59	10	51	2	43	75	35
47	7	39	80	31	72	23	55	15
36	68	19	60	11	52	3	44	76
16	48	8	40	81	32	64	24	56
77	28	69	20	61	12	53	4	45
57	17	49	9	41	73	33	65	25
37	78	29	70	21	62	13	54	5

Fig. 70

Remark. As already said (p. 15), it would seem that Moschopoulos relies
on a collection of squares found in some Arabic or Persian source.[51]

(2) We again start in the upper middle cell and again proceed with the
knight's move. But we pass from one sequence to the next by descending
one cell.[52] Here the resulting squares are symmetrical, and thus the magic
condition of the diagonals is automatically fulfilled (Fig. 71-73).

46	31	16	1	42	27	12
5	39	24	9	43	35	20
13	47	32	17	2	36	28
21	6	40	25	10	44	29
22	14	48	33	18	3	37
30	15	7	41	26	11	45
38	23	8	49	34	19	4

23	12	1	20	9
4	18	7	21	15
10	24	13	2	16
11	5	19	8	22
17	6	25	14	3

Fig. 71   Fig. 72

Such methods are described in various Arabic manuscrits, but not
always with due care. Thus the later (18th century) Kishnāwī, who pro-
poses a 'general' method for the construction of odd-order squares and
uses the same two moves as in the first method above, says that *one begins*

---

[51] See our *Les carrés magiques de Manuel Moschopoulos*, pp. 392-393.
[52] See MSS. Istanbul Ayasofya (Hagia Sophia, thus saint Wisdom) 2794, fol. 8ʳ,
Mashhad Āstān-i Quds 12167, fol. 167-168, and 14159, fol. 82ᵛ.

in any cell whatsoever of the square, without fixing any particular point of departure.[53] He then chooses to start in a corner cell and constructs the squares of orders 5 and 7. It was just as well that he stopped there: the square for $n = 9$ would not be magic; whereas, starting as above from the upper middle cell, we shall always obtain a magic square (but which will be pandiagonal only if the order is not divisible by 3). This must have been noted since we find in a Persian manuscript, on the same page, a 5×5 square constructed with the starting cell in the corner whereas the 9×9 follows the second method above.[54]

77	58	39	20	1	72	53	34	15
6	68	49	30	11	73	63	44	25
16	78	59	40	21	2	64	54	35
26	7	69	50	31	12	74	55	45
36	17	79	60	41	22	3	65	46
37	27	8	70	51	32	13	75	56
47	28	18	80	61	42	23	4	66
57	38	19	9	71	52	33	14	76
67	48	29	10	81	62	43	24	5

Fig. 73

Remark. The methods we have seen are often related. Thus the squares obtained by the first method above would also be reached starting with the magic squares of § 2 having the same order: putting the row with 1 as top row and having it followed by each $(k + 1)$th horizontal row if the order is $n = 2k + 1$. Likewise, if in the second of the two above methods the knight's move is replaced by a downward descent of $n - 1$ cells in the next column, we shall find the squares brought from India by la Loubère.

§ 6. Principles of these methods

We may now consider the common features of these various general methods (excluding the individual placings varying with the order considered, as performed by Abū'l-Wafā' Būzjānī and, for the time being, the separation by parity).

(1) When putting sequences of n numbers the move may vary in length,

---

[53] See our *Quelques méthodes*, p. 57.
[54] MS. Mashhad Āstān-i Quds 14159, fol. 82ᵛ.

but not its fundamental nature: it is diagonal, and jumps from one column to the next.

(2) The move from one sequence of n numbers to the next is proper to each method; but here too there is a common feature, namely the diagonal move made by the starting points of the sequences.

(3) In each square to be constructed the initial cell is fixed.

As a result of the diagonal moves of both the first movement and the starting points of the sequences in the second, a representative of each unit and of each multiple of the order considered will appear in each line and each column. This ensures that they will contain the required magic sum, as was also the case for the diagonals in the natural square (above, Fig. 46). The rôle of fixing the initial cell is to ensure the magic property of the two main diagonals.

Now the property of the main or broken diagonals of the natural square used by Ibn al-Hay<u>th</u>am for constructing the rows of his magic square may be generalized: supposing the natural square of a given order iterated in the plane, any diagonal route to n cells, parallel or not to a main diagonal, will produce the magic sum if it meets each unit and each multiple of the order. Such diagonal routes may thus, under certain conditions, be used to construct the horizontal and vertical rows of the magic square, for to determine their elements it suffices to advance with regular steps.

Let us thus consider the natural square of a certain order, iterated in the plane, and let us choose one of its cells as starting point. The element it contains will be, say, the first number of the square to be constructed, say its corner element.

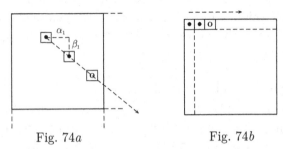

Fig. 74a   Fig. 74b

(1) Let the two natural numbers α_1, β_1 give the steps to pass in the natural square from this starting element to the next, which will thus be distant from α_1 cells horizontally and β_1 cells vertically; we shall take the element reached to be the horizontal neighbour of the first number in the square to be constructed. Repeating this same move we shall determine

successively the elements of the whole first line (Fig. 74a & b).

(2) Let α_2, β_2 be the steps to pass from the starting element in the natural square to another, thus distant from α_2 cells horizontally and β_2 cells vertically; the element reached will be, in the square to be constructed, the vertical neighbour of the initial number. With this same move we shall determine successively the elements of the whole first column.

Thus doing we shall have filled one line and one column of the magic square. Applying these same steps to any number already placed we may complete the square.

We had mentioned the existence of restrictive conditions. They have to do with the choice of our quantities α_1, β_1, α_2, β_2, which cannot be fully arbitrary if we are to obtain a magic square.

First, it is evident that, in order to have an equal distribution of units and order multiples, the two movements must follow in the natural square an oblique direction so as not to remain in a same line or column. So we must require:

Condition I: α_1, β_1, α_2, $\beta_2 \neq 0$.

It is likewise evident that, since the two moves start from the same point of departure in the natural square, they must be different, for otherwise we would find the same succession of numbers; so we must also have:

Condition II: $\alpha_1 \neq \alpha_2$ and $\beta_1 \neq \beta_2$.

But the two oblique lines will not be distinct and we shall still go through the same set of n numbers if their slopes $\frac{\alpha_1}{\beta_1}$ and $\frac{\alpha_2}{\beta_2}$ are not different. So we must further require:

Condition III: $\alpha_1\beta_2 - \alpha_2\beta_1 \neq 0$.

So much for the horizontal and vertical rows. Now the elements of the diagonal descending to the right ($\searrow$) and of the diagonal descending to the left ($\swarrow$) are obtained by combining the movements $\alpha_1 + \alpha_2$ and $\beta_1 + \beta_2$ for the first (which gives the element of the adjacent lower right-hand cell in the square to be constructed) and $\alpha_2 - \alpha_1$ and $\beta_2 - \beta_1$ for the other (which gives the element of the adjacent lower left-hand cell). Now here too we are not to remain in the same horizontal or vertical row of the natural square. So we are to require:

Condition IV: $\alpha_2 \pm \alpha_1$, $\beta_2 \pm \beta_1 \neq 0$.

The figures 75a and 75b ($\alpha_1 = 1$, $\beta_1 = 2$, $\alpha_2 = 3$, $\beta_2 = 4$) represent a square constructed in this manner. We could not, however, proceed thus for any order. Indeed, no further condition is necessary if the order is prime, but if the order is a composite number, the previous conditions do not ensure that we shall meet n different numbers during the movement. In order to avoid repetition in the lines, columns and diagonals being constructed, we are also to require that the above quantities be prime to the order. Hence we shall require:

Condition V: α_1, β_1, α_2, β_2, $\alpha_1\beta_2 - \alpha_2\beta_1$, $\alpha_2 \pm \alpha_1$, $\beta_2 \pm \beta_1$ prime to n.

1	2	3	4	5	6	7
8	9	10	11	12	13	14
15	16	17	18	19	20	21
22	23	24	25	26	27	28
29	30	31	32	33	34	35
36	37	38	39	40	41	42
43	44	45	46	47	48	49

Fig. 75a

1	16	31	46	12	27	42
32	47	13	28	36	2	17
14	22	37	3	18	33	48
38	4	19	34	49	8	23
20	35	43	9	24	39	5
44	10	25	40	6	21	29
26	41	7	15	30	45	11

Fig. 75b

If all these conditions are met, the square obtained will be magic, and even pandiagonal since Condition IV was not just valid for the main diagonals. But these same conditions also involve restrictions.

• First, the order considered must be odd. If n is even, Condition V will require the α_is and β_is to be odd; but then their sums and differences will be even, and thus the diagonals cannot be magic.

• Next, the odd order must not be divisible by 3. Indeed, if n is odd and divisible by 3, the α_is and β_is shall have, in order to be prime with n, one of the two forms $3t + 1$, $3t' + 2$; but then either their sum or their difference will be divisible by 3, and therefore the two diagonals cannot both be magic.

Since a generally applicable construction method for odd orders must also include those divisible by 3, we shall resort to another property of the natural square, and that is what the methods seen above do when they fix the initial cell: the two diagonals will thereby contain a sequence of numbers making the required magic sum.

Remark. Pandiagonal squares with odd orders divisible by 3 are possible. But if we use the natural square, it must first undergo transformations. This does not seem to have been considered at the time we are dealing with.

Application to the general methods seen above

1. Common method (method of diagonal placing, §2)

The values of the parameters are here, for any odd order $n = 2k + 1$ (Fig. 75a and 56; subtracting, if need be, $2k + 1$),

α_1	β_1	α_2	β_2	$\alpha_1 + \alpha_2$	$\beta_1 + \beta_2$	$\alpha_2 - \alpha_1$	$\beta_2 - \beta_1$
$k+1$	k	$k+1$	$k+1$	1	0	0	1.

As we see, the diagonals descending to the right will be the horizontal rows of the natural square (since $\beta_1 + \beta_2 = 0$), of which they will reproduce the terms in succession (since $\alpha_1 + \alpha_2 = 1$), whereas the diagonals descending to the left will be the columns of the natural square (since $\alpha_2 - \alpha_1 = 0$) with their terms taken in succession (since $\beta_2 - \beta_1 = 1$). The square will then be magic if the main diagonals descending to the right and to the left are, respectively, the median line and the median column of the natural square. It is in order to reach this situation that the initial cell was put below the central cell.

2. Method brought from India by la Loubère (§3)

We obtain the values (Fig. 75a and 59; adding, if need be, $2k + 1$)

α_1	β_1	α_2	β_2	$\alpha_1 + \alpha_2$	$\beta_1 + \beta_2$	$\alpha_2 - \alpha_1$	$\beta_2 - \beta_1$
2	1	1	1	3	2	$2k$	0.

All diagonals descending to the right will then fulfill (provided the order is not divisible by 3) the magic condition, and the diagonals descending to the left will comprise the elements of the horizontal rows of the natural square, but taken the other way round. For the main diagonal descending to the left also to be magic, we shall have here again to fix the initial cell.

3. Method of separation by parity (§4)

This corresponds to (Fig. 75a and 65)

α_1	β_1	α_2	β_2	$\alpha_1 + \alpha_2$	$\beta_1 + \beta_2$	$\alpha_2 - \alpha_1$	$\beta_2 - \beta_1$
1	$2k$	1	1	2	0	0	2.

For the diagonals, the situation is similar to that of the common method, except that we meet every other element of the natural row.

4. Methods using the knight's move (§5)

• In the first one, with the vertical jump of four cells between two sequences, we have (Fig. 75a and 69; adding, if need be, $2k + 1$)

α_1	β_1	α_2	β_2	$\alpha_1 + \alpha_2$	$\beta_1 + \beta_2$	$\alpha_2 - \alpha_1$	$\beta_2 - \beta_1$
$k+1$	k	k^2	k^2	$k^2 + k + 1$	$k^2 + k$	$k^2 + k$	$k^2 + k + 1.$

We shall then have magic, and pandiagonal, squares and may thus choose any initial cell if the order is not divisible by 3; indeed, each of the $2n + 2$ rows comprises all units and all multiples of the order. This is not the case if the order is divisible by 3. Thus, if $n = 9$ (therefore $k^2 + k + 1 = 21 \equiv 3$, $k^2 + k = 20 \equiv 2$), there will be in each third ($= \frac{n}{3}$) term repetition of the units for the diagonals descending to the right and repetition of the order's multiples for the diagonals descending to the left. So here we are to fix the initial cell so that the main diagonals will be magic. In our case ($n = 9$, Fig. 70), the two diagonals will then comprise the following multiples of the order and units: $0, 5$; $1, 2$; $2, 8$; $3, 5$; $4, 2$; $5, 8$; $6, 5$; $7, 2$; $8, 8$ and $1, 2$; $1, 5$; $1, 8$; $4, 1$; $4, 4$; $4, 7$; $7, 3$; $7, 6$; $7, 9$, respectively. Generally, for the first diagonal of an order divisible by 3, we find each multiple of the order, the sum of which gives $n \cdot \frac{n(n-1)}{2}$, and, for the units, three times the quantity $\frac{n}{3}\frac{n+1}{2}$, thus $\frac{n(n+1)}{2}$; and, for the second diagonal, each representative of the units, thus $\frac{n(n+1)}{2}$, and three times the multiple $\frac{n}{3}\frac{n-1}{2}$ of n, thus $n \cdot \frac{n(n-1)}{2}$. The sum of the two quantities will therefore give M_n.

• In the second method, the move between two sequences of n numbers is to the next cell. We have in this case (Fig. 75a and 72)

α_1	β_1	α_2	β_2	$\alpha_1 + \alpha_2$	$\beta_1 + \beta_2$	$\alpha_2 - \alpha_1$	$\beta_2 - \beta_1$
$2k$	$2k-1$	1	1	0	$2k$	2	3.

The diagonals descending to the right are the columns of the corresponding natural square, with their elements taken in reverse order. Therefore, the initial cell must be fixed in such a way that the main diagonal descending to the right will be the median column (in the case of Fig. 73 we find, beginning below, following multiples of the order $n = 9$ and units: $0, 5$; $1, 5$; $2, 5$; $3, 5$; $4, 5$; $5, 5$; $6, 5$; $7, 5$; $8, 5$). As for the other diagonal, it contains each of the units and three times the same multiples of the order as in the previous case (thus, in Fig. 73, we have $1, 3$; $1, 6$; $1, 9$; $4, 2$; $4, 5$; $4, 8$; $7, 1$; $7, 4$; $7, 7$).

B. SQUARES OF EVENLY-EVEN ORDERS

§7. The square of order 4

Whereas the square of order 3 displayed a single possibility of construction with the first nine natural numbers, the first sixteen natural numbers can be arranged in the 4×4 square in 880 different magic configurations, all represented by Bernard Frénicle de Bessy in his *Table generale des quarrez de quatre*, published in 1693. Thus arose a further problem, that of enumerating all the possible configurations of a magic square of given order, which is far from being solved.[55]

1	15	14	4
12	6	7	9
8	10	11	5
13	3	2	16

Fig. 76

1	14	15	4
12	7	6	9
8	11	10	5
13	2	3	16

Fig. 77

1	14	11	8
12	7	2	13
6	9	16	3
15	4	5	10

Fig. 78

Here are three examples of magic squares of order 4. As already noted (p. 5), figures arising from the rotation or inversion of a given square are not to be considered as different. However, squares obtained by other transformations, in particular rows exchanged in a symmetrical square (above, p. 8), count as being different. Thus, the square of Fig. 77 is regarded as different from that of Fig. 76, even if they display merely an exchange of the central columns; indeed, the elements forming each main diagonal differ. But the square of Fig. 78 is altogether different.

These three examples have not been chosen at random. They are the simplest cases of general methods to be explained now. Furthermore, they are of interest historically. The first is one of the squares transmitted to Europe at the end of the Middle Ages (above, Fig. 26), and it is frequently represented from that time on, while the second (but rotated by 180°) arrived in Europe at the same time, and is best known by its representation in Albrecht Dürer's *Melencolia* (above, p. 16). The third square is one of the oldest, since it is already seen in the earliest texts, and it displays in addition the pandiagonal property.

§8. Method of dotting

Take an empty square of order 4 and mark its main diagonals with dots (Fig. 79). Count then the cells from a corner and write in each marked cell the number thus attained. Upon reaching the last cell, start again from that one and, going through the cells in reverse, put the number reached in each unfilled place, thus in the cells without dots. The resulting square is the first one above (Fig. 76).

This method can easily be extended to higher evenly-even orders. Take (Fig. 80) an empty square of order $n = 4k$ and divide it into k^2 squares of order 4, where you are to arrange dots as in the square just

---

[55] A not entirely new problem since some attempts for particular categories seem to have been considered earlier; see *Magic squares in the tenth century*, pp. 44-45, 60.

seen. Here as before, we shall have constructed the corresponding magic square by counting the cells from one corner and then from the opposite corner (Fig. 81). The square of order 12 seen in Fig. 82, constructed in same manner, is also magic; thus it does not matter whether the axes of the main square cut or not the median squares of order 4.

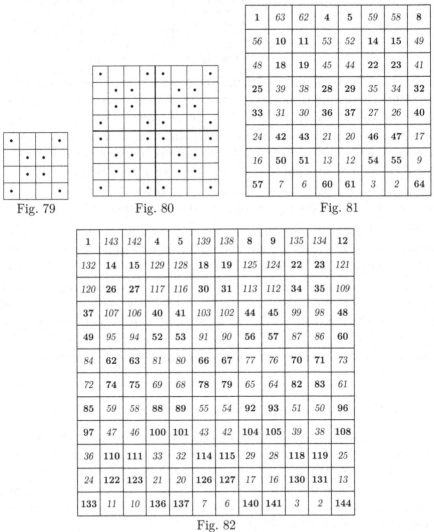

Fig. 79

Fig. 80

1	63	62	4	5	59	58	8
56	10	11	53	52	14	15	49
48	18	19	45	44	22	23	41
25	39	38	28	29	35	34	32
33	31	30	36	37	27	26	40
24	42	43	21	20	46	47	17
16	50	51	13	12	54	55	9
57	7	6	60	61	3	2	64

Fig. 81

1	143	142	4	5	139	138	8	9	135	134	12
132	14	15	129	128	18	19	125	124	22	23	121
120	26	27	117	116	30	31	113	112	34	35	109
37	107	106	40	41	103	102	44	45	99	98	48
49	95	94	52	53	91	90	56	57	87	86	60
84	62	63	81	80	66	67	77	76	70	71	73
72	74	75	69	68	78	79	65	64	82	83	61
85	59	58	88	89	55	54	92	93	51	50	96
97	47	46	100	101	43	42	104	105	39	38	108
36	110	111	33	32	114	115	29	28	118	119	25
24	122	123	21	20	126	127	17	16	130	131	13
133	11	10	136	137	7	6	140	141	3	2	144

Fig. 82

The author of the 12th-century compilation —already mentioned several times— takes credit for discovering this method.[56] That can, however, only be for the preliminary arrangement of dots since squares of this type occur earlier: the *Harmonious arrangement* teaches how to fill first

---

[56] *Une compilation arabe*, pp. 166-168 and lines 302-326 of the Arabic text.

the *diagonals of the squares of 4 contained in the square*, then, from the opposite corner, the empty cells.[57] Such squares were transmitted to the West: first in Moschopoulos' treatise (for $n = 4$ and $n = 16$) and then in late mediaeval Europe (besides $n = 4$, for $n = 8$, see p. 18).

§9. Exchange of subsquares

I	II	III	IV
V	VI	VII	VIII
IX	X	XI	XII
XIII	XIV	XV	XVI

Fig. 83

I	$\overline{XX}$	$\overline{XIX}$	IV
$\overline{IIX}$	VI	VII	$\overline{XI}$
$\overline{IIIV}$	X	XI	$\overline{X}$
XIII	$\overline{III}$	$\overline{II}$	XVI

Fig. 84

Considering now the natural square of order $4k$, we divide it into $4^2 = 16$ squares of order k, which we number (Fig. 83). Place these numbered squares according to the configuration seen for the 4×4 square in Fig. 76, but turning by 180° the subsquares displaced (Fig. 84).

1	2	3	4	5	6	7	8	9	10	11	12
13	14	15	16	17	18	19	20	21	22	23	24
25	26	27	28	29	30	31	32	33	34	35	36
37	38	39	40	41	42	43	44	45	46	47	48
49	50	51	52	53	54	55	56	57	58	59	60
61	62	63	64	65	66	67	68	69	70	71	72
73	74	75	76	77	78	79	80	81	82	83	84
85	86	87	88	89	90	91	92	93	94	95	96
97	98	99	100	101	102	103	104	105	106	107	108
109	110	111	112	113	114	115	116	117	118	119	120
121	122	123	124	125	126	127	128	129	130	131	132
133	134	135	136	137	138	139	140	141	142	143	144

Fig. 85

The natural square thus transformed is magic (Fig. 85-86). This method is attributed by our anonymous source to Asfizārī. As he tells us, he has chosen to present it because it was simpler than both that of Anṭākī and that of Ibn al-Haytham. Of Ibn al-Haytham's method we only know

---

[57] *Un traité médiéval*, pp. 40-42, lines 232-244 of the Arabic text.

that it proceeded using exchanges between opposite rows; the method of Anṭākī is known, and fills subsquares of order 4 (below, § 22).

1	2	3	141	140	139	138	137	136	10	11	12
13	14	15	129	128	127	126	125	124	22	23	24
25	26	27	117	116	115	114	113	112	34	35	36
108	107	106	40	41	42	43	44	45	99	98	97
96	95	94	52	53	54	55	56	57	87	86	85
84	83	82	64	65	66	67	68	69	75	74	73
72	71	70	76	77	78	79	80	81	63	62	61
60	59	58	88	89	90	91	92	93	51	50	49
48	47	46	100	101	102	103	104	105	39	38	37
109	110	111	33	32	31	30	29	28	118	119	120
121	122	123	21	20	19	18	17	16	130	131	132
133	134	135	9	8	7	6	5	4	142	143	144

Fig. 86

§ 10. Generalization

These two methods seem altogether different. As a matter of fact, they rely on a same principle, that of *diagonal exchanges in the natural square*. Let us divide the empty square into its quadrants. Since the order of the whole square is $4k$, that of the quadrant is $2k$; there is therefore an even number of cells in each of its lines and columns. So write in the first quadrant dots in such a way as to cover half of its lines' and columns' cells, thus with exactly k dots in each line and each column. Next turn down this first quadrant over the others successively, and carry over the dots to the cells covered by cells having dots. The result is a figure where the distribution of the dots displays a central symmetry: if a cell contains a dot, so will the diagonally opposite one, and the same holds for empty cells. Fig. 87-98 represent various such configurations for a square of order 8.

Remark. A notably easy way to obtain such a configuration appears from Fig. 88-91, 94-96, 98. We may simply write k dots in the first quadrant's horizontal row and fill with dots the diagonals, main or partial, starting with the cells thus marked and following the same oblique direction; half the quadrant being thus filled, it is then turned over its neighbours to complete the marking in the whole square.

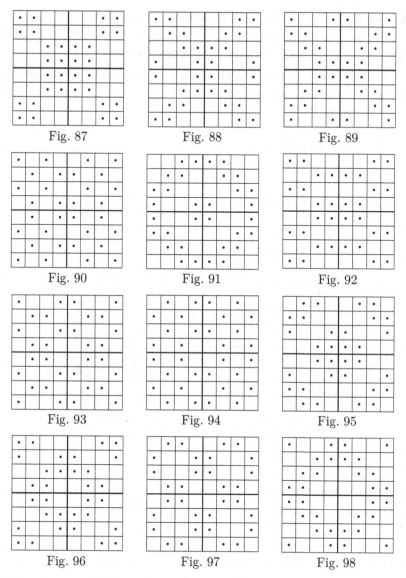

Fig. 87   Fig. 88   Fig. 89

Fig. 90   Fig. 91   Fig. 92

Fig. 93   Fig. 94   Fig. 95

Fig. 96   Fig. 97   Fig. 98

The square is then ready to receive the numbers. As seen before, fill first the cells containing a dot by enumerating them from a corner, then resume enumerating from the opposite corner to fill the empty cells. The square is thus completed. Since the configuration of the dots displays a central symmetry, no repetition of numbers will occur: each number is placed just once since diagonally opposite cells (which would give the same number with opposite enumerations) are filled during the same enumeration. Because of that central symmetry, the square obtained will be symmetrical, this symmetry being also present in the natural square.

We see now why the two previous methods were related: they correspond to the configurations of Fig. 80 and Fig. 87. Some of the configurations above also appear explicitly in Arabic sources.[58]

Remark. The larger the order, the greater the number of possible configurations for the dots and thus the greater the number of possible arrangements.

§ 11. Continuous filling

1	134	135	4	5	138	139	8	9	142	143	12
132	23	22	129	128	19	18	125	124	15	14	121
25	110	111	28	29	114	115	32	33	118	119	36
108	47	46	105	104	43	42	101	100	39	38	97
49	86	87	52	53	90	91	56	57	94	95	60
84	71	70	81	80	67	66	77	76	63	62	73
72	83	82	69	68	79	78	65	64	75	74	61
85	50	51	88	89	54	55	92	93	58	59	96
48	107	106	45	44	103	102	41	40	99	98	37
109	26	27	112	113	30	31	116	117	34	35	120
24	131	130	21	20	127	126	17	16	123	122	13
133	2	3	136	137	6	7	140	141	10	11	144

Fig. 99

We place (Fig. 99) the numbers from 1 to $\frac{n^2}{2}$ alternately in opposite lines by pairs (except at the beginning and end of lines), going to and fro then passing to the next pair of opposite rows. After covering in this way the two median rows, we resume the movement from these two rows in reverse. In order to be sure about the starting cell of the second half, we may note that the final square will, like the previous ones, be symmetrical; therefore, the cell with $\frac{n^2}{2} + 1$ (here 73) will be diagonally opposite that of $\frac{n^2}{2}$ (here 72).

Our second square of 4 (Fig. 77) has been constructed in that way.

---

[58] That of Fig. 89 in Kharaqī (*Herstellungsverfahren III*, p. 201), those of the figures 80, 87, 88 in Shabrāmallisī (MS. Paris BNF Arabe 2698, fol. 24ᵛ - 25ʳ), those of the figures 80, 89-91, 94-98 in the MS. London BL Delhi Arabic 110 (fol. 43ʳ), the configuration of Fig. 90 in Persian texts (MS. Mashhad Āstān-i Quds 12167, fol. 170ᵛ - 171ʳ, or 12235, p. 411).

Use of this method is attested at the beginning of the 11th century, for the *Harmonious arrangement* has a similar construction.[59] This method of continuous filling is one of the easiest for constructing magic squares of evenly-even orders, for it does not require any preliminaries, such as dotting.

Remark. The method of §9 leads to the 8 × 8 square of Fig. 100*a*. Now the (not symmetrical) 8 × 8 square of Fig. 100*b* occurs in some manuscripts.[60] This notably simple method for filling continuously opposite lines, which evidently meets the conditions for lines and columns, is not generally applicable since it will not make the diagonals magic for the next evenly-even order (but will for the 16 × 16 square).

1	2	62	61	60	59	7	8
9	10	54	53	52	51	15	16
48	47	19	20	21	22	42	41
40	39	27	28	29	30	34	33
32	31	35	36	37	38	26	25
24	23	43	44	45	46	18	17
49	50	14	13	12	11	55	56
57	58	6	5	4	3	63	64

Fig. 100*a*

1	2	62	61	60	59	7	8
16	15	51	52	53	54	10	9
17	18	46	45	44	43	23	24
32	31	35	36	37	38	26	25
33	34	30	29	28	27	39	40
48	47	19	20	21	22	42	41
49	50	14	13	12	11	55	56
64	63	3	4	5	6	58	57

Fig. 100*b*

§ 12. Principles of these methods

Considering a natural square of *even* order (Fig. 101), Ibn al-Haytham observed that it had, besides the property of the diagonals —which is not peculiar to odd orders— the property that *the sum of the content of two complementary halves of horizontal or vertical rows equidistant from the centre equals the content of either one of the two (main) diagonals.*[61]

Generally, the natural square of even order has the following two properties:

I. As in the case of the natural square of odd order, the sum in each diagonal, main or broken, equals the magic sum for the order considered.

---

[59] *Un traité médiéval*, p. 43 (and n. 71 there).

[60] MSS. Istanbul Ayasofya 2794, fol. 15ʳ, Mashhad Āstān-i Quds 12235, p. 437, and 14159, fol. 85ᵛ.

[61] *Une compilation arabe*, p. 165, lines 283-287 of the Arabic text; or below, Appendix 8. He illustrated it *at least* with Fig. 76, attributed to him in MS. London BL Delhi Arabic 110, fol. 58ʳ.

II. The sum of half the elements of any horizontal row and half the elements, not aligned with them, of the row placed symmetrically equals the magic sum, and the same holds for the vertical rows.

This second property was also, *mutatis mutandis*, verified in the natural square of odd order: one row and its symmetrical differed from the magic sum by the same quantity, and therefore the median horizontal and vertical rows made exactly the magic sum. This was then used to obtain the magic condition for the diagonals, while the first property was the basis for obtaining magic horizontal and vertical rows. In the case of even squares it is the second property which will play the main rôle since it will be used for the magic sums in the rows, the number of cells of which is even, while the first property will occasionally serve for the diagonals.

1	2	3	4	5	6	7	8
9	10	11	12	13	14	15	16
17	18	19	20	21	22	23	24
25	26	27	28	29	30	31	32
33	34	35	36	37	38	39	40
41	42	43	44	45	46	47	48
49	50	51	52	53	54	55	56
57	58	59	60	61	62	63	64

Fig. 101

Let us now consider mathematically this second property (we have already examined the first for the natural square of odd order, see p. 26). As for the horizontal rows, the elements of line L_i placed in the upper half of the square will be

$$(i-1)n+1, \ (i-1)n+2, \ \ldots, \ (i-1)n+n$$

the sum of which is

$$(i-1)n^2 + \frac{n(n+1)}{2} = \frac{n}{2}[2ni - 2n + n + 1]$$

$$= \frac{n}{2}[n^2 + 1 - n^2 + 2ni - n] = M_n - \frac{n}{2}[n(n-2i+1)].$$

The opposite row L_{n-i+1} contains the larger numbers

$$(n-i)n+1, \ (n-i)n+2, \ \ldots, \ (n-i)n+n$$

the sum of which is

$$(n-i)n^2 + \frac{n(n+1)}{2} = M_n + \frac{n}{2}[n(n-2i+1)].$$

Now the difference between two of their elements vertically aligned is uniformly $n(n-2i+1)$, and the upper row is in deficit of $\frac{n}{2}$ times this quantity while its opposite has exactly that in excess; therefore the vertical exchange of $\frac{n}{2}$ elements between two opposite horizontal rows of the natural square will produce the magic sum in these two rows.

The same holds for the columns. Column C_j of the left-hand half contains the numbers

$$j, \ j+n, \ j+2n, \ \ldots, \ j+(n-1)n$$

with the sum

$$nj + n\frac{(n-1)n}{2} = \frac{n}{2}[2j+n^2-n]$$

$$= \frac{n}{2}[n^2 + 1 + 2j - n - 1] = M_n - \frac{n}{2}[n-2j+1],$$

and we have likewise for the numbers of C_{n-j+1}

$$n-j+1, \ (n-j+1)+n, \ \ldots, \ (n-j+1)+(n-1)n$$

with the sum

$$n(n-j+1) + n\frac{(n-1)n}{2} = M_n + \frac{n}{2}[n-2j+1].$$

The difference between two elements horizontally aligned being uniformly $n-2j+1$, and the deficit of a column and the excess of its conjugate being equal to $\frac{n}{2}$ times this quantity, we shall have equalized two columns by exchanging between them half their elements, just as before.

Application to the above methods

The principle of equalization is now clear, but its application requires some care: if we exchange half the elements between conjugate lines, opposite elements of the *columns* involved may no longer display the constant difference, and their horizontal exchange will then not lead to the desired equalization. This difficulty may be avoided in two ways.

1. Diagonal exchanges only

Consider four elements placed symmetrically relative to the two axes in the natural square (Fig. 102a). We first exchange the upper elements with the corresponding ones in the opposite horizontal row (Fig. 102b). We then exchange the left-hand elements with the corresponding ones in the opposite column (Fig. 102c). These four exchanges, two vertical and two horizontal, in fact correspond to a pair of diagonal exchanges, for in the end the four elements have been exchanged with their diagonally opposites. Doing this in the natural square, we shall have eliminated the part $\frac{4}{n}$ of the deficit and excess for the horizontal and vertical rows involved.

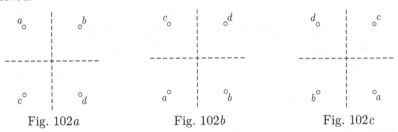

Fig. 102a Fig. 102b Fig. 102c

Therefore, operating $\frac{n}{4}$ such pairs of diagonal exchanges between each pair of opposite horizontal rows in an evenly-even-order natural square, that is, moving $\frac{n}{2}$ numbers in each horizontal row, thus $\frac{n^2}{2}$ numbers in the whole natural square, we shall have equalized completely and simultaneously the horizontal as well as vertical rows (the sums in the diagonals remaining unchanged). The square obtained will then be magic and also, because of the diagonal exchanges, symmetrical. That was the purpose of our $\frac{n^2}{2}$ dots to mark (say) the unchanged cells, while counting from the opposite corner enabled us to fill the still empty $\frac{n^2}{2}$ cells whilst operating the required diagonal exchanges.

2. With horizontal and vertical exchanges

Equalizing solely by means of diagonal exchanges is particularly easy, but not necessary. We may use horizontal and vertical exchanges, together with diagonal exchanges, as long as the fundamental principle (having $\frac{n}{2}$ exchanges of elements for each pair of opposite horizontal and vertical rows in the natural square) holds. But if the diagonals are involved, we must keep the central symmetry for the elements exchanged: in order to hold their sum, two complementary elements either are to remain in their diagonal or move together to the other diagonal.

Consider this for the natural square of order 8 (Fig. 103a). We imagine it divided into its 4×4 quadrants.

(a) Take, in the upper half, the lines of odd rank and exchange the median elements of the 4×4 squares with their opposite vertically in the lower half (Fig. 103b); these lines are thus equalized.

(b) In the columns already modified, exchange the remaining elements with their opposite horizontally (Fig. 103c); these columns are thus equalized.

1	2	3	4	5	6	7	8
9	10	11	12	13	14	15	16
17	18	19	20	21	22	23	24
25	26	27	28	29	30	31	32
33	34	35	36	37	38	39	40
41	42	43	44	45	46	47	48
49	50	51	52	53	54	55	56
57	58	59	60	61	62	63	64

Fig. 103a

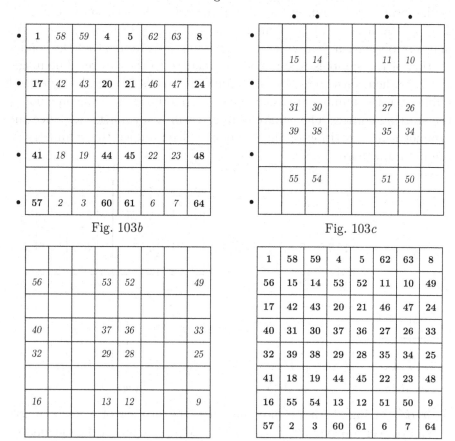

Fig. 103b

1	58	59	4	5	62	63	8
17	42	43	20	21	46	47	24
41	18	19	44	45	22	23	48
57	2	3	60	61	6	7	64

Fig. 103c

	15	14			11	10	
	31	30			27	26	
	39	38			35	34	
	55	54			51	50	

Fig. 103d

56			53	52			49
40			37	36			33
32			29	28			25
16			13	12			9

Fig. 103e

1	58	59	4	5	62	63	8
56	15	14	53	52	11	10	49
17	42	43	20	21	46	47	24
40	31	30	37	36	27	26	33
32	39	38	29	28	35	34	25
41	18	19	44	45	22	23	48
16	55	54	13	12	51	50	9
57	2	3	60	61	6	7	64

(c) We are left with equalizing the lines of the last exchanges and the columns left unmodified in the first exchanges; we shall do that simultaneously using diagonal exchanges (Fig. 103d). The resulting square (Fig. 103e) is magic and takes the form (particularly easy to remember) of the continuous filling (§ 11).

These various kinds of exchange may be extended to all other squares of evenly-even order, including the particular case of order 4 where there is just one row of odd rank in the upper half. These exchanges for the orders 4, 8, 12 are symbolized in the figures 104a-c, where • indicates that the numbers of the natural square keep their place, | means a vertical exchange between opposite lines, − stands for a horizontal exchange and × for a diagonal one. Thus the | and the × serve to equalize the lines, while − and × do so for the columns (× plays the two rôles since a diagonal exchange effects both a horizontal and vertical exchange). Therefore, in order for the lines to be equalized, each must contain altogether $\frac{n}{2}$ signs of the form | or ×, and, in the case of the columns, each must contain $\frac{n}{2}$ signs of the form − or ×. This is indeed the case in these figures. We may also observe, if we divide the squares into quadrants, that the configuration of the symbols is symmetrical relative to the centre. The magic square is then constructed by enumerating the cells from each corner and filling with the numbers reached the cells containing the same symbol as the starting corner. The square obtained will be symmetrical, since the symmetrical repartition of the symbols will reproduce the symmetry of the natural square.

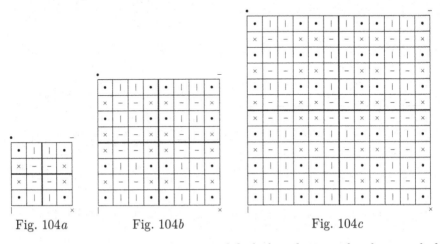

Fig. 104a Fig. 104b Fig. 104c

Remark. In Fig. 105a we have modified the placing of a few symbols and the central symmetry is lost in some places (not in the diagonals); the corresponding square (Fig. 105b) is magic, since the conditions of exchange are respected, but is no longer symmetrical where indicated (numbers underlined).

1	58	59	4	_61_	_6_	63	8
56	15	14	53	52	11	10	49
_41_	_18_	43	20	21	46	47	24
40	31	30	37	36	27	26	33
32	39	38	29	28	35	34	25
_17_	_42_	19	44	45	22	23	48
16	55	54	13	12	51	50	9
57	2	3	60	_5_	_62_	7	64

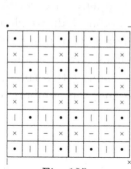

Fig. 105a   Fig. 105b

§ 13. Filling according to parity

The methods we have seen can be used to fill the squares not only with the sequence of numbers taken in natural order, but also with the sequence of even and odd numbers taken separately. The *Harmonious arrangement* has a whole section devoted to these cases.[62] We need only place the sequence of odd numbers, or even ones, just as before we had placed the natural sequence. Since the squares obtained are again symmetrical, the place of the even numbers (in particular 2, which starts the second enumeration) is immediately determined.

1	62	60	7	9	54	52	15
48	19	21	42	40	27	29	34
32	35	37	26	24	43	45	18
49	14	12	55	57	6	4	63
2	61	59	8	10	53	51	16
47	20	22	41	39	28	30	33
31	36	38	25	23	44	46	17
50	13	11	56	58	5	3	64

Fig. 106

1	62	60	7	9	54	52	15
48	19	21	42	40	27	29	34
33	30	28	39	41	22	20	47
16	51	53	10	8	59	61	2
63	4	6	57	55	12	14	49
18	45	43	24	26	37	35	32
31	36	38	25	23	44	46	17
50	13	11	56	58	5	3	64

Fig. 107

We recognize here the method of placing dots seen in Fig. 80 and 93 (Fig. 106 and 107); the configuration of Fig. 87 (Fig. 108); and the continuous placing which did not result from solely diagonal placings (Fig. 109). As for the placing of even numbers first, that will merely invert the

---

[62] *Un traité médiéval*, pp. 59-62 of the translation, lines 402-434 of the Arabic text.

two halves of the square (Fig. 109 & Fig. 110). We may also observe that with the methods of enumerating the cells from two opposite corners (Fig. 106-108), we shall end in the median line of the square, after writing half the numbers of the parity considered. For the second half, we shall start from the corner opposite, and resume the enumeration but filling this time the cells of the other type (thus without dots if we have previously filled those with dots); indeed, the cells to be filled cannot this time display central symmetry since this would mean repeating the numbers already written.

1	3	60	58	56	54	13	15
17	19	44	42	40	38	29	31
32	30	37	39	41	43	20	18
16	14	53	55	57	59	4	2
63	61	6	8	10	12	51	49
47	45	22	24	26	28	35	33
34	36	27	25	23	21	46	48
50	52	11	9	7	5	62	64

Fig. 108

1	52	54	7	9	60	62	15
48	29	27	42	40	21	19	34
33	20	22	39	41	28	30	47
16	61	59	10	8	53	51	2
63	14	12	57	55	6	4	49
18	35	37	24	26	43	45	32
31	46	44	25	23	38	36	17
50	3	5	56	58	11	13	64

Fig. 109

2	51	53	8	10	59	61	16
47	30	28	41	39	22	20	33
34	19	21	40	42	27	29	48
15	62	60	9	7	54	52	1
64	13	11	58	56	5	3	50
17	36	38	23	25	44	46	31
32	45	43	26	24	37	35	18
49	4	6	55	57	12	14	63

Fig. 110

§ 14. An older method

Abū'l-Wafā' Būzjānī fills a square of order 8 using exchanges in the natural square, which he describes as follows (Fig. 111a and 111b).[63] We

---

[63] *Magic squares in the tenth century*, B.11.*ii-iii*; or below, Appendix 9. Our figures, though reading from left to right, still reflect the instructions given.

begin their displacement with the numbers in the square of 4. We move 20 to the cell of 21, 21 to the cell of 44, 44 to the cell of 45, and 45 to the cell of 20. Next, we move likewise 35 to the cell of 27, 27 to the cell of 38, 38 to the cell of 30, and 30 to the cell of 35. Next, we move likewise 12 to the cell of 53, 53 to the cell of 13, 13 to the cell of 52, and 52 to the cell of 12. We also move 26 to the cell of 31, 31 to the cell of 34, 34 to the cell of 39, and 39 to the cell of 26. Next, we move 11 to the cell of 54, 54 to the cell of 51, 51 to the cell of 14, and 14 to the cell of 11, and we move 18 to the cell of 42, 42 to the cell of 23, 23 to the cell of 47, and 47 to the cell of 18. Then the quantities of the square of 6 within the square of 8 are all the same, the sum of the content of each row being 195. (...)

1	2	3	4	5	6	7	8
9	10	11	12	13	14	15	16
17	18	19	20	21	22	23	24
25	26	27	28	29	30	31	32
33	34	35	36	37	38	39	40
41	42	43	44	45	46	47	48
49	50	51	52	53	54	55	56
57	58	59	60	61	62	63	64

Fig. 111a

1	58	59	5	4	62	63	8
16	10	14	52	53	51	15	49
24	47	19	45	20	22	42	41
33	39	35	28	29	38	26	32
25	31	30	36	37	27	34	40
48	18	43	21	44	46	23	17
56	50	54	13	12	11	55	9
57	7	6	60	61	3	2	64

Fig. 111b

Next, we move 5 to the cell of 4 and 4 to the cell of 5, (but) leave 60 and 61 in their places. Next we move 2 to the cell of 63, 63 to the cell of 7, 7 to the cell of 58, and 58 to the cell of 2. Next we move, by analogy, 3 to the cell of 62, 62 to the cell of 6, 6 to the cell of 59, and 59 to the cell of 3. Then the sum in all inner vertical rows is the same, and in accordance with the quantity to be found in each row, namely 260, and the first and eighth horizontal rows also make the same sum, (for) they contain the quantity 260. Next, we move 25 to the cell of 33 and 33 to the cell of 25, (but) leave 32 and 40 in their cells. Next we displace the (contents of the) cells of the first and eighth vertical rows by analogy with the displacement performed for the (contents of the) cells in the first and eighth horizontal rows. For we move 49 to the cell of 16, 16 to the cell of 9, 9 to the cell of 56, and 56 to the cell of 49. Next, we move likewise 41 to the cell of 24, 24 to the cell of 17, 17 to the cell of 48, and 48 to the cell of 41. Then the quantities of all rows of this square are equalized.

All these moves are represented in Fig. 112.

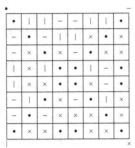

Fig. 112

Remarks.

(1) It appears from the text that the moves in the squares of 4, 6 and 8
are performed separately and independently; Abū'l-Wafā' repeats for
the first two squares the instructions given when he considered these
orders (our figures 113*a* and 113*b* represent his construction for order
6, the inner part of which is filled according to what he has taught for
order 4).

(2) His square of order 8 is not symmetrical, and its author is evidently
unaware of the use of diagonal exchanges, which is so easy and characteristic for evenly-even orders. He does not make a distinction here
between the two types of even order, and, as a matter of fact, he just
surrounds his 6 × 6 square with a border. Thus he does not know any
general treatment for ordinary magic squares of evenly-even orders,
and his example cannot be extended to higher orders.

1	5	33	34	32	6
30	8	28	9	11	25
24	20	15	16	23	13
18	17	21	22	14	19
7	26	10	27	29	12
31	35	4	3	2	36

Fig. 113*a*

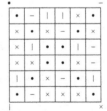

Fig. 113*b*

§ 15. Crossing the quadrants

The general methods seen so far were easy to apply and to memorize.
This is not always the case, and more complicated methods are seen being
used in later times, along with the others. Such is the method reported
by Shabrāmallisī for the order $n = 4k$, here $n = 12$.[64]

---

[64] MS. Paris BNF Arabe 2698, fol. 34ᵛ - 36ʳ.

Place (Fig. 114) 1 in the first cell and descend cell by cell diagonally towards the left (thus jumping to the opposite column of the square); when reaching the median line, jump to the same cell in the upper row of the diagonally opposed quadrant (thus, in this case, the second of its upper row), and proceed diagonally as before, but changing direction. We repeat the same movement for the three (generally k) sequences of twelve (respectively n) numbers, starting in the same quadrant but shifting each time the point of departure by two cells. We shall thus have placed the numbers 1 to 36 —generally a quarter of the numbers to be placed. Putting now 37 in the second bottom cell of the third quadrant, we go up diagonally towards the right, then towards the left after jumping to the upper half and starting in the first bottom cell of the diagonally opposite quadrant. Doing the same with the next two sequences, omitting one cell for the new departure, we shall have placed the numbers to 72 ($= \frac{n^2}{2}$). The third group of k sequences (starting with 73, 85, 97 in Fig. 109) starts in the lower part of the second quadrant, goes up to the left and changes direction when resuming on the bottom of the opposite quadrant, which is the fourth. The last group (starting with 109 in Fig. 114) starts at the top of the fourth quadrant, in its second cell, proceeds down to the right, then to the left when resuming on the top of the quadrant opposite to that of the cell last reached.

1	78	13	90	25	102	115	48	127	60	139	72
71	14	77	26	89	116	101	128	47	140	59	2
15	70	27	76	117	88	129	100	141	46	3	58
57	28	69	118	75	130	87	142	99	4	45	16
29	56	119	68	131	74	143	86	5	98	17	44
43	120	55	132	67	144	73	6	85	18	97	30
84	7	96	19	108	31	42	109	54	121	66	133
134	95	8	107	20	41	32	53	110	65	122	83
94	135	106	9	40	21	52	33	64	111	82	123
124	105	136	39	10	51	22	63	34	81	112	93
104	125	38	137	50	11	62	23	80	35	92	113
114	37	126	49	138	61	12	79	24	91	36	103

Fig. 114

In short, we place each time a quarter of the numbers. We start for the first group from the first cell at the top of the first quadrant and proceed to the left; for the second, from the second cell at the bottom of the third quadrant and proceed to the right; for the third, from the first cell at the bottom of the second quadrant and proceed to the left; for the fourth, from the second cell at the top of the fourth quadrant and proceed to the right. By the way, the initial direction of the movement is easy to remember: if we start at the top of a quadrant, we first go down away from the horizontally adjacent quadrant, if at the bottom we first go up towards the horizontally adjacent quadrant.

1	**136**	17	152	33	168	49	184	**201**	**80**	217	96	233	112	249	128
127	18	**135**	34	151	50	167	**202**	183	218	**79**	234	95	250	111	2
19	126	35	**134**	51	150	**203**	166	219	182	235	**78**	251	94	**3**	110
109	36	125	52	**133**	**204**	149	220	165	236	181	252	**77**	**4**	93	20
37	108	53	124	**205**	**132**	221	148	237	164	253	180	**5**	**76**	21	92
91	54	107	**206**	123	222	**131**	238	147	254	163	**6**	179	22	**75**	38
55	90	**207**	106	223	122	239	**130**	255	146	**7**	162	23	178	39	**74**
73	**208**	89	224	105	240	121	256	**129**	**8**	145	24	161	40	177	56
144	**9**	160	25	176	41	192	57	**72**	**193**	88	209	104	225	120	241
242	159	**10**	175	26	191	42	**71**	58	87	**194**	103	210	119	226	143
158	243	174	**11**	190	27	**70**	43	86	59	102	**195**	118	211	**142**	227
228	173	244	189	**12**	**69**	28	85	44	101	60	117	**196**	**141**	212	157
172	229	188	245	**68**	**13**	84	29	100	45	116	61	**140**	**197**	156	213
214	187	230	**67**	246	83	**14**	99	30	115	46	**139**	62	155	**198**	171
186	215	**66**	231	82	247	98	**15**	114	31	**138**	47	154	63	170	199
200	**65**	216	81	232	97	248	113	**16**	**137**	32	153	48	169	64	185

Fig. 115

We have represented a square of order 16 in order to better illustrate this method (Fig. 115); the first sequences of the four groups (beginning with 1, 65, 129, 193) are in bold.

Remark. As a matter of fact, one needs only to remember how to place the first set: those of the second, which are their complements to $\frac{n^2}{2}+1$, will be in the horizontally opposite cells; those of the fourth and third sets are, respectively, their complements to $n^2 + 1$ and will be placed in the diagonally opposite cells of the same quadrant.

0,1	6,6	1,1	7,6	2,1	8,6	9,7	3,12	10,7	4,12	11,7	5,12
5,11	1,2	6,5	2,2	7,5	9,8	8,5	10,8	3,11	11,8	4,11	0,2
1,3	5,10	2,3	6,4	9,9	7,4	10,9	8,4	11,9	3,10	0,3	4,10
4,9	2,4	5,9	9,10	6,3	10,10	7,3	11,10	8,3	0,4	3,9	1,4
2,5	4,8	9,11	5,8	10,11	6,2	11,11	7,2	0,5	8,2	1,5	3,8
3,7	9,12	4,7	10,12	5,7	11,12	6,1	0,6	7,1	1,6	8,1	2,6
6,12	0,7	7,12	1,7	8,12	2,7	3,6	9,1	4,6	10,1	5,6	11,1
11,2	7,11	0,8	8,11	1,8	3,5	2,8	4,5	9,2	5,5	10,2	6,11
7,10	11,3	8,10	0,9	3,4	1,9	4,4	2,9	5,4	9,3	6,10	10,3
10,4	8,9	11,4	3,3	0,10	4,3	1,10	5,3	2,10	6,9	9,4	7,9
8,8	10,5	3,2	11,5	4,2	0,11	5,2	1,11	6,8	2,11	7,8	9,5
9,6	3,1	10,6	4,1	11,6	5,1	0,12	6,7	1,12	7,7	2,12	8,7

Fig. 116

This method may not be easy to remember, but its magic property is easy to explain. Each column and each main diagonal contains exactly once each unit from 1 to n and each multiple of n, from $0 \cdot n$ to $(n-1)n$ (Fig. 116, corresponding to the square of Fig. 114). From the above *remark* we infer that each line consists of $2k$ pairs of elements placed symmetrically, k pairs adding up to $\frac{n^2}{2} + 1$ and k pairs to $\frac{3n^2}{2} + 1$, so on the whole $k \left(2n^2 + 2\right) = \frac{n(n^2+1)}{2}$.

§ 16. Descent by the knight's move

1. Square of order 4

Just as for odd-order squares, we may use the knight's move to construct evenly-even squares. Some of these, too, will be pandiagonal.

Let us examine the two squares of order 4 found in the figures 117a and 118a, both of which are pandiagonal (the first has already been given as example, see Fig. 78). Their method of construction is similar: descending first by the knight's move, then moving towards or away from the side by a queen's move, then resuming with the knight's move. With 8 $(= \frac{n^2}{2})$ we descend symmetrically with decreasing numbers. It then appears that each line contains the same sum, 9, and two alternate columns also the same sum, 7 or 11. It also appears that each bishop's cell of a cell filled has remained empty (Fig. 117b and 118b). We shall fill them in such a way that the sum of two conjugate cells —that is, joined by a bishop's move— makes $n^2 + 1 = 17$. Since all lines contain, after the preliminary filling, the same sum, the complements of any line will produce the magic sum in any other, in particular in its conjugate. As for the columns, the sums in them are not always the same but they are in two conjugate columns, so when filled they will display the magic sum. The case of the diagonals is clear since they contain pairs of complements.

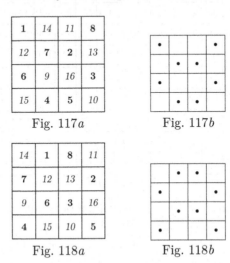

1	14	11	8
12	7	2	13
6	9	16	3
15	4	5	10

Fig. 117a

Fig. 117b

14	1	8	11
7	12	13	2
9	6	3	16
4	15	10	5

Fig. 118a

Fig. 118b

The same holds for the square of Fig. 119a and 119b, filled by the same moves, and with the second sequence in adjacent cells: after the preliminary filling, the sums in lines and in conjugate columns are the same, and the bishop's cells are left empty for complements.

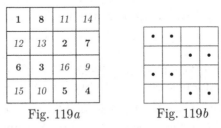

1	8	11	14
12	13	2	7
6	3	16	9
15	10	5	4

Fig. 119a

Fig. 119b

All subsequent fillings will rely on the same principle. One should also keep in mind that the choice of the point of departure is arbitrary provided that the moves from it are as above. Indeed, the resulting square is pandiagonal, thus not affected by any cyclical permutation of rows (see the figures 5 and 6, p. 4). And with such permutations formerly conjugate columns remain conjugate.

Remark. These three squares differ from one another only by the position of the columns, for median ones become lateral or conversely, in such a way that conjugate columns remain conjugate.

2. Squares of higher orders

This type of construction may be generalized as follows. Take an empty square of order $n = 4k$, and divide it into subsquares of order 4. As before, start in the first horizontal row and choose in each subsquare two like cells, either median or lateral. They will be the points of departure for $\frac{n}{2}$ continuous sequences of n numbers covering the first half of the numbers to be placed, of which the first $\frac{n}{4}$ sequences will be ascending and the last $\frac{n}{4}$ descending (Fig. 120). The direction of the movement is easy to determine, for the first move is a knight's and it must be in the same subsquare as the starting cell. Then, descend through the given square by knight's moves, except near its border, which is either reached by a queen's move if we are one cell away from it (case of departure from a lateral cell), or from which we move away with such a move if we attained it by a knight's move (departure from a median cell). With this last step the sequences are distributed in a regular manner in all columns. In Fig. 121, there will be $\frac{n}{4} = 3$ ascending sequences and as many descending ones, namely from 1 to 12, 13 to 24, 25 to 36, and 48 to 37, 60 to 49, 72 to 61. As in the previous example, the points of departure of opposite sequences are located in cells placed symmetrically in the top line.

	1	9			24	32	
10		2	31				23
	11	30			3	22	
29			12	21			4
	28	20			13	5	
19			27	6			14
	18	7			26	15	
8			17	16			25

Fig. 120

35	1	9	54	43	24	32	62
10	53	36	2	31	61	44	23
56	11	30	64	33	3	22	41
29	63	55	12	21	42	34	4
58	28	20	47	50	13	5	39
19	48	57	27	6	40	49	14
45	18	7	37	60	26	15	52
8	38	46	17	16	51	59	25

Fig. 122

The remaining cells are easy to fill. According to the configuration, each bishop's cell of a cell filled within the same 4×4 square has remained empty. It will receive the complement to $n^2 + 1$ of the number already placed. Thus, in our examples (Fig. 120 and Fig. 121), if i occupies a cell, its bishop's cell in the same subsquare will contain $65 - i$ and $145 - i$, respectively. This will give the squares of Fig. 122 and 123.

The preliminary filling of half the cells leads to the same situation as for the above 4×4 squares. In the half-filled square, the lines contain

$\frac{n}{4}$ pairs making each the same sum as the first two terms of symmetrically placed sequences; this sum being $\frac{n^2}{2}+1$, each line contains the sum $\frac{n}{4}(\frac{n^2}{2}+1)$. The sum in the columns is not uniformly the same, but it is in conjugate columns, thus those separated by one column and within the same vertical strip of 4×4 squares; indeed, because of the knight's move and initial fixing of direction, two conjugate columns contain $\frac{n}{2}$ pairs

1			13	25			48	60			72
	14	2			47	26			71	59	
15			46	3			70	27			58
	45	16			69	4			57	28	
44			68	17			56	5			29
	67	43			55	18			30	6	
66			54	42			31	19			7
	53	65			32	41			8	20	
52			33	64			9	40			21
	34	51			10	63			22	39	
35			11	50			23	62			38
	12	36			24	49			37	61	

Fig. 121

1	99	130	13	25	75	142	48	60	87	118	72
129	14	2	100	141	47	26	76	117	71	59	88
15	132	144	46	3	97	120	70	27	73	85	58
143	45	16	131	119	69	4	98	86	57	28	74
44	91	79	68	17	114	103	56	5	138	126	29
80	67	43	92	104	55	18	113	125	30	6	137
66	77	101	54	42	89	128	31	19	116	140	7
102	53	65	78	127	32	41	90	139	8	20	115
52	134	110	33	64	122	95	9	40	107	83	21
109	34	51	133	96	10	63	121	84	22	39	108
35	112	93	11	50	136	81	23	62	124	105	38
94	12	36	111	82	24	49	135	106	37	61	123

Fig. 123

of consecutive numbers, half in ascending and half in descending order, so that one of these columns displays, relative to the other, $\frac{n}{4}$ times an excess of $+1$ and $\frac{n}{4}$ times a deficit of -1. The two characteristics of the initial placing, namely equality of the sum in each line and equality of the sum in conjugate columns, may be used to verify that it has been done correctly.

Supposing thus $\frac{n}{2}$ numbers α_i to have been initially placed in a row, line or column, and $\frac{n}{2}$ numbers β_i in the conjugate row, we have $\sum \alpha_i = \sum \beta_i$, and with the $\frac{n}{2}$ complements in place, the conjugate row will contain

$$\sum \beta_i + \sum [(n^2 + 1) - \alpha_i] = \frac{n}{2}(n^2 + 1),$$

which is the required sum. As for the main diagonals, it is evident that they contain the magic sum since they consist in the diagonals of $\frac{n}{4}$ sub-squares containing each two pairs of complements, thus $2(n^2 + 1)$.

This method was known in the first half of the 11th century; it is described in the *Brief treatise*, where the author seems to take credit for its discovery. In any case he must have recognized the origin of the magic property since he mentions the above criterion for the correctness of the initial placing. Here are his explanations, with the departures in the upper row as in our Fig. 121.[65]

You put the small extreme and the small middle in the two ends of the upper row, and the two subsequent —ascending (in the numbers) from the extreme and descending from the middle— in the two corresponding knight's cells of the second row, then of the third row, then of the fourth row (and so on). When the passage from one row to that underneath is not possible with the knight's move, you pass from one to the other with the queen's move. Then you return to the knight's move by inverting the oblique direction, (thus passing) from the right (direction) to the left and from the left to the right. Having reached with these two (movements) the last row, you jump in both cases to the first row, (namely to) the second diagonal of the square of 4 the first diagonal of which contained the point of departure.[66] You proceed with these two (movements), following the oblique which is inverse to that of the first point of departure, to the last

---

[65] *L'Abrégé*, pp. 118-121 & 125-126, p. 149 lines 4-18 & p. 150 lines 11-16 of the Arabic text; or below, Appendix 10. The vocabulary needs no explanation except perhaps for *extreme terms* and *middle terms*: of n^2 numbers to be placed, 1 and n^2 are the extremes, while $\frac{n^2}{2}$ and $\frac{n^2}{2} + 1$ are the middle terms of the sequence. These latter numbers separate the 'small numbers' (from 1 to $\frac{n^2}{2}$) from their complements, which form the 'large numbers' ($\frac{n^2}{2} + 1$ to n^2). Consequently, 1 is the 'small extreme' and $\frac{n^2}{2}$ the 'small middle'. Words in brackets are our additions.

[66] Here and in what follows the 'first' diagonal is that of the initial movement, thus

row. *If the small numbers are completed (case of the 8 × 8 square), that
is it. Otherwise, you return for both to the first diagonal of the second
square of 4 —that following the first square (of 4)— and you proceed with
both (movements) following the first oblique (direction), to the last row.
If the small numbers are completed (case of the 12 × 12 square), that is
it. Otherwise, you return for both (movements) to the second diagonal of
the second square of 4 (...). You continue in this way until you are done
with half the square's cells. The (number of these pairs of) jumps from
the bottom to the top will grow (by one) with each increment of the size
of the square by four cells. Afterwards, you will place the complements
in the corresponding bishop's cells of the (corresponding) squares of 4.
So the magic square will be completed.*

118	1	48	99	87	72	25	142	130	13	60	75
47	100	117	2	26	141	88	71	59	76	129	14
97	46	27	144	120	3	58	73	85	70	15	132
28	143	98	45	57	74	119	4	16	131	86	69
126	29	56	91	79	44	17	114	138	5	68	103
55	92	125	30	18	113	80	43	67	104	137	6
89	54	19	116	128	31	66	101	77	42	7	140
20	115	90	53	65	102	127	32	8	139	78	41
134	21	64	83	107	52	9	122	110	33	40	95
63	84	133	22	10	121	108	51	39	96	109	34
81	62	11	124	136	23	38	93	105	50	35	112
12	123	82	61	37	94	135	24	36	111	106	49

Fig. 124

This method may be generalized in the sense that the initial row may
be any one, and the starting cells either lateral or median; furthermore,
the pairs adding up to $\frac{n^2}{2} + 1$ do not need to occupy cells placed symmetrically. But the manner of placing the $\frac{n^2}{2}$ first numbers and their
complements follows the previous rules (Fig. 124-125). Our source, the
Brief treatise, explains a few such cases. It closes the subject with the
aforementioned criterion of validity for the preliminary placement: *When
you have placed the small numbers and wish to ascertain that the plac-*

(for us) the descending diagonal for the increasing sequence and the ascending diagonal
for the decreasing sequence.

ing of the small numbers contains no error, so as to place with certainty their complementary large numbers in the corresponding bishop's cells, consider the horizontal and vertical rows. If the (sums in all) horizontal rows are equal to one another, and if (the sum in) each vertical row is equal to (that of) the (vertical) row linked with a bishop's move, then the placing of the small numbers is correct, otherwise not.

89	46	27	116	128	3	58	101	77	70	15	140
28	115	90	45	57	102	127	4	16	139	78	69
118	29	56	99	87	44	17	142	130	5	68	75
55	100	117	30	18	141	88	43	67	76	129	6
81	54	19	124	136	31	66	93	105	42	7	112
20	123	82	53	65	94	135	32	8	111	106	41
126	21	64	91	79	52	9	114	138	33	40	103
63	92	125	22	10	113	80	51	39	104	137	34
97	62	11	144	120	23	38	73	85	50	35	132
12	143	98	61	37	74	119	24	36	131	86	49
134	1	48	83	107	72	25	122	110	13	60	95
47	84	133	2	26	121	108	71	59	96	109	14

Fig. 125

§ 17. Filling pairs of horizontal rows

Instead of descending with the knight's move from line to line, as in the previous method, we may remain in a pair of lines, and use the queen's move in passing from one pair of lines to the next, which also marks the change in direction (Fig. 126). Having arrived at the bottom of the square —or at the line preceding the initial one if the initial line was not the first— we shall start the inverse, upward movement from the cell horizontally symmetrical to that reached. (We may also start simultaneously from 1 with the ascending sequence and from $\frac{n^2}{2}$ with the descending sequence, as in the previous method.) We shall thus have placed half the numbers. The second half will be placed by putting, as before, the complements in the bishop's cells in each subsquare (Fig. 127).

The origin of the magic property is the same as before: each line contains, before placing of the complements, pairs of numbers adding up to $\frac{n^2}{2} + 1$, thus $\frac{n}{4}\left(\frac{n^2}{2} + 1\right)$ altogether, and two conjugate columns contain the same sum; the complements will make the magic sum. Here too, the diagonals contain pairs of complements.

1			68	3			70	5			72
	67	2			69	4			71	6	
66			11	64			9	62			7
	12	65			10	63			8	61	
13			56	15			58	17			60
	55	14			57	16			59	18	
54			23	52			21	50			19
	24	53			22	51			20	49	
25			44	27			46	29			48
	43	26			45	28			47	30	
42			35	40			33	38			31
	36	41			34	39			32	37	

Fig. 126

1	134	79	68	3	136	81	70	5	138	83	72
80	67	2	133	82	69	4	135	84	71	6	137
66	77	144	11	64	75	142	9	62	73	140	7
143	12	65	78	141	10	63	76	139	8	61	74
13	122	91	56	15	124	93	58	17	126	95	60
92	55	14	121	94	57	16	123	96	59	18	125
54	89	132	23	52	87	130	21	50	85	128	19
131	24	53	90	129	22	51	88	127	20	49	86
25	110	103	44	27	112	105	46	29	114	107	48
104	43	26	109	106	45	28	111	108	47	30	113
42	101	120	35	40	99	118	33	38	97	116	31
119	36	41	102	117	34	39	100	115	32	37	98

Fig. 127

This elegant method is reported by S̲h̲abrāmallisī, but it is much older since it is already found in the *Harmonious arrangement*.[67] This earlier

---

[67] MS. Paris BNF Arabe 2698, fol. 30ᵛ - 31ᵛ & 32ᵛ; *Un traité médiéval*, pp. 44-47 in the translation, lines 261-291 of the Arabic text.

text considers the various places the initial cell may occupy in the first row. That is a banal consequence of the pandiagonality of the squares thus formed; but, as already said (p. 3), it was mainly the possibility of choosing any point of departure which seems to have been of interest at that time.

Our figure 4 (p. 3), found in one version of the 11th-century astronomer Zarqālī's text,[68] results from a less obvious placing, but which follows the same principle (equality of the sums in the lines and the conjugate columns before insertion of the complements): the pairs of associated horizontal rows are, successively, 5–2, 7–4, 1–6, 3–8; the sequences are placed from right to left with the knight's move (1 to 16), then in inverse order and direction (17 to 32), whereby in the second half pairs of numbers in conjugate columns differ by 3, and the pairs i and $33 - i$ are put side by side as in the subsequent method (§ 18). Indeed, we may at will vary the point of departure, even if texts usually set the starting cell in the top row. For, as the author of the *Harmonious arrangement* puts it when dealing with this method and starting in the first line, *it naturally comes to mind to associate the first row with the second and to move between them with the knight's move, as we have seen. It will not immediately come to mind to associate the first row with the last and imagine them placed side by side, to move between them with the knight's move, then to associate the row following the first with that preceding the last, using the knight's move to move between them and the queen's move to pass from each pair of rows to the following pair. (...) If you begin in the second row, it will naturally come to mind to associate it with the third one. It will not immediately come to mind to associate it with the first row, to move between them by the knight's move, then to associate the fourth with the third, and so on, passing from one pair to the other —imagining the row of departure and the row of arrival as associated— by the queen's move.*[69]

§ 18. Four knight's routes

Shabrāmallisī calls this method 'knight's move in four cycles (*adwār*)'; indeed, here we place each group of $\frac{n^2}{4}$ numbers by repeated knight's moves.[70] We begin by placing the first group by putting, as before, $\frac{n}{2}$ numbers in the first pair of rows; we do the same for each next pair of

---

[68] MS. London BL Arabic 977, fol. 133^r - 145^v.

[69] *Un traité médiéval*, pp. 64-66 of the translation, lines 451-462 of the Arabic text; or below, Appendix 11.

[70] MS. Paris BNF Arabe 2698, fol. 25^v - 26^v.

rows, progressing in the same direction but beginning alternately in the first or second cell (from 1 to 4 in Fig. 128, from 1 to 16 in Fig. 129, from 1 to 36 in Fig. 130).

1	32	47	50	3	30	45	52
48	49	2	31	46	51	4	29
28	5	54	43	26	7	56	41
53	44	27	6	55	42	25	8
9	24	39	58	11	22	37	60
40	57	10	23	38	59	12	21
20	13	62	35	18	15	64	33
61	36	19	14	63	34	17	16

1	8	11	14
12	13	2	7
6	3	16	9
15	10	5	4

Fig. 128   Fig. 129

1	72	107	110	3	70	105	112	5	68	103	114
108	109	2	71	106	111	4	69	104	113	6	67
66	7	116	101	64	9	118	99	62	11	120	97
115	102	65	8	117	100	63	10	119	98	61	12
13	60	95	122	15	58	93	124	17	56	91	126
96	121	14	59	94	123	16	57	92	125	18	55
54	19	128	89	52	21	130	87	50	23	132	85
127	90	53	20	129	88	51	22	131	86	49	24
25	48	83	134	27	46	81	136	29	44	79	138
84	133	26	47	82	135	28	45	80	137	30	43
42	31	140	77	40	33	142	75	38	35	144	73
139	78	41	32	141	76	39	34	143	74	37	36

Fig. 130

This being done, we place the second sequence of $\frac{n^2}{4}$ numbers starting from the cell next to that last reached, then going in the opposite direction and upwards, beginning alternately in the first or second cell, as before. There is a break at the end of the second movement: the first number of the third cycle must be written at the other end of the broken diagonal last reached, thus above the last cell (9 in Fig. 128, 33 in Fig. 129, 73 in Fig. 130); we shall place (again with the knight's move) the numbers of the third cycle going up the pairs of rows. From the cell filled with the last number of the third cycle, which is the first cell of the second line, we shall shift to the next horizontally. The last quarter of numbers will be placed as before, but this time moving to the right and downwards. The square thus constructed is pandiagonal.

The magic property of the horizontal rows is justified as in the previous method: indeed, the $\frac{n^2}{2}$ elements put after the first two cycles are the same, merely arranged differently. Consequently, some of the columns have changed their position: the odd-rank columns are the same, whereas the even-rank columns $2i$ and $n - 2i + 2$ have exchanged their places. As a consequence, the sums in all odd-rank columns are the same after the first two cycles, and the same holds for the even-rank ones (with another sum). Then, the numbers placed during the last two cycles complete them, and the same applies to the diagonals.[71]

§ 19. Knight's shuttle

This time we shall need only two cycles, as mentioned by Shabrāmallisī, who presents this method next. See Fig. 131 and Fig. 132.[72]

It is possible to fill a magic square of this kind (= the evenly-even order) by means of two cycles only. This is an elegant method, which proceeds as follows. You put 1 in one of the square's corners, and advance by the knight's move, (putting) the numbers one after the other in the cells in succession in the row beginning at the corner and in the adjacent row until the movement is stopped. You put then the number following the one reached in the cell next to that reached and in the same row (…). Return then within the pair of rows, again by the knight's move, till the movement comes to an end. Put then the number following the one reached in the cell aligned with the cell reached, belonging to the first of the next two rows, adjacent to the two previous ones. Advance in them as in the previous pair back and forth till the end of the movement, after which you are to pass to the next two rows just as you did to these. In this way, in each pair of rows the return begins in the cell next to that reached, alternately before and after. This is done until you reach the end of the figure. This concludes the first cycle and the filling of half the cells of the magic square; half the cells of each row and diagonal are filled, and the number reached equals half the number of cells of the square.

---

[71] A similar method, but with the second and fourth cycles starting in the vertical neighbour of the cell reached, in MS. London BL Delhi Arabic 110, fol. 54ᵛ.

[72] MS. Paris BNF Arabe 2698, fol. 27ʳ, 1 - 27ᵛ, 2; or below, Appendix 12. Several Persian manuscripts contain the same square as our figure 132 (Istanbul Ayasofya 2794, fol. 16ᵛ, Mashhad Āstān-i Quds 12235, p. 438, and 14159, fol. 86ʳ).

1	12	134	143	3	10	136	141	5	8	138	139
133	144	2	11	135	142	4	9	137	140	6	7
24	13	131	122	22	15	129	124	20	17	127	126
132	121	23	14	130	123	21	16	128	125	19	18
25	36	110	119	27	34	112	117	29	32	114	115
109	120	26	35	111	118	28	33	113	116	30	31
48	37	107	98	46	39	105	100	44	41	103	102
108	97	47	38	106	99	45	40	104	101	43	42
49	60	86	95	51	58	88	93	53	56	90	91
85	96	50	59	87	94	52	57	89	92	54	55
72	61	83	74	70	63	81	76	68	65	79	78
84	73	71	62	82	75	69	64	80	77	67	66

1	4	14	15
13	16	2	3
8	5	11	10
12	9	7	6

Fig. 131   Fig. 132

After that, you are to begin the placement of the second cycle by putting the next number in the queen's cell of the last number placed, within the pair of rows filled at the end of the first cycle. You are then to move in them as for the first movement, back and forth, until they are filled completely. Put then the number following the one reached in the cell aligned with it, belonging to the first of the next two rows adjacent to those we have completed the filling of, and advance in them in the same manner until they are completed. Pass then from them to the next pair, and so on until the magic square is completed.

This method is as elegant as its justification is easy: the diagonals of the magic square contain pairs of complements, while the same sum is found, after the first cycle, in all columns and in pairs of adjacent lines, each of which will receive the complements of the other. We may also observe that if we divide this square into subsquares of order 4, each will contain the same sum $2\left(n^2+1\right)$ in all its rows and diagonals, and this sum is also found in each of their own subsquares since their diagonals are occupied by pairs of complements.

Remark. If, after the first cycle, the complements are put in the bishop's cells of each 4×4 square, the resulting square will be pandiagonal.

§ 20. Filling by knight's and bishop's moves

Placing initially 1 and n^2 at the ends of (say) the first horizontal row, we remain in the first pair of rows where we place by knight's moves the numbers in succession, ascending and descending respectively. When we reach a side, thus after placing $\frac{n}{2}$ numbers for each sequence, we pass to the next pair of rows, and this time begin in the second cell of the upper row whilst keeping the direction of the two previous movements (as in the method of §18). We do the same with each pair of rows, starting alternately from the first and the second cell. We shall have thus placed the first $\frac{n^2}{4}$ numbers and the last $\frac{n^2}{4}$ ones (1 to 4 and 16 to 13 in Fig. 133, 1 to 36 and 144 to 109 in Fig. 134).

1			140	3			142	5			144
	139	2			141	4			143	6	
	7	134			9	136			11	138	
133			8	135			10	137			12
13			128	15			130	17			132
	127	14			129	16			131	18	
	19	122			21	124			23	126	
121			20	123			22	125			24
25			116	27			118	29			120
	115	26			117	28			119	30	
	31	110			33	112			35	114	
109			32	111			34	113			36

1			16
	15	2	
	3	14	
13			4

Fig. 133   Fig. 134

The second step consists in filling the broken diagonals with bishop's moves starting from the vertical sides. We first put the next small number (5 and 37, respectively) in the same column as the last small number, in the third cell from the top, and do the same for the next large number on the other side (thus 12 and 108). Then we start filling diagonally sequences of $\frac{n}{2}$ numbers, going up to the left for the ascending sequences, and to the right for the descending ones; see the arrows in Fig. 135 and Fig. 136, which indicate the beginning of the broken diagonal for each sequence, the first terms of which are, from top to bottom on the left sides, 12, 10 and 108, 102, 96, 90, 84, 78, and, on the right sides, again from top to bottom, 5, 7 and 37, 43, 49, 55, 61, 67). We shall have thus placed twice a further $\frac{n^2}{4}$ numbers, which completes the filling. Although elaborate, this method appears already in the *Harmonious arrangement*.[73]

---

[73] *Un traité médiéval*, pp. 66-67 of the translation, lines 466-491 of the Arabic text.

As with the method seen in § 15 (not easy to grasp and memorize either) we have added an example (Fig. 137).

1	66	107	140	3	52	93	142	5	38	79	144
72	139	2	101	58	141	4	87	44	143	6	73
108	7	134	65	94	9	136	51	80	11	138	37
133	102	71	8	135	88	57	10	137	74	43	12
13	42	95	128	15	64	81	130	17	50	103	132
48	127	14	89	70	129	16	75	56	131	18	97
96	19	122	41	82	21	124	63	104	23	126	49
121	90	47	20	123	76	69	22	125	98	55	24
25	54	83	116	27	40	105	118	29	62	91	120
60	115	26	77	46	117	28	99	68	119	30	85
84	31	110	53	106	33	112	39	92	35	114	61
109	78	59	32	111	100	45	34	113	86	67	36

1	6	11	16
8	15	2	9
12	3	14	5
13	10	7	4

Fig. 135

Fig. 136

The magic property of these squares may be explained as follows.

— The horizontal rows comprise pairs of complements, altogether the sum $\frac{n}{2}(n^2+1)$.

— For the columns, we may verify that in each the sum of the smaller numbers ($\leq \frac{n^2}{2}$) is the same, and so also for the sum of the larger numbers. This has to do with the regular distribution of the numbers. First, each of the n columns contains in its first four rows one, and only one, of the first n numbers. Let this smallest element i (thus $i = 1, \ldots, n$) be taken to designate its column. Then column i, when filled, will contain $\frac{n}{2}$ smaller numbers and $\frac{n}{2}$ larger numbers belonging to one of four arithmetical progressions to $\frac{n}{4}$ terms; namely, with t taking, in each column, the values $1, 2, \ldots, \frac{n}{4}$,

$$i + (t-1) \cdot n,$$

$$\left(\frac{n^2}{2} - i + 1\right) - (t-1) \cdot n,$$

for the smaller numbers, and, for the larger numbers,

$$\left(\frac{n^2}{2} + n - i + 1\right) + (t-1) \cdot n = \left(\frac{n^2}{2} - i + 1\right) + t \cdot n$$

$$\left(n^2 - n + i\right) - (t-1) \cdot n = \left(n^2 + i\right) - t \cdot n.$$

1	120	191	250	3	102	173	252	5	84	155	254	7	66	137	256
128	249	2	183	110	251	4	165	92	253	6	147	74	255	8	129
192	9	242	119	174	11	244	101	156	13	246	83	138	15	248	65
241	184	127	10	243	166	109	12	245	148	91	14	247	130	73	16
17	72	175	234	19	118	157	236	21	100	139	238	23	82	185	240
80	233	18	167	126	235	20	149	108	237	22	131	90	239	24	177
176	25	226	71	158	27	228	117	140	29	230	99	186	31	232	81
225	168	79	26	227	150	125	28	229	132	107	30	231	178	89	32
33	88	159	218	35	70	141	220	37	116	187	222	39	98	169	224
96	217	34	151	78	219	36	133	124	221	38	179	106	223	40	161
160	41	210	87	142	43	212	69	188	45	214	115	170	47	216	97
209	152	95	42	211	134	77	44	213	180	123	46	215	162	105	48
49	104	143	202	51	86	189	204	53	68	171	206	55	114	153	208
112	201	50	135	94	203	52	181	76	205	54	163	122	207	56	145
144	57	194	103	190	59	196	85	172	61	198	67	154	63	200	113
193	136	111	58	195	182	93	60	197	164	75	62	199	146	121	64

Fig. 137

In each column i there are thus $\frac{n}{4}$ pairs of smaller numbers, altogether adding up to $\frac{n}{4}\left(\frac{n^2}{2}+1\right)$, and larger numbers adding up to the complementary quantity $\frac{n}{4}\left(\frac{3n^2}{2}+1\right)$.

— The descending diagonal is made up of numbers of the form

$$1 + s\left(\frac{n}{2}+1\right)$$
$$n^2 - s'\left(\frac{n}{2}-1\right)$$

with $s = 0, 1, \ldots, \frac{n}{2}-1$ and $s' = 1, 2, \ldots, \frac{n}{2}$; the sum of these n numbers will then be

$$\frac{n}{2}\left[1 + \left(\frac{n}{2}-1\right)\left(\frac{n}{2}+1\right)\right] + \frac{n}{2}\left[n^2 - \left(\frac{n}{2}+1\right)\left(\frac{n}{2}-1\right)\right] = \frac{n}{2}(n^2+1).$$

— Since the other diagonal contains their complements, it too gives the magic sum.

§ 21. Filling according to parity

We have seen that the simple filling methods made it possible to place numbers by parity (§ 13). This may also be done with constructions using the knight's move. Thus we recognize the method of § 16 in Fig. 138, of

§ 17 in Fig. 139 and 140, the method of § 18 in Fig. 141 (the third of the four cycles starts placing the even numbers), the method of § 19 in Fig. 142 (the second of the two cycles places the numbers of the other parity), finally the method of § 20 in Fig. 143 (completing first the placement of the odd numbers in the broken diagonals from the right-hand side and writing then the even numbers as complements to n^2+1 in the horizontally symmetrical cells).

1	6	44	17	47	22	60	63
42	19	3	8	58	61	45	24
21	48	64	59	5	2	18	43
62	57	23	46	20	41	7	4
55	52	30	39	25	36	14	9
32	37	53	50	16	11	27	34
35	26	10	13	51	56	40	29
12	15	33	28	38	31	49	54

Fig. 138

1	52	10	59	5	56	14	63
12	57	3	50	16	61	7	54
55	6	64	13	51	2	60	9
62	15	53	8	58	11	49	4
17	36	26	43	21	40	30	47
28	41	19	34	32	45	23	38
39	22	48	29	35	18	44	25
46	31	37	24	42	27	33	20

Fig. 139

2	51	9	60	6	55	13	64
11	58	4	49	15	62	8	53
56	5	63	14	52	1	59	10
61	16	54	7	57	12	50	3
18	35	25	44	22	39	29	48
27	42	20	33	31	46	24	37
40	21	47	30	36	17	43	26
45	32	38	23	41	28	34	19

Fig. 140

1	63	30	36	5	59	26	40
32	34	3	61	28	38	7	57
55	9	44	22	51	13	48	18
42	24	53	11	46	20	49	15
17	47	14	52	21	43	10	56
16	50	19	45	12	54	23	41
39	25	60	6	35	29	64	2
58	8	37	27	62	4	33	31

Fig. 141

These placings preserve the original features: the squares of the figures 138 to 140 are pandiagonal, in Fig. 142 the rows of the 4×4 squares and their quadrants make the sum $2(n^2 + 1)$.

The *Harmonious arrangement* presents the examples of our figures 139 and 140. In § 17 we said that it considered various positions for the initial cell; the author finds it unnecessary to repeat that for placing by parity, since he concludes by saying:[74] *All the methods we have explained*

---

[74] *Un traité médiéval*, pp. 64 and 68, lines 446-448 and 497-498 of the Arabic text;

for placing consecutive numbers in this way apply here for placing the odd numbers and the complementary even numbers in the bishop's cells, and conversely, without any difference; thus there is no purpose in protracting (the discourse about it in) this book. The other example he constructs is that of our figure 143. Having completed it, he observes that the quadrants display a particular arrangement: *A careful look will enable you to note that the odd numbers are separated from the even ones inside each 4 × 4 square, the position of their cells being like a hexagon having two of its opposite sides elongated* (see the numbers underlined by us). This seemingly insignificant remark may remind us of the importance attached at that time to the separation of numbers by parity, either whilst placing them or in the result.[75]

1	15	52	62	5	11	56	58
50	64	3	13	54	60	7	9
31	17	46	36	27	21	42	40
48	34	29	19	44	38	25	23
33	47	20	30	37	43	24	26
18	32	35	45	22	28	39	41
63	49	14	4	59	53	10	8
16	2	61	51	12	6	57	55

Fig. 142

1	55	_30_	_60_	5	35	10	64
63	_58_	3	_22_	43	62	7	2
_32_	9	_52_	53	12	13	56	33
_50_	_24_	61	11	54	4	41	15
17	39	14	44	21	51	26	48
47	42	19	6	59	46	23	18
16	25	36	37	28	29	40	49
34	8	45	27	38	20	57	31

Fig. 143

§ 22. Filling the subsquares of order 4

Let us divide the square of order $n = 4k$ into subsquares of order 4. We have already seen easy methods for constructing a pandiagonal square of order 4 in § 16 (Fig. 117a-119a, p. 64). Using such squares of order 4, we can obtain, for any square of larger order $n = 4k$, magic squares. What they will all have in common is that we first fill half the cells in each subsquare, thus placing the smaller numbers $1, \ldots, \frac{n^2}{2}$, and then fill their bishop's cells with the complements to $n^2 + 1$. According to how the eight smaller numbers are placed in the 4 × 4 squares, the resulting magic squares will differ in aspect. Furthermore, depending on the succession chosen for filling the 4 × 4 subsquares, the resulting magic square may even be pandiagonal. We shall consider here squares of order

---

or below, Appendix 13.

[75] See above, pp. 34-36 & 57-58, and below, pp. 146-148 & Ch. V.

12, all constructed using the last of the pandiagonal 4×4 squares seen in §16, 1, reproduced here (Fig. 144).

1	8	139	142	9	16	131	134	17	24	123	126
140	141	2	7	132	133	10	15	124	125	18	23
6	3	144	137	14	11	136	129	22	19	128	121
143	138	5	4	135	130	13	12	127	122	21	20
25	32	115	118	33	40	107	110	41	48	99	102
116	117	26	31	108	109	34	39	100	101	42	47
30	27	120	113	38	35	112	105	46	43	104	97
119	114	29	28	111	106	37	36	103	98	45	44
49	56	91	94	57	64	83	86	65	72	75	78
92	93	50	55	84	85	58	63	76	77	66	71
54	51	96	89	62	59	88	81	70	67	80	73
95	90	53	52	87	82	61	60	79	74	69	68

1	8	11	14
12	13	2	7
6	3	16	9
15	10	5	4

Fig. 144   Fig. 145

1	40	107	142	5	44	103	138	9	48	99	134
108	141	2	39	104	137	6	43	100	133	10	47
38	3	144	105	42	7	140	101	46	11	136	97
143	106	37	4	139	102	41	8	135	98	45	12
13	52	95	130	17	56	91	126	21	60	87	122
96	129	14	51	92	125	18	55	88	121	22	59
50	15	132	93	54	19	128	89	58	23	124	85
131	94	49	16	127	90	53	20	123	86	57	24
25	64	83	118	29	68	79	114	33	72	75	110
84	117	26	63	80	113	30	67	76	109	34	71
62	27	120	81	66	31	116	77	70	35	112	73
119	82	61	28	115	78	65	32	111	74	69	36

Fig. 146

(1) Filling eight cells of each 4×4 square taken successively with the natural sequence of numbers, we shall have placed the $\frac{n^2}{2}$ first numbers. Completing each smaller square with the complements to $n^2 + 1 = 145$,

we obtain the square of Fig. 145.

1	72	75	142	5	68	79	138	9	64	83	134
76	141	2	71	80	137	6	67	84	133	10	63
70	3	144	73	66	7	140	77	62	11	136	81
143	74	69	4	139	78	65	8	135	82	61	12
13	60	87	130	17	56	91	126	21	52	95	122
88	129	14	59	92	125	18	55	96	121	22	51
58	15	132	85	54	19	128	89	50	23	124	93
131	86	57	16	127	90	53	20	123	94	49	24
25	48	99	118	29	44	103	114	33	40	107	110
100	117	26	47	104	113	30	43	108	109	34	39
46	27	120	97	42	31	116	101	38	35	112	105
119	98	45	28	115	102	41	32	111	106	37	36

Fig. 147

1	16	131	142	5	20	127	138	9	24	123	134
132	141	2	15	128	137	6	19	124	133	10	23
14	3	144	129	18	7	140	125	22	11	136	121
143	130	13	4	139	126	17	8	135	122	21	12
25	40	107	118	29	44	103	114	33	48	99	110
108	117	26	39	104	113	30	43	100	109	34	47
38	27	120	105	42	31	116	101	46	35	112	97
119	106	37	28	115	102	41	32	111	98	45	36
49	64	83	94	53	68	79	90	57	72	75	86
84	93	50	63	80	89	54	67	76	85	58	71
62	51	96	81	66	55	92	77	70	59	88	73
95	82	61	52	91	78	65	56	87	74	69	60

Fig. 148

(2) We may also fill the subsquares' cells four by four. There are then various ways of placing the second set of four cells in each subsquare: we may return to the first subsquare (Fig. 146) or start from the last (Fig. 147); we may also use these two ways separately for each horizontal strip of subsquares, namely by placing the second set of four cells in our

starting subsquare (Fig. 148) or in that of arrival (Fig. 149).

1	24	123	142	5	20	127	138	9	16	131	134
124	141	2	23	128	137	6	19	132	133	10	15
22	3	144	121	18	7	140	125	14	11	136	129
143	122	21	4	139	126	17	8	135	130	13	12
25	48	99	118	29	44	103	114	33	40	107	110
100	117	26	47	104	113	30	43	108	109	34	39
46	27	120	97	42	31	116	101	38	35	112	105
119	98	45	28	115	102	41	32	111	106	37	36
49	72	75	94	53	68	79	90	57	64	83	86
76	93	50	71	80	89	54	67	84	85	58	63
70	51	96	73	66	55	92	77	62	59	88	81
95	74	69	52	91	78	65	56	87	82	61	60

Fig. 149

(3) We may also fill the cells of each subsquare pair by pair, starting for each new cycle from the point of departure (Fig. 150) or arrival (Fig. 151).

1	56	107	126	3	58	105	124	5	60	103	122
108	125	2	55	106	123	4	57	104	121	6	59
38	19	144	89	40	21	142	87	42	23	140	85
143	90	37	20	141	88	39	22	139	86	41	24
7	62	101	120	9	64	99	118	11	66	97	116
102	119	8	61	100	117	10	63	98	115	12	65
44	25	138	83	46	27	136	81	48	29	134	79
137	84	43	26	135	82	45	28	133	80	47	30
13	68	95	114	15	70	93	112	17	72	91	110
96	113	14	67	94	111	16	69	92	109	18	71
50	31	132	77	52	33	130	75	54	35	128	73
131	78	49	32	129	76	51	34	127	74	53	36

Fig. 150

(4) Finally, we may write each time a single number in each subsquare (Fig. 152).

1	72	107	110	3	70	105	112	5	68	103	114
108	109	2	71	106	111	4	69	104	113	6	67
38	35	144	73	40	33	142	75	42	31	140	77
143	74	37	36	141	76	39	34	139	78	41	32
7	66	101	116	9	64	99	118	11	62	97	120
102	115	8	65	100	117	10	63	98	119	12	61
44	29	138	79	46	27	136	81	48	25	134	83
137	80	43	30	135	82	45	28	133	84	47	26
13	60	95	122	15	58	93	124	17	56	91	126
96	121	14	59	94	123	16	57	92	125	18	55
50	23	132	85	52	21	130	87	54	19	128	89
131	86	49	24	129	88	51	22	127	90	53	20

Fig. 151

1	64	99	126	2	65	98	125	3	66	97	124
108	117	10	55	107	116	11	56	106	115	12	57
46	19	144	81	47	20	143	80	48	21	142	79
135	90	37	28	134	89	38	29	133	88	39	30
4	67	96	123	5	68	95	122	6	69	94	121
105	114	13	58	104	113	14	59	103	112	15	60
49	22	141	78	50	23	140	77	51	24	139	76
132	87	40	31	131	86	41	32	130	85	42	33
7	70	93	120	8	71	92	119	9	72	91	118
102	111	16	61	101	110	17	62	100	109	18	63
52	25	138	75	53	26	137	74	54	27	136	73
129	84	43	34	128	83	44	35	127	82	45	36

Fig. 152

Remarks.

(1) How many numbers are placed simultaneously in each subsquare is
determined by the powers of 2 (thus 1, 2, 4, 8).

(2) When with these methods the subsquares are filled in natural order,
as here above, the resulting squares are pandiagonal; but, as in the
case of squares displaying the same feature, this property seems to

have remained mostly unnoticed or not worth mentioning.

It is certainly due to the similarity of these methods that they do not occur all in the same work; thus the *Harmonious arrangement* presents the first and the last, while Shabrāmallisī has only the first.[76] Furthermore, later authors might prefer the 'newer' methods which cover the whole square with regular sequences of moves without requiring computation of the complements (§§ 18-20). Indeed, to our knowledge, global constructions of (non-bordered, non-composite) evenly-even squares are not attested before the 11th century.

On the contrary, the construction by means of 4×4 squares was held in high esteem in the 10th century. It is also particularly convenient: we may start with any cell of any subsquare and continue with any other subsquare. (This is a notable difference to odd-order composite squares, which cannot be filled in any order since the magic sums in the subsquares cannot be equal.) Thus, Anṭākī has an example of a 12×12 square with the subsquares filled in arbitrary order —by the way reproduced from Thābit's translation of a Greek text.[77] The convenience of this construction by means of subsquares filled in any order is confirmed in the 10th century by the instructions in Abū'l-Wafā' Būzjānī's text:[78] *It is possible to set up the magic arrangement in squares with sides having a fourth in a manner more elegant than that we have explained.[79] It consists in dividing the square (considered) into a (certain) number of squares in all rows of which the sums will be uniformly the same. If we wish that, we divide the square (of the order) by 16. The quotient will be the number of squares in which the considered square is to be divided, each of these squares having 16 cells. We arrange in half (of the cells) of one of these squares, whichever it is, the consecutive numbers taken in natural order from 1 to 8 following one of the arrangements established in the second chapter for the square of 4,[80] and we complete them by means of the equalizing number for the (large) square.[81] We move then to any other square and arrange in half of its (cells) the numbers from 9 to 16, and we complete them by means of the equalizing number. We proceed in this way until we have finished with all its squares. At this point, the*

---

[76] *Un traité médiéval*, pp. 78-80; MS. Paris BNF Arabe 2698, fol. 31ᵛ - 32ᵛ.

[77] *Magic squares in the tenth century*, A.II.48. See below, Fig. 217.

[78] *Magic squares in the tenth century*, B.26.*i-ii*; or below, Appendix 14.

[79] Namely the arrangement of composite squares where each subsquare displays a different magic sum (see below, Ch. III, § 1).

[80] Equivalent to our figures 117 to 119.

[81] Completing the bishop's cells to $n^2 + 1$.

quantities in all the rows of the (large) square make the same sum, while the quantities in all the rows of each (small) square are also the same.[82]

It is thus clear that knowing a single construction for a (pandiagonal) square of 4 one could easily obtain ordinary evenly-even magic squares, with their subsquares arranged in any order since each is a magic entity with the same magic sum. This method was thus in common use in the oldest texts.

Remark. It is not mentioned, or not considered worth mentioning, that these subsquares, when filled, may be individually turned or inverted.

Particular cases

(1) These arrangements are also applicable to the square of order 8 since a square of order 2 can be magic only if its four elements are equal; and here the four 4×4 squares display equal magic sums. In the particular example below (Fig. 153) the numbers are placed according to an arrangement as above (Fig. 117b), but with each number placed in the next quadrant; each 2×2 square makes the same sum 130, and the pairs of complements are in opposite 2×2 squares belonging to the same quadrant.[83]

1	62	51	16	17	46	35	32
36	31	18	45	52	15	2	61
14	49	64	3	30	33	48	19
47	20	29	34	63	4	13	50
28	39	42	21	12	55	58	5
57	6	11	56	41	22	27	40
23	44	37	26	7	60	53	10
54	9	8	59	38	25	24	43

Fig. 153

(2) The construction of the square in Fig. 154 is entirely based on the configuration seen in Fig. 117b, both for the succession within the subsquares as for the subsquares' own.[84] We first fill a quarter of eight subsquares; this leads us to 32. The numbers 33 to 64 are put in the same cells of the subsquares distant from the previous ones by a bishop's move and taken in inverse order. Starting with 65 from the subsquare just reached, we put

---

[82] Namely $2\,(n^2 + 1)$.

[83] MS. Mashhad Āstān-i Quds 14159, fol. 87ᵛ.

[84] MS. Mashhad Āstān-i Quds 12235, p. 439.

the third and fourth sequence, four cells by four, in the same subsquares taken in inverse order. This leads us to 128, in the initial subsquare. This determines the remainder since all bishop's cells are to be filled with complements. The result is interesting: both the whole square and the subsquares are pandiagonal.

1	254	131	128	53	202	183	76	41	214	171	88	29	226	159	100
132	127	2	253	184	75	54	201	172	87	42	213	160	99	30	225
126	129	256	3	74	181	204	55	86	169	216	43	98	157	228	31
255	4	125	130	203	56	73	182	215	44	85	170	227	32	97	158
45	210	175	84	25	230	155	104	5	250	135	124	49	206	179	80
176	83	46	209	156	103	26	229	136	123	6	249	180	79	50	205
82	173	212	47	102	153	232	27	122	133	252	7	78	177	208	51
211	48	81	174	231	28	101	154	251	8	121	134	207	52	77	178
21	234	151	108	33	222	163	96	61	194	191	68	9	246	139	120
152	107	22	233	164	95	34	221	192	67	62	193	140	119	10	245
106	149	236	23	94	161	224	35	66	189	196	63	118	137	248	11
235	24	105	150	223	36	93	162	195	64	65	190	247	12	117	138
57	198	187	72	13	242	143	116	17	238	147	112	37	218	167	92
188	71	58	197	144	115	14	241	148	111	18	237	168	91	38	217
70	185	200	59	114	141	244	15	110	145	240	19	90	165	220	39
199	60	69	186	243	16	113	142	239	20	109	146	219	40	89	166

Fig. 154

(3) The same Persian manuscript has (Fig. 155) an ordinary 12×12 magic square divided into subsquares of order 3, which are once again filled successively according to the moves in the 4×4 pandiagonal square.[85] To begin with, we fill the first four subsquares, three cells by three in a to-and-fro movement, the cells being taken according to the succession in the 3×3 square (numbers from 1 to 36). We do the same for the four next subsquares (37 to 72). We proceed with the bishop's subsquares of the second sequence (73 to 108) and conclude with those associated with the first sequence, taken in reverse order (109 to 144). As a result, any subsquare has the same sum in its lines and columns, and these sums

---

[85] MS. Mashhad Āstān-i Quds 12235, p. 440.

follow an arithmetical progression of difference 3 in each sequence; then, adding them up in two complementary subsquares makes each time the same quantity.

2	25	24	113	136	129	80	103	90	47	70	51
27	23	1	138	128	112	105	89	79	72	50	46
22	3	26	127	114	137	88	81	104	49	48	71
83	106	87	44	67	54	5	28	21	110	133	132
108	86	82	69	53	43	30	20	4	135	131	109
85	84	107	52	45	68	19	6	29	130	111	134
41	64	57	74	97	96	119	142	123	8	31	18
66	56	40	99	95	73	144	122	118	33	17	7
55	42	65	94	75	98	121	120	143	16	9	32
116	139	126	11	34	15	38	61	60	77	100	93
141	125	115	36	14	10	63	59	37	102	92	76
124	117	140	13	12	35	58	39	62	91	78	101

Fig. 155

31	18	74	67	54	2	22	63	38
21	59	43	30	14	79	66	50	7
71	46	6	26	55	42	35	10	78
4	72	47	40	27	56	76	36	11
75	32	16	3	68	52	39	23	61
44	19	60	80	28	15	8	64	51
58	45	20	13	81	29	49	9	65
48	5	70	57	41	25	12	77	34
17	73	33	53	1	69	62	37	24

Fig. 156

(4) Unlike with 4 × 4 subsquares, we cannot find an odd-order magic square with each part displaying the same magic sum. But two Persian manuscripts have a 9 × 9 magic square where each 3 × 3 subsquare makes in its lines and columns 123, thus a third of the magic sum (Fig. 156).[86] These subsquares of order 3 are filled each by the same moves as for the 3 × 3 square. But, first, the two numbers following 1 are placed in

---

[86] MSS. Āstān-i Quds 12235, p. 430, and (with slight changes) 14159, fol. 83ᵛ.

subsquares within the same vertical strip. The next three numbers are in the next strip, beginning from the middle subsquare, the next three in the remaining strip, starting with the top subsquare. The next sequence of nine numbers (10 to 18) follows the same movement, starting with the top subsquare of the same strip of subsquares. Generally, to pass from one sequence of nine numbers to the next, we move one subsquare down (or from bottom to top) in the same strip, except if the subsquare is in the median strip, in which case we are to remain in the same subsquare (from 27 to 28 and from 54 to 55).

C. SQUARES OF EVENLY-ODD ORDERS

The squares of evenly-even orders were, as a category, particularly easy to fill with just diagonal exchanges. This is no longer the case with even orders not divisible by 4, which is probably why the method of constructing them seems to have appeared later, after the technique of performing horizontal and vertical exchanges separately was fully mastered. In any event, whereas the methods we have seen are attested in the early 11th century, or even in the late tenth, the first general construction of ordinary magic squares of evenly-odd order is seen towards the end of the 11th century or, if not, the early twelfth. This is not to say that such squares could not be constructed before. They were obtained either by applying exchanges to particular classes of evenly-odd orders, or by reducing their construction to that of evenly-even orders by filling two lines and two columns, or a small inner square of evenly-odd order, separately. We shall first consider the general method and then earlier attempts.

§ 23. Exchanges in the natural square

The general method for any square of evenly-odd order is clearly explained by Kharaqī, around or soon after 1100.[87]

If the square is evenly-odd, like (those) of six, ten, fourteen and so on, we divide the square into quadrants and set a reference quadrant.

We put in its first cell a dot. Then, if the square is that of six, we put in its second cell a zero and in its third cell, a cross. We complete with dots the cells of the (main) diagonal, and with zeros and crosses (the broken diagonals) on either side of the (main) diagonal having altogether as many cells as the (main) diagonal. The square of six will then have

---

[87] *Herstellungsverfahren III*, pp. 202-204 of the summary, lines 298-331 of the Arabic text; or below, Appendix 15. Our additions are, as usual, in brackets. Keep in mind that our figures here read left to right.

the following aspect (Fig. 157). Then, we turn this quadrant, thus the reference one, over its left-hand neighbour for (reproducing there) the dots and the crosses, over its lower neighbour for (reproducing) the dots and the zeros, and over the fourth quadrant for (reproducing) the dots only. The square has then the following aspect (Fig. 158).

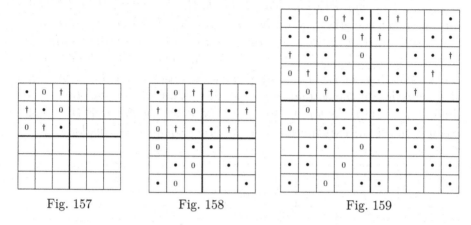

Fig. 157   Fig. 158   Fig. 159

If the square is that of ten, we put in the first cell of the (quadrant of) reference a dot, leave then a cell empty and put a zero in the third cell, a cross in the fourth and a dot in the fifth. We complete (with dots the main diagonal), with dots, zeros and crosses (the broken diagonals) on either part of the first diagonal of the quadrant of reference and having altogether the same number of cells as the first diagonal. Next we turn the reference quadrant over the remaining quadrants, (to reproduce) the zeros in the adjacent lower quadrant, the crosses in the neighbouring left-hand quadrant, and the dots in all three quadrants. The square will then have the following aspect (Fig. 159).

If the square is that of fourteen, we put a dot in the first cell, leave the second empty, put a zero in the third cell, a cross in the fourth, a dot in the fifth, leave the sixth empty and put a dot in the seventh cell. You are to know that one writes a zero and a cross only once in the first horizontal row of the first quadrant; as to the (number of) dots, it increases together with the side. Next we complete this (quadrant for its main diagonal and) on either side of it, then turn it over the other quadrants.

When we wish to place the numbers, we begin in the first cell, which we shall call 'start of the dots', and proceed to the left and going down, writing the numbers in each cell containing a dot, beginning with 1 and continuing to the end of the first diagonal. Next, we return from there, and call this end 'start of the blanks'; we advance along the last row *towards the right and going up, beginning with 1, and write in each blank cell the number reached, until we get to the first cell of the square. Next we begin at the first of the two ends of the second diagonal, which we call 'start of the zeros'. We advance along the first horizontal row towards the right, writing in each cell containing a zero the number (reached), beginning (the enumeration) with 1; we do the same with the other rows, until we reach the other end of the second diagonal. We return from there, and call it 'start of the crosses', and go forward in the last horizontal row towards the left and going up, and write in each cell containing a cross the number (reached), beginning (the enumeration) with 1, till we reach the beginning of the second diagonal. The square of ten will then be as follows* (Fig. 160). *As for the square of six, it will be like that* (Fig. 161). *Keep in mind our explanations for the other squares (of this kind).*

• → ← 0

1	99	8	94	5	6	97	93	92	10
11	12	88	17	85	86	84	83	19	20
71	22	23	77	26	75	74	28	29	80
40	62	33	34	66	65	37	38	69	61
60	49	53	44	45	46	47	58	52	51
50	59	48	54	55	56	57	43	42	41
70	39	63	64	36	35	67	68	32	31
30	72	73	27	76	25	24	78	79	21
81	82	18	87	16	15	14	13	89	90
91	9	98	7	95	96	4	3	2	100

† →

Fig. 160

• → ← 0

1	5	33	34	32	6
25	8	10	27	11	30
18	20	15	16	23	19
24	17	21	22	14	13
12	26	28	9	29	7
31	35	4	3	2	36

† →

Fig. 161

In short, we divide the square of the order considered $4k+2$ into its quadrants and put in the first row of the first one, with $2k+1$ cells, k dots, one zero, one cross; their place is arbitrary (which the text does not mention) as long as the zero and the cross do not fall in the main diagonal. We then complete the arrangement by reproducing these same symbols along the quadrant's diagonals, main and broken. We then turn this quadrant over the others, and reproduce the dots in all the corresponding cells, but the crosses only in the horizontal neighbour and the zeros, in the vertical neighbour. We fill the cells by enumerating them in turn from the four corners, as we did for the exchanges which were not solely diagonal (p. 56): the cells with dots from the corner of the first quadrant, the blank cells from the opposite corner, the cells with crosses from the corner opposite to the quadrant with the carried-over crosses, the cells with zeros from the remaining corner, which is the corner opposite to the quadrant of the carried-over zeros.

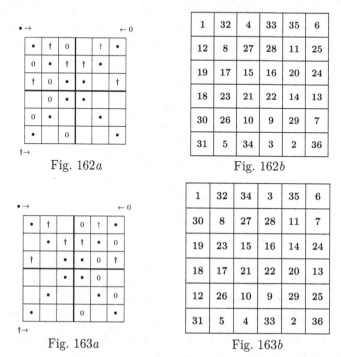

Fig. 162a

1	32	4	33	35	6
12	8	27	28	11	25
19	17	15	16	20	24
18	23	21	22	14	13
30	26	10	9	29	7
31	5	34	3	2	36

Fig. 162b

Fig. 163a

1	32	34	3	35	6
30	8	27	28	11	7
19	23	15	16	14	24
18	17	21	22	20	13
12	26	10	9	29	25
31	5	4	33	2	36

Fig. 163b

Remark. This, even for the smallest order, admits of a certain number of variations. Thus that of Fig. 162a (exchanging the places of † and 0 in the first quadrant) will produce the square of Fig. 162b.[88] The square of Fig. 163a (reversing Fig. 162a) will produce Fig. 163b.[89] Still another configuration (Fig. 163a turned this time around its horizontal axis) gives one of the two 6 × 6 squares transmitted to Europe (Fig. 30, slightly different from Fig. 163b).

§ 24. An older method

The generally applicable method we have just seen for orders of the form $n = 4k + 2$ was taught by Kharaqī among others he presents, thus without any claim to originality. Its first appearance is not known. An author from the close of the 11th century (probably Asfizārī, p. 14 above) knew a method using exchanges in the natural square, but this method turns out to be valid only for an evenly-odd order $n = 4k + 2$ with k

[88] MSS. Mashhad Āstān-i Quds 12167, fol. 172ᵛ, and 12235, p. 412.
[89] MS. Mashhad Āstān-i Quds 14159, fol. 87ᵛ - 88ʳ.

even, thus of the form $8t + 2$.[90] Since the anonymous author of the 12th century reporting it says that he chose, from the various treatises seen by him, the more general and elegant methods of construction, he cannot have encountered a method valid for all evenly-odd orders in his sources, which include Ibn al-Ha$\underline{\text{y}}$tham and Asfizārī. Indeed, the older method expounded below, apart from being of limited use, is not very convenient, at least in comparison with the one we have seen, which therefore cannot yet have been known or widespread.

I	II		III	IV
V	VI		VII	VIII
IX	X		XI	XII
XIII	XIV		XV	XVI

Fig. 164

We first separate, in the square of the order considered, thus $8t + 2$, the pairs of median horizontal and vertical rows and divide each of the four quadrants thus defined into four squares, each with order $2t$ (Fig. 164). We then perform the following exchanges (see Fig. 164-166):

(1) We exchange diagonally the squares I and XVI, VI and XI, VII and X, IV and XIII, turning them by 180° as was done in the method of exchanging quadrants (§ 9[91]).

(2) We exchange diagonally the content of the four central cells.

(3) We switch $4t - 1$ elements between adjacent rows in the branches of the cross.

(4) We exchange the diagonals of the two remaining subsquares of the first quadrant, II and V, with those of their vertical or horizontal opposite; this is performed in such a way that the lateral elements remain so and the diagonal orientation is changed. Considering first II, we exchange its first diagonal ($\diagdown$) with the second diagonal ($\diagup$) of III and its second diagonal with the first diagonal of XIV; considering then V, we exchange its first diagonal with the second diagonal of IX, and its second diagonal with the first diagonal of VIII. See Fig. 166, where the numbers moved

---

[90] *Une compilation arabe*, pp. 170-172 of the translation, lines 364-408 of the Arabic text.

   [91] This method was expounded by Asfizārī and the present one just follows it; whence its presumed authorship.

are in bold.[92]

1	2	3	4	5	6	7	8	9	10
11	12	13	14	15	16	17	18	19	20
21	22	23	24	25	26	27	28	29	30
31	32	33	34	35	36	37	38	39	40
41	42	43	44	45	46	47	48	49	50
51	52	53	54	55	56	57	58	59	60
61	62	63	64	65	66	67	68	69	70
71	72	73	74	75	76	77	78	79	80
81	82	83	84	85	86	87	88	89	90
91	92	93	94	95	96	97	98	99	100

Fig. 165

100	99	8	94	5	6	7	3	92	91
90	89	83	17	16	15	14	18	82	81
71	29	78	77	26	25	74	73	22	30
40	62	68	67	36	35	64	63	39	31
41	52	53	54	56	55	47	48	49	50
51	42	43	44	46	45	57	58	59	60
61	32	38	37	65	66	34	33	69	70
21	72	28	27	75	76	24	23	79	80
20	19	13	84	85	86	87	88	12	11
10	9	93	4	95	96	97	98	2	1

Fig. 166

Remark. If the order is not of the form $8t + 2$, and is thus of the form $8t + 6$, it will not be possible to perform the exchanges of diagonals for the squares II and V since their two diagonals have one element in common. The diagonals of the subsquares, thus also their rows, must therefore contain an even number of cells.

§ 25. Principles of these methods

In order to understand what these two methods have in common, we need only remember the origin of the exchange methods in the natural square for the case of evenly-even squares. Two principles were used: either exchanging separately half the elements of lines and half the elements of columns with their respective opposite, preferably without touching the diagonals, or exchanging half the elements of half-lines with the corresponding ones in the diagonally opposite quadrant. Now in the case of an evenly-odd order, where the quadrant has order $2k + 1$, it is no longer possible to make diagonal exchanges only. Therefore we must either perform separate exchanges for lines and columns or add to a number $2k$ of diagonal exchanges a single exchange between corresponding horizontal and vertical rows. This will indeed give the required number of $2k + 1$ horizontal and vertical exchanges, that is, the number equal to half the order.

Remark. Because of these single exchanges it is not possible to construct a symmetrical square of evenly-odd order.

---

[92] A similar figure is found in MS. London BL Delhi Arabic 110, fol. 49ʳ (the main difference is the exchange of ╲ of V with ╱ of VIII and ╱ of V with ╲ of IX).

Application to the methods seen

1. Older method (§ 24)

We shall again use the symbols seen on p. 56, with • keeping the
place of the numbers in the natural square, | marking a vertical exchange
between opposite lines, − indicating a horizontal one and × a diagonal
one. Fig. 167 shows the configuration of the symbols and where the
enumeration of the cells is to begin; this will give Fig. 166. We may
verify that the required number of exchanges appears in each vertical
and horizontal row: there are indeed five symbols × or | in each line and
five symbols × or − in each column. (The diagonals keep their elements.)

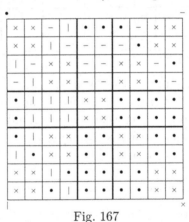

Fig. 167

2. General method (§ 23)

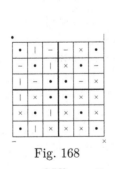

Fig. 168

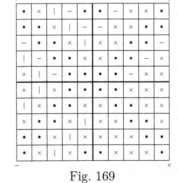

Fig. 169

The squares of <u>Kh</u>araqī would be obtained from the configurations of
the figures 168 and 169. Note, however, that, relative to the previous
enumeration from the corners, | and − have changed place. Indeed, −
(which corresponds to <u>Kh</u>araqī's †) means here a horizontal exchange *after*
a diagonal one, and likewise | (which is <u>Kh</u>araqī's 0) a vertical exchange

following a diagonal one, of which it eliminates the vertical component. Thus, in Fig. 161, the numbers 5 and 35, which had first been exchanged diagonally and thus reached the second column, have then changed places vertically. This anomaly, if we might consider it as such, probably originated with the discovery of the method: first there were $2k + 2$ diagonal exchanges for each row, thus one too many; one vertical exchange per line was then eliminated, with the numbers concerned moving back up or down, and likewise one diagonal exchange for each column was reduced by its horizontal component, the numbers involved moving back right or left.

Remark. The very curious arrangement of Fig. 170 is found in some treatises.[93] S̲h̲abrāmallisī attempts, in vain, to extend it to the next order, 10.

1	31	22	15	30	12
2	32	14	23	29	11
3	33	24	13	28	10
34	4	16	21	9	27
35	5	17	20	8	26
36	6	18	19	7	25

Fig. 170

§ 26. Method of the cross

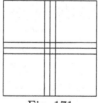

Fig. 171

Probably because there was no general method yet known, people tried at first to reduce the construction of evenly-odd squares to the simpler case of evenly-even order. One possibility was to eliminate a pair of horizontal and vertical rows —in other words, to fill them separately. Such is the purpose of the 10th-century (as a matter of fact, ancient) method of the cross (Fig. 171), which divides the square into four squares of even order $2k$ and two pairs of median rows in between (the method of

---

[93] MSS. Paris BNF Arabe 2698, fol. 39ᵛ - 40ᵛ; Āstān-i Quds 12235, p. 442.

§ 24 may represent the transition from this idea to the general method). Without the cross, the four squares are, singly or combined, of evenly-even order, according to whether k is even or odd.

Since the order of the whole square is $4k + 2$, it will contain the numbers from 1 to $n^2 = 16k^2 + 16k + 4$. We shall use $16k + 4$ of these numbers for the cross, for example those in the middle of the sequence, while the $8k^2$ first and the $8k^2$ last will fill the subsquares.

Filling the cross will require the use of a fundamental notion (already mentioned, p. 2) which we shall often meet later, that of the 'sum due' (*ḥaqq*). For an even square of order n filled with the first n^2 numbers, the magic sum is

$$M_n = \frac{n(n^2 + 1)}{2}.$$

Since there are n numbers in each row, thus $\frac{n}{2}$ pairs of numbers, the ideal contribution of each pair of numbers would be n^2+1. Now this is precisely the sum displayed by a pair of complements, namely two numbers of the form i and $n^2 + 1 - i$. For equalizing the cross, the branches of which are two cells wide, we shall put in each pair of such cells a pair of complements. This will give the sum due for two cells and we shall no longer need to take them into account when filling the subsquares.

We are still to obtain the sum due longitudinally in the branches and diametrically in the four central cells. Now it is already possible to attain the sum due with a length of six cells, comprising the cells in the middle of the cross, thus altogether twenty cells, by using ten consecutive pairs among the $8k + 2$ pairs to be placed in the cross. Consider first these $16k + 4$ numbers, aligned vertically by pairs of complements with sum $n^2 + 1 = 16k^2 + 16k + 5$:

$$8k^2 + 1 \qquad 8k^2 + 2 \qquad \ldots \qquad 8k^2 + 8k + 1 \qquad 8k^2 + 8k + 2$$
$$8k^2 + 16k + 4 \quad 8k^2 + 16k + 3 \quad \ldots \quad 8k^2 + 8k + 4 \quad 8k^2 + 8k + 3.$$

Let us for instance take the last ten pairs, which may also be expressed as

$\frac{n^2}{2}-9$	$\frac{n^2}{2}-8$	$\frac{n^2}{2}-7$	$\frac{n^2}{2}-6$	$\frac{n^2}{2}-5$	$\frac{n^2}{2}-4$	$\frac{n^2}{2}-3$	$\frac{n^2}{2}-2$	$\frac{n^2}{2}-1$	$\frac{n^2}{2}$
$\frac{n^2}{2}+10$	$\frac{n^2}{2}+9$	$\frac{n^2}{2}+8$	$\frac{n^2}{2}+7$	$\frac{n^2}{2}+6$	$\frac{n^2}{2}+5$	$\frac{n^2}{2}+4$	$\frac{n^2}{2}+3$	$\frac{n^2}{2}+2$	$\frac{n^2}{2}+1$
X	IX	VIII	VII	VI	V	IV	III	II	I

We shall place I and III in the centre, with the elements of each pair diagonally opposite. Next, the elements of the pairs II, IV, V, VI will occupy the vertical branch and the others the horizontal branch, as represented in Fig. 172. With this arrangement, we shall have the following sums:

— diagonally: $n^2 + 1$ for two cells,

— vertically: $3(n^2 + 1)$ for six cells,

— horizontally: $3(n^2 + 1)$ for six cells.

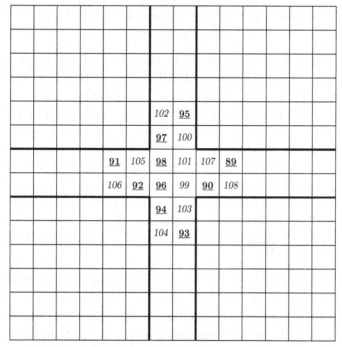

Fig. 172

Fig. 173 applies that to order 14.

Fig. 173

All the occupied cells now display the required sum in all three directions. We are left with placing in the branches of the cross the $16k - 16$ remaining numbers, thus two sequences of $8(k-1)$ complementary numbers, each sequence being formed by consecutive numbers since the centre of the cross has been filled with the last ten pairs.

This is where we meet another fundamental principle, that of the placing which we shall call *neutral*. Let us consider four consecutive numbers, or in arithmetical progression,

$$a, \; a+s, \; a+2s, \; a+3s.$$

Since the sum of the extremes equals the sum of the means, namely $2a + 3s$, we put in a row two extremes, in the opposite row two means, and write their complements opposite (Fig. 174). Vertically we shall have the sum of a pair of complements, thus the sum due for two cells, $n^2 + 1$, and horizontally twice this sum, thus the sum due for four cells.

	a	$n^2+1-(a+s)$	$n^2+1-(a+2s)$	$a+3s$		$\Longrightarrow$	$2(n^2+1)$
	n^2+1-a	$a+s$	$a+2s$	$n^2+1-(a+3s)$		$\Longrightarrow$	$2(n^2+1)$

Fig. 174

With that it is easy to complete the filling of the branches of the cross. If $n = 6$, they are already filled. For $n = 4k + 2$ with $k > 1$, we still have to place $8(k-1)$ pairs of complements, that is, $8(k-1)$ smaller consecutive numbers and their $8(k-1)$ complements, thus $4(k-1)$ pairs horizontally and as much vertically. Since $4(k-1)$ is divisible by 4, we have two possible placings:
– putting the smaller numbers by tetrads of consecutive numbers, the extremes on one side of a branch and the means on the other;
– putting $k - 1$ consecutive smaller numbers each time, a first group on one side, the next two opposite, the last on the first side.
After placing the complements, the branches will display the sum due (Fig. 175).

Filling the part outside the cross is easy, and it was our purpose to end there. We shall fill the four squares with the remaining numbers, dismissing those already used for the cross. There are various ways of doing that, all the more with increasing order. Fig. 176 divides these four squares into subsquares of order 4, some of them cut by the cross. In the figures 177-179, they are filled as a whole using various methods seen for filling squares of evenly-even orders. Finally, Fig. 180 fills a square of order 6.

						81	116						
						115	82						
						114	83						
						84	113						
						102	95						
						97	100						
73	123	122	76	91	105	98	101	107	89	77	119	118	80
124	74	75	121	106	92	96	99	90	108	120	78	79	117
						94	103						
						104	93						
						85	112						
						111	86						
						110	87						
						88	109						

Fig. 175

1	194	191	8	9	186	81	116	183	16	17	178	175	24
192	7	2	193	184	15	115	82	10	185	176	23	18	177
6	189	196	3	14	181	114	83	188	11	22	173	180	19
195	4	5	190	187	12	84	113	13	182	179	20	21	174
25	170	167	32	33	162	102	95	159	40	41	154	151	48
168	31	26	169	160	39	97	100	34	161	152	47	42	153
73	123	122	76	91	105	98	101	107	89	77	119	118	80
124	74	75	121	106	92	96	99	90	108	120	78	79	117
30	165	172	27	38	157	94	103	164	35	46	149	156	43
171	28	29	166	163	36	104	93	37	158	155	44	45	150
49	146	143	56	57	138	85	112	135	64	65	130	127	72
144	55	50	145	136	63	111	86	58	137	128	71	66	129
54	141	148	51	62	133	110	87	140	59	70	125	132	67
147	52	53	142	139	60	88	109	61	134	131	68	69	126

Fig. 176

1	99	98	4	33	68	5	95	94	8
92	10	11	89	36	65	88	14	15	85
84	18	19	81	54	47	80	22	23	77
25	75	74	28	49	52	29	71	70	32
37	40	43	57	50	53	59	41	63	62
64	61	58	44	48	51	42	60	38	39
69	31	30	72	46	55	73	27	26	76
24	78	79	21	56	45	20	82	83	17
16	86	87	13	67	34	12	90	91	9
93	7	6	96	66	35	97	3	2	100

Fig. 177

This method of the cross goes back to antiquity, for it is fully described in the ninth century by Thābit ibn Qurra, whose text is reproduced a century later by Anṭākī.[94] It is also mentioned that equalization of the cross may be performed, for orders $n = 4k + 2$ with k even, on just one side of the branch, the remaining cells being filled with neutral placings. Thābit's text has the example of Fig. 181.[95]

1	2	98	97	33	68	96	95	7	8
9	10	90	89	36	65	88	87	15	16
84	83	19	20	54	47	21	22	78	77
76	75	27	28	49	52	29	30	70	69
37	40	43	57	50	53	59	41	63	62
64	61	58	44	48	51	42	60	38	39
32	31	71	72	46	55	73	74	26	25
24	23	79	80	56	45	81	82	18	17
85	86	14	13	67	34	12	11	91	92
93	94	6	5	66	35	4	3	99	100

Fig. 178

---

[94] *Magic squares in the tenth century*, A.II.50-54, and MS. London BL Delhi Arabic 110, fol. 100ʳ - 100ᵛ.

[95] MS. London BL Delhi Arabic 110, fol. 102ʳ. With k even there will be an integral number of neutral placings on each side.

1	71	90	9	33	68	24	79	98	32
89	10	2	72	36	65	97	31	23	80
11	92	100	30	54	47	3	69	77	22
99	29	12	91	49	52	78	21	4	70
37	40	43	57	50	53	59	41	63	62
64	61	58	44	48	51	42	60	38	39
28	94	83	20	46	55	13	86	75	5
84	19	27	93	56	45	76	6	14	85
18	81	73	7	67	34	26	96	88	15
74	8	17	82	66	35	87	16	25	95

Fig. 179

1	35	22	15	34	4
32	6	17	20	7	29
11	25	18	21	27	9
26	12	16	19	10	28
8	30	14	23	31	5
33	3	24	13	2	36

Fig. 180

28	75	70	29	40	61	20	83	78	21
74	25	32	71	62	39	82	17	24	79
31	72	73	26	63	38	23	80	81	18
69	30	27	76	37	64	77	22	19	84
41	59	43	57	50	53	36	66	67	33
60	42	58	44	48	51	65	35	34	68
4	99	94	5	49	52	12	91	86	13
98	1	8	95	54	47	90	9	16	67
7	96	97	2	46	55	15	88	89	10
93	6	3	100	56	45	85	14	11	92

Fig. 181

Thābit's translation includes several examples of larger squares with a central cross. We shall examine them in § 4 of the next chapter, on composite squares.

Remark. Some authors, instead of filling the cross with the numbers in the middle, do it with the first and the last ones; see Fig. 182.[96]

---

[96] MSS. Istanbul Ayasofya 2794, fol. 23ᵛ, Mashhad Āstān-i Quds 12235, p. 444 (incomplete), or 14159, fol. 89ʳ. Shabrāmallisī has an analogous construction (MS. Paris BNF Arabe 2698, fol. 50ʳ).

19	80	77	26	89	12	27	72	69	34
78	25	20	79	11	90	70	33	28	71
24	75	82	21	92	9	32	67	74	29
81	22	23	76	99	2	73	30	31	68
84	85	15	10	1	4	98	96	94	18
17	16	86	91	97	100	3	5	7	83
35	64	61	42	6	95	43	56	53	50
62	41	36	63	8	93	54	49	44	55
40	59	66	37	14	87	48	51	58	45
65	38	39	60	88	13	57	46	47	52

Fig. 182

§ 27. Construction of a border

As a matter of fact, the method of the cross is just a particular case, namely that in which the pairs of rows separated are the median ones. Other pairs of parallel rows may be selected, for instance so as to leave a 4×4 square in the centre. Fig. 184 illustrates that for order 14: a border of order 6, filled with the numbers from 33 to 42 and their complements (according to the arrangement of Fig. 183) surrounds a square of order 4 filled with the subsequent smaller numbers and their complements, and the remainder of the rows selected is filled with neutral placings. This elegant construction occurs in the three Persian manuscripts mentioned in the previous note, at the same place.[97]

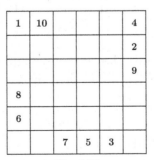

Fig. 183

---

[97] Also, but incomplete, on p. 374 of MS. Āstān-i Quds 12235. Similar example in MS. London BL Delhi Arabic 110, fol. 102^{r} (see below Fig. 225, p. 132).

1	194	191	8	83	9	186	183	16	114	17	178	175	24
192	7	2	193	113	184	15	10	185	84	176	23	18	177
6	189	196	3	112	14	181	188	11	85	22	173	180	19
195	4	5	190	86	187	12	13	182	111	179	20	21	174
87	88	108	107	33	42	158	160	162	36	106	105	93	94
25	170	167	32	163	43	152	149	50	34	51	144	141	58
168	31	26	169	156	150	49	44	151	41	142	57	52	143
30	165	172	27	40	48	147	154	45	157	56	139	146	53
171	28	29	166	38	153	46	47	148	159	145	54	55	140
110	109	89	90	161	155	39	37	35	164	91	92	104	103
59	136	133	66	95	67	128	125	74	102	75	120	117	82
134	65	60	135	101	126	73	68	127	96	118	81	76	119
64	131	138	61	100	72	123	130	69	97	80	115	122	77
137	62	63	132	98	129	70	71	124	99	121	78	79	116

Fig. 184

1	10	15	85	84	18	94	96	98	4
99	19	80	77	26	27	72	69	34	2
92	78	25	20	79	70	33	28	71	9
11	24	75	82	21	32	67	74	29	90
89	81	22	23	76	73	30	31	68	12
88	35	64	61	42	43	56	53	50	13
14	62	41	36	63	54	49	44	55	87
8	40	59	66	37	48	51	58	45	93
6	65	38	39	60	57	46	47	52	95
97	91	86	16	17	83	7	5	3	100

Fig. 185

Instead of surrounding a central square, we can also have the square itself surrounded by a border, and that too will leave an evenly-even square, as taught for instance by the *Harmonious arrangement*.[98] Thus, in Fig. 185, we have taken the initial numbers from 1 to 18 and the final

---

[98] *Un traité médiéval*, pp. 68-69 of the translation, lines 503-517 of the Arabic text.

ones from 83 to 100, and placed them in a border as in Fig. 183, the border
then being completed with neutral placings. The remaining numbers have
been used for the 4 × 4 squares, each arranged according to one of the
pandiagonal configurations. Comparing the result with that of Fig. 182, it
will appear that the branches of the cross are now simply in the border,
with the central cells in the corners —as is to be expected since they
are common to two rows. (Indeed, the three Persian manuscripts which
report these two constructions have them together.)

Knowing thus how to construct an evenly-odd border, we are brought
back to filling an evenly-even square, which in turn can be reduced to
filling 4 × 4 squares. Such separation of two pairs of rows, in one or the
other configuration, was already in use in antiquity, together with the
method we shall see now.

§ 28. Method of the central square

This other possibility of reverting to an evenly-even square is suc-
cinctly explained in Thābit's (and therefore Anṭākī's) text.[99] Suppose
that the square to be constructed has the order $n = 4k + 2$. Draw in
its centre a square of order $4t + 2$ $(k > t \geq 1$, thus $k \geq 2)$. The whole
square is then broken up as follows (Fig. 186): in its centre, the square of
order $4t + 2$, distant by $2(k - t)$ from the sides; in each corner, one square
of order $2(k - t)$, which is either evenly-even or the quadrant of such a
square; finally, four $2(k - t) \times (4t + 2)$ lateral rectangles, not containing
any part of the main diagonals and with one of their dimensions, or that
of two opposite rectangles, evenly-even.

Fig. 186

For the filling, we first pair up vertically the complementary numbers
to be placed:

$$\begin{array}{cccc} 1 & 2 & \cdots & \frac{n^2}{2} \\[6pt] n^2 & n^2 - 1 & \cdots & \frac{n^2}{2} + 1 \end{array}$$

We then fill, say, the corner squares with the first smaller numbers and their complements (here in a pandiagonal arrangement); their magic sum will thus be the same and equal to the sum due. Next we may fill the central square with the subsequent smaller numbers and their complements. We are left with the rectangles, which we already know how to fill: we may either fill them with neutral placings only (since their length, $4t+2$, is even while their width $2(k-t)$, or twice it, is divisible by 4); or we can consider in them subsquares (which can always be 4×4), which we shall fill in such a way that the sum due appears, this leaving us again with filling smaller rectangular strips using neutral placings.

1	194	191	8	51	55	59	138	142	146	9	186	183	16
192	7	2	193	145	141	137	60	56	52	184	15	10	185
6	189	196	3	144	140	136	61	57	53	14	181	188	11
195	4	5	190	54	58	62	135	139	143	187	12	13	182
107	89	88	110	33	37	161	162	160	38	63	133	132	66
103	93	92	106	153	40	42	155	43	158	67	129	128	70
99	97	96	102	50	148	47	48	151	147	71	125	124	74
98	100	101	95	152	49	149	150	46	45	126	72	73	123
94	104	105	91	44	154	156	41	157	39	130	68	69	127
90	108	109	87	159	163	36	35	34	164	134	64	65	131
17	178	175	24	122	118	114	83	79	75	25	170	167	32
176	23	18	177	76	80	84	113	117	121	168	31	26	169
22	173	180	19	77	81	85	112	116	120	30	165	172	27
179	20	21	174	119	115	111	86	82	78	171	28	29	166

Fig. 187

Our figures 187 and 188 illustrate these two possibilities. The square in the centre has been treated in the manner of Fig. 161, that is, according to the general method of exchanges. Thus in Fig. 187 the sequences from 33 to 50 and from 147 to 164 take the place of 1 to 18 and 19 to 36, respectively, while in Fig. 188 the central square is filled with the continuous sequence 81 to 116.

Thābit ibn Qurra's text (and thus Anṭākī's) mentions several examples of division, some of which are actually constructed in Thābit's text. We shall examine them in § 5 of the next chapter. (But the case seen in Fig. 188, of a square part cut by a strip, does not occur.)

1	194	191	8	9	186	73	124	183	16	17	178	175	24
192	7	2	193	184	15	123	74	10	185	176	23	18	177
6	189	196	3	14	181	122	75	188	11	22	173	180	19
195	4	5	190	187	12	76	121	13	182	179	20	21	174
57	138	135	64	81	85	113	114	112	86	25	170	167	32
136	63	58	137	105	88	90	107	91	110	168	31	26	169
65	131	130	68	98	100	95	96	103	99	69	127	126	72
132	66	67	129	104	97	101	102	94	93	128	70	71	125
62	133	140	59	92	106	108	89	109	87	30	165	172	27
139	60	61	134	111	115	84	83	82	116	171	28	29	166
49	146	143	56	41	154	77	120	151	48	33	162	159	40
144	55	50	145	152	47	119	78	42	153	160	39	34	161
54	141	148	51	46	149	118	79	156	43	38	157	164	35
147	52	53	142	155	44	80	117	45	150	163	36	37	158

Fig. 188

Chapter III

Composite magic squares

Here the construction of a larger square is once again reduced to that of its individual parts, mostly squares, each of which is itself magic. The possibility of constructing a square in such a way depends of course on the divisibility of its order.

A. What is usually meant by composite square is a square divided into even or odd subsquares each of the same order, *filled one after the other completely by a continuous sequence of numbers in magic arrangement.* Since the sums in the subsquares are different, but they form an arithmetical progression, these subsquares must be arranged in such a way that the main square will itself be magic.

B. If the main square can be divided into subsquares of the same *even* order ≥ 4, they can be *filled individually with pairs of complements in magic arrangement.* Since these subsquares are all of the same size and display the same sum, they can be arranged in any manner.

C. A less conventional possibility is to divide the main square into *unequal parts*, square or even rectangular, but here again filled with pairs of complements. Each part is to display its sum due, which of course will vary according to its dimensions.

Such are the three variants we shall examine in turn. The last two, of which we have already seen a few cases, are studied in the text translated by Thābit ibn Qurra and reproduced by Anṭākī. As to the first kind, it is explained in the tenth century by Abū'l-Wafā' Būzjānī, but the London manuscript also refers to Thābit's translation in this context (below, p. 111). We may thus say that the construction of composite squares in these three forms was well established in the tenth century and probably inherited altogether from antiquity.

A. *EQUAL SUBSQUARES DISPLAYING DIFFERENT SUMS*

§1. Composition using squares of orders larger than 2

Suppose our main square to have a composite order of the form $n = r \cdot s$. We can therefore divide our square into r^2 squares of order s which will be filled each with a continuous sequence of s^2 numbers arranged according to any method suitable for the order s, the r^2 squares thus filled

J. Sesiano, *Magic Squares*, Sources and Studies in the History of Mathematics and Physical Sciences, https://doi.org/10.1007/978-3-030-17993-9_3

being themselves placed according to any arrangement suitable for the order r. If $r \neq s$, we shall obtain a different arrangement by exchanging the rôles of r and s. Likewise, another factorization of n will lead to another arrangement or two arrangements.

There are restrictions to the construction of a composite square. First, obviously, the order of the main square must not be prime. But it cannot be twice a prime either: since a square of order 2 can only be magic if its elements are equal, we could not place four squares of unequal magic sums in a magic arrangement. Since that excludes both order 6 and order 8, we shall suppose that $n = r \cdot s$ with r and s larger than 2. Therefore, the smallest possible composite magic square is that of order 9, broken up into nine squares of order 3 (Fig. 189 & Fig. 190).

31	36	29	76	81	74	13	18	11
30	32	34	75	77	79	12	14	16
35	28	33	80	73	78	17	10	15
22	27	20	40	45	38	58	63	56
21	23	25	39	41	43	57	59	61
26	19	24	44	37	42	62	55	60
67	72	65	4	9	2	49	54	47
66	68	70	3	5	7	48	50	52
71	64	69	8	1	6	53	46	51

Fig. 189   Fig. 190

For once, we may give an example of explanation in a late text. Kishnāwī describes the above construction as follows.[100] *You are to know that the odd (in this section he deals only with odd-order squares) cannot be constructed as composite if its side is not composite relative to multiplication; such are the square of 9, the square of 15, the square of 21, and other similar ones. The method is as follows. You divide the main square into small squares, each having uniformly the same number of parts on the side and their global quantity being equal to the square of the number by which the side of these small squares measures the side of the large square. Thus, the square of 9 is divided into 9 squares of 3, the square of 15 into 9 squares of 5 or, if you will, into 25 squares of 3, the square of 21*

---

[100] *Quelques méthodes*, p. 68; MS. London School of Oriental and African studies 65496, fol. 96^r, 14-96^v, 4, or below, Appendix 16. Text almost identical to that of Shabrāmallisī, MS. Paris BNF Arabe 2698, fol. 10^v-11^v.

into 9 squares of 7 or into 49 squares of 3, and so on. Then you put in the first of the small squares the natural numbers from 1 to the number of cells of the small square, performing the placing according to one of the methods for filling this category of square. Then you put, in the second of the small squares, the natural numbers beginning with that following the last of the numbers placed in the first square. Then you put, in the third of the small squares, the natural numbers beginning with that following the last of the numbers placed in the second square. Then you put the natural numbers in the fourth square, beginning with that following the last of the numbers placed in the third square. You continue filling the squares, one after the other, in this same manner, till you have filled the main square. And the first of the small squares is that which would take the place of 1 if you were to consider the small squares as cells, the second that which would take the place of 2, the third that which would take the place of 3, the fourth that which would take the place of 4, and so on in a similar way. After following this method for dealing with the small magic squares and arranging them, the main square will also be magic. Here is an example of that for the 9×9 square, filled completely to serve as a model in other cases (Fig. 190 above).

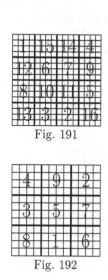

Fig. 191

Fig. 192

49	63	62	52	129	143	142	132	17	31	30	20
60	54	55	57	140	134	135	137	28	22	23	25
56	58	59	53	136	138	139	133	24	26	27	21
61	51	50	64	141	131	130	144	29	19	18	32
33	47	46	36	65	79	78	68	97	111	110	100
44	38	39	41	76	70	71	73	108	102	103	105
40	42	43	37	72	74	75	69	104	106	107	101
45	35	34	48	77	67	66	80	109	99	98	112
113	127	126	116	1	15	14	4	81	95	94	84
124	118	119	121	12	6	7	9	92	86	87	89
120	122	123	117	8	10	11	5	88	90	91	85
125	115	114	128	13	3	2	16	93	83	82	96

Fig. 193

The next possible order is $n = 12$, which may be divided into sixteen squares of order 3 arranged according to one of the ways for order 4 (Fig. 191), or into nine squares of order 4 placed according to the magic arrangement for order 3 (Fig. 192-193). We have added the construction of a composite square of order 15 (Fig. 194), and a variant with a central bordered square (Fig. 195; see, for the border, Fig. 239).[101]

The magic property of a composite square is easy to explain. We have r^2 subsquares of order s. The tth of them ($t = 1, \ldots, r^2$) is made up of the elements $i + (t-1)s^2$ ($i = 1, \ldots, s^2$); that is, it will have the magic sum

$$M^{(t)} = M_s + (t-1)s^3$$

and thus the sum on the whole tth square will be

$$s \cdot M^{(t)} = s \cdot M_s + (t-1)s^4.$$

The problem is now to arrange magically r^2 quantities in arithmetical progression with constant difference s^4, the first term of which is $s \cdot M_s$. Any magical arrangement for order r will do: indeed, whether the numbers are consecutive or in arithmetical progression with a constant difference other than 1, their arrangement remains the same. Thus, our example of Fig. 194 corresponds to Fig. 196, where $s \cdot M_s + (t-1)s^4 = 325 + (t-1) \cdot 625$.

198	187	176	195	184	23	12	1	20	9	148	137	126	145	134
179	193	182	196	190	4	18	7	21	15	129	143	132	146	140
185	199	188	177	191	10	24	13	2	16	135	149	138	127	141
186	180	194	183	197	11	5	19	8	22	136	130	144	133	147
192	181	200	189	178	17	6	25	14	3	142	131	150	139	128
73	62	51	70	59	123	112	101	120	109	173	162	151	170	159
54	68	57	71	65	104	118	107	121	115	154	168	157	171	165
60	74	63	52	66	110	124	113	102	116	160	174	163	152	166
61	55	69	58	72	111	105	119	108	122	161	155	169	158	172
67	56	75	64	53	117	106	125	114	103	167	156	175	164	153
98	87	76	95	84	223	212	201	220	209	48	37	26	45	34
79	93	82	96	90	204	218	207	221	215	29	43	32	46	40
85	99	88	77	91	210	224	213	202	216	35	49	38	27	41
86	80	94	83	97	211	205	219	208	222	36	30	44	33	47
92	81	100	89	78	217	206	225	214	203	42	31	50	39	28

Fig. 194

[101] In both, the move between sequences is to an adjacent cell. That with the bordered square in the centre and the starting elements in the upper corner cells (see p. 39) is found in MSS. Istanbul Ayasofya 2794, fol. 12^r, and Mashhad Āstān-i Quds 12235, p. 434; same in MS. Āstān-i Quds 14159, fol. 84^v, but the central square is not bordered.

126	150	144	138	132	151	175	169	163	157	26	50	44	38	32
139	133	127	146	145	164	158	152	171	170	39	33	27	46	45
147	141	140	134	128	172	166	165	159	153	47	41	40	34	28
135	129	148	142	136	160	154	173	167	161	35	29	48	42	36
143	137	131	130	149	168	162	156	155	174	43	37	31	30	49
1	25	19	13	7	123	108	105	122	107	201	225	219	213	207
14	8	2	21	20	120	116	111	112	106	214	208	202	221	220
22	16	15	9	3	101	109	113	117	125	222	216	215	209	203
10	4	23	17	11	102	114	115	110	124	210	204	223	217	211
18	12	6	5	24	119	118	121	104	103	218	212	206	205	224
176	200	194	188	182	51	75	69	63	57	76	100	94	88	82
189	183	177	196	195	64	58	52	71	70	89	83	77	96	95
197	191	190	184	178	72	66	65	59	53	97	91	90	84	78
185	179	198	192	186	60	54	73	67	61	85	79	98	92	86
193	187	181	180	199	68	62	56	55	74	93	87	81	80	99

4700	325	3450
1575	2825	4075
2200	5325	950

Fig. 195   Fig. 196

We may of course apply to each subsquare the elementary transformations seen in I, §3. Furthermore, if one of the factors is itself composite, a finer break-up is possible: $n = r \cdot s \cdot t = (r \cdot s) \cdot t$. In Fig. 197, a square of order 27 is divided into nine squares arranged magically as for the order 3, each of these nine squares being in turn divided into nine squares of order 3.

As said in the beginning, the construction of higher-order squares by division into smaller ones displaying different sums was well known in the tenth century. For instance, for odd squares, Abū'l-Wafā' Būzjānī constructs the square of order 9. That is the only one; of further odd orders, he mentions the cases of $15 = 3 \cdot 5$ and $21 = 3 \cdot 7$, and, for even squares, $12 = 3 \cdot 4$ and $20 = 4 \cdot 5$, each time with the two variants. He does not construct any of them: the extant manuscript of his work, without a single figure but leaving each time a blank space, has none here and the text does not refer to any.[102] Now the London manuscript reporting some of Thābit ibn Qurra's translation refers here to both Abū'l-Wafā' and Thābit, and has five examples of squares, all illustrating the cases $12 = 3 \cdot 4$ and $15 = 3 \cdot 5$.[103] A Greek origin of

---

[102] *Magic squares in the tenth century*, B.24-25.
[103] MS. London BL Delhi Arabic 110, fol. 94^r-95^v.

274	279	272	319	324	317	256	261	254	679	684	677	724	729	722	661	666	659	112	117	110	157	162	155	94	99	92
273	275	277	318	320	322	255	257	259	678	680	682	723	725	727	660	662	664	111	113	115	156	158	160	93	95	97
278	271	276	323	316	321	260	253	258	683	676	681	728	721	726	665	658	663	116	109	114	161	154	159	98	91	96
265	270	263	283	288	281	301	306	299	670	675	668	688	693	686	706	711	704	103	108	101	121	126	119	139	144	137
264	266	268	282	284	286	300	302	304	669	671	673	687	689	691	705	707	709	102	104	106	120	122	124	138	140	142
269	262	267	287	280	285	305	298	303	674	667	672	692	685	690	710	703	708	107	100	105	125	118	123	143	136	141
310	315	308	247	252	245	292	297	290	715	720	713	652	657	650	697	702	695	148	153	146	85	90	83	130	135	128
309	311	313	246	248	250	291	293	295	714	716	718	651	653	655	696	698	700	147	149	151	84	86	88	129	131	133
314	307	312	251	244	249	296	289	294	719	712	717	656	649	654	701	694	699	152	145	150	89	82	87	134	127	132
193	198	191	238	243	236	175	180	173	355	360	353	400	405	398	337	342	335	517	522	515	562	567	560	499	504	497
192	194	196	237	239	241	174	176	178	354	356	358	399	401	403	336	338	340	516	518	520	561	563	565	498	500	502
197	190	195	242	235	240	179	172	177	359	352	357	404	397	402	341	334	339	521	514	519	566	559	564	503	496	501
184	189	182	202	207	200	220	225	218	346	351	344	364	369	362	382	387	380	508	513	506	526	531	524	544	549	542
183	185	187	201	203	205	219	221	223	345	347	349	363	365	367	381	383	385	507	509	511	525	527	529	543	545	547
188	181	186	206	199	204	224	217	222	350	343	348	368	361	366	386	379	384	512	505	510	530	523	528	548	541	546
229	234	227	166	171	164	211	216	209	391	396	389	328	333	326	373	378	371	553	558	551	490	495	488	535	540	533
228	230	232	165	167	169	210	212	214	390	392	394	327	329	331	372	374	376	552	554	556	489	491	493	534	536	538
233	226	231	170	163	168	215	208	213	395	388	393	332	325	330	377	370	375	557	550	555	494	487	492	539	532	537
598	603	596	643	648	641	580	585	578	31	36	29	76	81	74	13	18	11	436	441	434	481	486	479	418	423	416
597	599	601	642	644	646	579	581	583	30	32	34	75	77	79	12	14	16	435	437	439	480	482	484	417	419	421
602	595	600	647	640	645	584	577	582	35	28	33	80	73	78	17	10	15	440	433	438	485	478	483	422	415	420
589	594	587	607	612	605	625	630	623	22	27	20	40	45	38	58	63	56	427	432	425	445	450	443	463	468	461
588	590	592	606	608	610	624	626	628	21	23	25	39	41	43	57	59	61	426	428	430	444	446	448	462	464	466
593	586	591	611	604	609	629	622	627	26	19	24	44	37	42	62	55	60	431	424	429	449	442	447	467	460	465
634	639	632	571	576	569	616	621	614	67	72	65	4	9	2	49	54	47	472	477	470	409	414	407	454	459	452
633	635	637	570	572	574	615	617	619	66	68	70	3	5	7	48	50	52	471	473	475	408	410	412	453	455	457
638	631	636	575	568	573	620	613	618	71	64	69	8	1	6	53	46	51	476	469	474	413	406	411	458	451	456

Fig. 197

such, or even these, squares is therefore not unlikely. We thus find successively (here reading left to right):

— A 12 × 12 square divided into sixteen 3 × 3 subsquares (Fig. 198a), arranged as in Fig. 198b.

— A 12 × 12 square divided into sixteen 3 × 3 subsquares, filled alternately by changing parity (Fig. 199a), arranged as in Fig. 199b.

— A 12 × 12 square divided into nine 4 × 4 subsquares (Fig. 200; see Fig. 193).

— A 15 × 15 square divided into nine 5 × 5 (bordered) subsquares (Fig. 201; cf. our Fig. 194, and Fig. 202b for the borders).

— A 15 × 15 square divided into twenty-five 3 × 3 subsquares, arranged as for a 5 × 5 bordered square (Fig. 202a). We have also indicated the sequence in which its parts are filled (Fig. 202b).

33	28	35	128	135	130	83	90	85	42	37	44
34	32	30	133	131	129	88	86	84	43	41	39
29	36	31	132	127	134	87	82	89	38	45	40
119	126	121	2	9	4	65	72	67	92	99	94
124	122	120	7	5	3	70	68	66	97	95	93
123	118	125	6	1	8	69	64	71	96	91	98
56	63	58	101	108	103	110	117	112	11	18	13
61	59	57	106	104	102	115	113	111	16	14	12
60	55	62	105	100	107	114	109	116	15	10	17
74	81	76	47	54	49	20	27	22	137	144	139
79	77	75	52	50	48	25	23	21	142	140	138
78	73	80	51	46	53	24	19	26	141	136	143

Fig. 198a

Fig. 198b

124	110	120	21	35	25	4	18	8	141	127	137
114	118	122	31	27	23	14	10	6	131	135	139
116	126	112	29	19	33	12	2	16	133	143	129
58	72	62	87	73	83	106	92	102	39	53	43
68	64	60	77	81	85	96	100	104	49	45	41
66	56	70	79	89	75	98	108	94	47	37	51
105	91	101	40	54	44	57	71	61	88	74	84
95	99	103	50	46	42	67	63	59	78	82	86
97	107	93	48	38	52	65	55	69	80	90	76
3	17	7	142	128	138	123	109	119	22	36	26
13	9	5	132	136	140	113	117	121	32	28	24
11	1	15	134	144	130	115	125	111	30	20	34

Fig. 199a

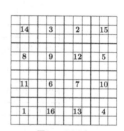

Fig. 199b

20	31	26	21	132	143	138	133	52	63	58	53
30	17	24	27	142	129	136	139	62	49	56	59
23	28	29	18	135	140	141	130	55	60	61	50
25	22	19	32	137	134	131	144	57	54	51	64
100	111	106	101	68	79	74	69	36	47	42	37
110	97	104	107	78	65	72	75	46	33	40	43
103	108	109	98	71	76	77	66	39	44	45	34
105	102	99	112	73	70	67	80	41	38	35	48
84	95	90	85	4	15	10	5	116	127	122	117
94	81	88	91	14	1	8	11	126	113	120	123
87	92	93	82	7	12	13	2	119	124	125	114
89	86	83	96	9	6	3	16	121	118	115	128

Fig. 200

129	149	148	133	131	20	1	21	19	4	179	199	198	183	181
144	135	142	137	132	2	10	17	12	24	194	185	192	187	182
146	140	138	136	130	3	15	13	11	23	196	190	188	186	180
126	139	134	141	150	18	14	9	16	8	176	189	184	191	200
145	127	128	143	147	22	25	5	7	6	195	177	178	193	197
154	174	173	158	156	120	101	121	119	104	54	74	73	58	56
169	160	167	162	157	102	110	117	112	124	69	60	67	62	57
171	165	163	161	155	103	115	113	111	123	71	65	63	61	55
151	164	159	166	175	118	114	109	116	108	51	64	59	66	75
170	152	153	168	172	122	125	105	107	106	70	52	53	68	72
29	49	48	33	31	204	224	223	208	206	79	99	98	83	81
44	35	42	37	32	219	210	217	212	207	94	85	92	87	82
46	40	38	36	30	221	215	213	211	205	96	90	88	86	80
26	39	34	41	50	201	214	209	216	225	76	89	84	91	100
45	27	28	43	47	220	202	203	218	222	95	77	78	93	97

Fig. 201

173	180	175	6	7	2	182	189	184	164	171	166	29	36	31
178	176	174	1	5	9	187	185	183	169	167	165	34	32	30
177	172	179	8	3	4	186	181	188	168	163	170	33	28	35
11	18	13	83	90	85	146	153	148	101	108	103	209	216	211
16	14	12	88	86	84	151	149	147	106	104	102	214	212	210
15	10	17	87	82	89	150	145	152	105	100	107	213	208	215
20	27	22	128	135	130	110	117	112	92	99	94	200	207	202
25	23	21	133	131	129	115	113	111	97	95	93	205	203	201
24	19	26	132	127	134	114	109	116	96	91	98	204	199	206
155	162	157	119	126	121	74	81	76	137	144	139	65	72	67
160	158	156	124	122	120	79	77	75	142	140	138	70	68	66
159	154	161	123	118	125	78	73	80	141	136	143	69	64	71
191	198	193	218	225	220	38	45	40	56	63	58	47	54	49
196	194	192	223	221	219	43	41	39	61	59	57	52	50	48
195	190	197	222	217	224	42	37	44	60	55	62	51	46	53

Fig. 202a

20	1	21	19	4
2	10	17	12	24
3	15	13	11	23
18	14	9	16	8
22	25	5	7	6

Fig. 202b

1	62	35	32	197	250	231	220	137	182	171	152	77	114	111	84
36	31	2	61	232	219	198	249	172	151	138	181	112	83	78	113
30	33	64	3	218	229	252	199	150	169	184	139	82	109	116	79
63	4	29	34	251	200	217	230	183	140	149	170	115	80	81	110
141	178	175	148	73	118	107	88	5	58	39	28	193	254	227	224
176	147	142	177	108	87	74	117	40	27	6	57	228	223	194	253
146	173	180	143	86	105	120	75	26	37	60	7	222	225	256	195
179	144	145	174	119	76	85	106	59	8	25	38	255	196	221	226
69	122	103	92	129	190	163	160	205	242	239	212	9	54	43	24
104	91	70	121	164	159	130	189	240	211	206	241	44	23	10	53
90	101	124	71	158	161	192	131	210	237	244	207	22	41	56	11
123	72	89	102	191	132	157	162	243	208	209	238	55	12	21	42
201	246	235	216	13	50	47	20	65	126	99	96	133	186	167	156
236	215	202	245	48	19	14	49	100	95	66	125	168	155	134	185
214	233	248	203	18	45	52	15	94	97	128	67	154	165	188	135
247	204	213	234	51	16	17	46	127	68	93	98	187	136	153	166

Fig. 203

Let us conclude this section with one particular example taken from a later source. The square of Fig. 203, pandiagonal, has been constructed with four sequences of four pandiagonal subsquares, the magic sums of which are, within each sequence, equal: 130, 386, 642, 898 (whereas in Fig. 154 the individual sums are all equal).[104]

§2. Composition using squares of order 2

In the above paragraph, we supposed that $n = r \cdot s$ with $r, s > 2$. Another type of composition is applicable if one of the factors is 2, so that n need only be even (≥ 6). Consider thus an empty square of even order $n = 2r$ (r even or odd), which we divide into smaller squares of order 2.

Imagine filling a square of order 2 with the first four natural numbers. We already know that a magic arrangement is not possible since we cannot obtain uniformly the (theoretical) magic sum $M_2 = 5$. We shall, however, suppose that we can arrange in the square of order $2r$ such small squares, all filled with the first four numbers, in such a way that their individual excesses and deficits cancel one another out in each of the lines, columns, and main diagonals. In that case, we would obtain the constant sum $M_2 \cdot r = 5 \cdot r$ in all the rows (lines, columns, diagonals) of the larger square. We shall then, taking these r^2 small squares according to a magic arrangement for order r (so we must suppose $r \geq 3$), add the successive multiples of 4 to the four numbers already present. The main square of order $2r$ will then be magic.

Indeed, the tth small square (according to the magic arrangement), with each of its cells increased by $(t-1)4$, will comprise the four numbers $1 + (t-1)4$, $2 + (t-1)4$, $3 + (t-1)4$, $4 + (t-1)4$. Its *global* sum will then be

$$10 + (t-1) \cdot 16 = 2 \cdot M_2 + (t-1) \cdot 2^4,$$

a situation we have already seen (p. 110); thus these quantities, placed in the main square according to an arrangement appropriate for the order r, will indeed make in it the required sum M_{2r}.

Shabrāmallisī explains quite clearly the more difficult part, which is the preliminary placing of the smaller squares filled with the first four numbers in the main square in order to eliminate their individual differences with M_5.[105] He has thus first to consider these individual differences in the twenty-four configurations those smaller squares may display. As he observes, in each of these squares the differences between $M_2 = 5$ and

---

[104] MS. Istanbul Ayasofya 2794, fol. 17ᵛ, or Mashhad Āstān-i Quds 14159, fol. 86ᵛ.
[105] MS. Paris BNF Arabe 2698, fol. 47ʳ - 48ᵛ.

the sum in each line, column, diagonal are all unequal, but equal with opposite signs for the same kind of row; that is, the differences with M_2 in a single 2×2 square will be 0, ± 1, ± 2 while two rows of the same kind will differ by 0, 2, or 4 (*You are to know that each of these cells which are squares of 2 has two diagonals, two lines and two columns, with always two (homonymous) being equal, two differing by 2, two differing by 4*). This enables Shabrāmallisī to divide the twenty-four configurations into three groups, according to the difference (in absolute value) between their diagonals and M_2, which he does because of the particular rôle to be played by the diagonals. We shall designate here the difference between M_2 and the diagonals by Δ_d, between M_2 and the horizontal rows (lines) by Δ_h, between M_2 and the columns by Δ_v.

- First group (Fig. 204): the two diagonals contain M_2, thus either

$$\Delta_d^{(1)} = 0, \quad \Delta_h^{(1)} = \pm 1, \quad \Delta_v^{(1)} = \pm 2 \quad \text{(Fig. 204, } a\text{-}d\text{)}$$

or

$$\Delta_d^{(1)} = 0, \quad \Delta_h^{(1)} = \pm 2, \quad \Delta_v^{(1)} = \pm 1 \quad \text{(Fig. 204, } e\text{-}h\text{)}.$$

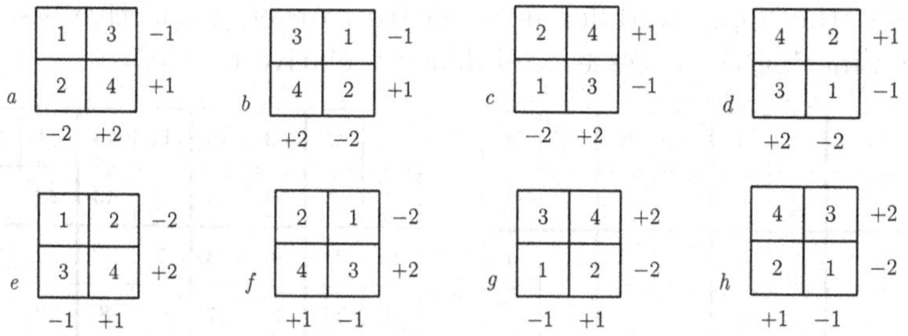

Fig. 204

- Second group (Fig. 205): the diagonals differ from M_2 by ± 1, thus either

$$\Delta_d^{(2)} = \pm 1, \quad \Delta_h^{(2)} = 0, \quad \Delta_v^{(2)} = \pm 2 \quad \text{(Fig. 205, } a\text{-}d\text{)}$$

or

$$\Delta_d^{(2)} = \pm 1, \quad \Delta_h^{(2)} = \pm 2, \quad \Delta_v^{(2)} = 0 \quad \text{(Fig. 205, } e\text{-}h\text{)}.$$

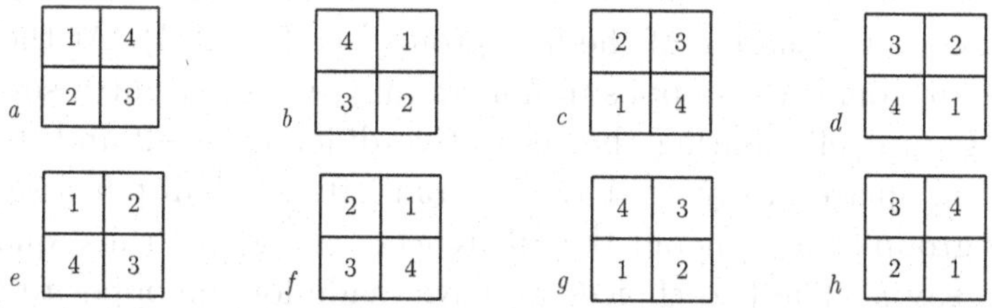

Fig. 205

- Third group (Fig. 206): the diagonals differ from M_2 by ± 2, thus either

$$\Delta_d^{(3)} = \pm 2, \quad \Delta_h^{(3)} = \pm 1, \quad \Delta_v^{(3)} = 0 \quad \text{(Fig. 206, } a\text{-}d\text{)}$$

or

$$\Delta_d^{(3)} = \pm 2, \quad \Delta_h^{(3)} = 0, \quad \Delta_v^{(3)} = \pm 1 \quad \text{(Fig. 206, } e\text{-}h\text{)}.$$

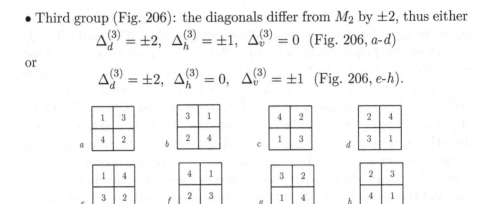

Fig. 206

We are now to place these small squares in an appropriate manner. Since it is always the filling of the diagonals which is most problematic, we shall deal with them first. For one diagonal we use squares of the first group, which meet the condition of the diagonal sum; for the other, squares of the first group as well, but ones which display, horizontally or vertically, the opposite difference relative to their respective associates, thus eliminating the excesses and deficits relative to $M_2 = 5$.

1	2	1	4	1	4	3	4
3	4	2	3	2	3	1	2
1	4	2	4	1	3	1	4
2	3	1	3	2	4	2	3
4	1	4	2	3	1	3	2
3	2	3	1	4	2	4	1
2	1	3	2	4	1	4	3
4	3	4	1	3	2	2	1

Fig. 207

1	2	53	56	41	44	31	32
3	4	54	55	42	43	29	30
45	48	26	28	5	7	49	52
46	47	25	27	6	8	50	51
24	21	36	34	63	61	11	10
23	22	35	33	64	62	12	9
58	57	15	14	20	17	40	39
60	59	16	13	19	18	38	37

Fig. 208

Suppose for instance that, as in Fig. 207, we have placed, in the upper left-hand corner, square e of the first group (see Fig. 204). At the top of the other diagonal, thus opposite horizontally, we shall put a square (of the first group) eliminating the respective differences -2 and $+2$, thus square g or square h, say g; at the bottom left, we shall put a square of the same group cancelling out the differences -1 and $+1$, thus square f or square h, say f. The two choices we have made for the opposite corners will determine which square to put in the fourth corner, in this case h.

Thus far we have remained in the lower row of the squares in Fig. 204. Had we chosen a square from the upper row, the other three would also be there. For instance, square c would have square a on the right and squares d and b below. That is what has been done with the four squares in the centre.

The two diagonals are thus covered by small squares eliminating the differences. The remainder of the horizontal and vertical rows, which do not interfere with the diagonals, will be filled by placing opposite to one another small squares displaying horizontally and vertically opposite differences (if any).

Such a procedure is valid for both evenly-even and evenly-odd orders. In the second case, however, we cannot complete the remainder of the square, after filling the diagonals, as simply as before, with squares of the same groups vertically or horizontally opposite: filling the central cross requires particular attention.

1	2	2	3	3	4
3	4	4	1	1	2
1	3	1	3	4	3
4	2	2	4	1	2
2	1	2	3	4	3
4	3	4	1	2	1

Fig. 209

13	14	34	35	7	8
15	16	36	33	5	6
9	11	17	19	28	27
12	10	18	20	25	26
30	29	2	3	24	23
32	31	4	1	22	21

Fig. 210

1	2	1	2	4	3	4	3
3	4	3	4	2	1	2	1
1	2	1	2	4	3	4	3
3	4	3	4	2	1	2	1
4	3	4	3	1	2	1	2
2	1	2	1	3	4	3	4
4	3	4	3	1	2	1	2
2	1	2	1	3	4	3	4

Fig. 211

To Fig. 207 and 209 of the initial configuration correspond the magic squares of order 8 and 6 of Fig. 208 and 210 (in the first one we recognize the usual configuration of the pandiagonal 4×4 square). These two examples are those of Shabrāmallisī. We have added Fig. 211, which uses only the squares e and h of Fig. 204; it enabled us to fill the two squares of Fig. 212 and 213, constructed on the model of the 4×4 of Fig. 76 and 78, respectively.

1	2	57	58	56	55	16	15
3	4	59	60	54	53	14	13
45	46	21	22	28	27	36	35
47	48	23	24	26	25	34	33
32	31	40	39	41	42	17	18
30	29	38	37	43	44	19	20
52	51	12	11	5	6	61	62
50	49	10	9	7	8	63	64

Fig. 212

1	2	53	54	44	43	32	31
3	4	55	56	42	41	30	29
45	46	25	26	8	7	52	51
47	48	27	28	6	5	50	49
24	23	36	35	61	62	9	10
22	21	34	33	63	64	11	12
60	59	16	15	17	18	37	38
58	57	14	13	19	20	39	40

Fig. 213

B. EQUAL SUBSQUARES DISPLAYING EQUAL SUMS

In what follows we shall examine the composition of a given square with same subsquares displaying their sum due *with respect to the whole square*, thus $\frac{m}{2} \cdot (n^2 + 1)$ for $m \times m$ subsquares within a square of order n.

§3. Examples

We have already seen (II, §22) squares of order 12 broken up into smaller, even-order squares which were filled individually with pairs of complements, each of these squares becoming a magic entity displaying its sum due, and these subsquares being filled in any succession we wish. This is extended to higher-order squares already in Greek times. Indeed, Thābit ibn Qurra's text (and thus Antākī's) mentions such examples of division. For evenly-even orders, we may always divide the square into four or sixteen subsquares; if the order (then also evenly-odd, see below) is divisible by 3, we may have nine even-order subsquares. As Thābit ibn Qurra (and after him Antākī) expresses it: *You draw the main square, which arises from the multiplication of an even number by itself. Then its inner part is divided into four squares such that the side of each equals half the side of the main square; or it is divided into sixteen squares such that the side of each of these squares equals a fourth of the side of the* main square; or it is divided into nine squares such that the side of each equals a third of the side of the main one (...). Then you will arrange the numbers in all the cells of the main square from 1 to the end of their quantity in such manner that in the (main) square the resulting sums are everywhere the same, and that in each inner square considered by itself the resulting sums are everywhere the same.[106] Whereas Anṭākī has only two constructed examples, namely four and nine subsquares for order 12, there are one smaller and three larger examples in the manuscript reporting Thābit's translation. The source is said to be Abū'l-Wafā' and Thābit; since Abū'l-Wafā' only *mentions* two small examples (8 × 8 and 12 × 12 with four and nine subsquares, respectively), these examples are probably taken from Thābit's translation. (We have indicated for the three larger squares the sequence in which their parts are filled: remember that for equal subsquares displaying equal sums any arrangement of the subsquares may be considered.)

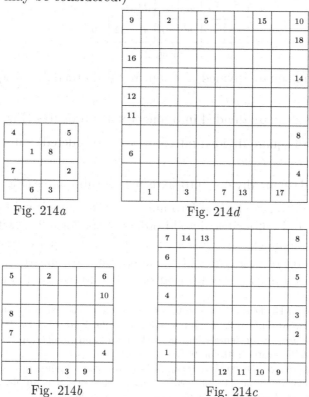

Fig. 214a

Fig. 214d

Fig. 214b

Fig. 214c

Remark. In all the examples we shall see, the arrangements adopted for

[106] *Magic squares in the tenth century*, A.II.44, and MS. London BL Delhi Arabic 110, fol. 99^r, 7-12.

the inner 4×4 square and the surrounding borders of orders 6, 8, 10, are the same, namely (omitting the complements) the above ones (Fig. 214a-d).

Thus we find successively:

— An 8×8 square divided into four 4×4 subsquares (Fig. 215).[107]

28	39	34	29	12	55	50	13
38	25	32	35	54	9	16	51
31	36	37	26	15	52	53	10
33	30	27	40	49	14	11	56
4	63	58	5	20	47	42	21
62	1	8	59	46	17	24	43
7	60	61	2	23	44	45	18
57	6	3	64	41	22	19	48

Fig. 215

— A 12×12 square divided into four 6×6 (bordered) subsquares (Fig. 216).[108]

— A 12×12 square divided into nine 4×4 subsquares (Fig. 217).[109]

— A 16×16 square divided into four 8×8 (bordered) subsquares (Fig. 218).[110]

— A 16×16 square divided into sixteen 4×4 subsquares (Fig. 219a-b).[111]

— A 18×18 square divided into nine 6×6 (bordered) subsquares (Fig. 220a-b).[112] Indeed, for evenly-odd orders of the form $2 \cdot r$ with r a composite (odd) number, say $r = s \cdot t$ with $s, t \geq 3$, the square may be divided into s^2 squares of even order $2t$. Thābit's (and Antākī's) text mentions, in addition to the case $n = 18$, that of a 30×30 square divided into nine

---

[107] MS. London BL Delhi Arabic 110, fol. 96ᵛ. One of the examples mentioned (not constructed) by Abū'l-Wafā', see *Magic squares in the tenth century*, B.26.*iv*.

[108] *Magic squares in the tenth century*, A.II.44 & 48 and (figure) pp. 174 & 314; MS. London BL Delhi Arabic 110, fol. 96ᵛ.

[109] *Magic squares in the tenth century*, A.II.44 & 48 and (figure) pp. 174 & 315; MS. London BL Delhi Arabic 110, fol. 96ᵛ. This is the second example mentioned by Abū'l-Wafā', see *Magic squares in the tenth century*, B.26.*iii*.

[110] MS. London BL Delhi Arabic 110, fol. 97ʳ.

[111] MS. London BL Delhi Arabic 110, fol. 97ᵛ.

[112] *Magic squares in the tenth century*, A.II.45, and MS. London BL Delhi Arabic 110, fol. 99ʳ, 19-20, (mention), and fol. 98ʳ (figure).

10 × 10 subsquares or twenty-five 6 × 6 subsquares.[113]

— A 20 × 20 square divided into twenty-five 4 × 4 subsquares (Fig. 221a-b).[114]

23	126	20	124	118	24	59	90	56	88	82	60
117	32	115	110	33	28	81	68	79	74	69	64
26	114	29	36	111	119	62	78	65	72	75	83
25	35	112	113	30	120	61	71	76	77	66	84
123	109	34	31	116	22	87	73	70	67	80	58
121	19	125	21	27	122	85	55	89	57	63	86
41	108	38	106	100	42	5	144	2	142	136	6
99	50	97	92	51	46	135	14	133	128	15	10
44	96	47	54	93	101	8	132	11	18	129	137
43	53	94	95	48	102	7	17	130	131	12	138
105	91	52	49	98	40	141	127	16	13	134	4
103	37	107	39	45	104	139	1	143	3	9	140

Fig. 216

60	87	82	61	28	119	114	29	68	79	74	69
86	57	64	83	118	25	32	115	78	65	72	75
63	84	85	58	31	116	117	26	71	76	77	66
81	62	59	88	113	30	27	120	73	70	67	80
12	135	130	13	52	95	90	53	20	127	122	21
134	9	16	131	94	49	56	91	126	17	24	123
15	132	133	10	55	92	93	50	23	124	125	18
129	14	11	136	89	54	51	96	121	22	19	128
36	111	106	37	4	143	138	5	44	103	98	45
110	33	40	107	142	1	8	139	102	41	48	99
39	108	109	34	7	140	141	2	47	100	101	42
105	38	35	112	137	6	3	144	97	46	43	104

Fig. 217

---

[113] *Magic squares in the tenth century*, A.II.45, and MS. London BL Delhi Arabic 110, fol. 99ʳ, 17 & 20-22.

[114] MS. London BL Delhi Arabic 110, fol. 98ᵛ.

103	110	109	149	150	151	152	104	39	46	45	213	214	215	216	40
102	115	146	112	144	138	116	155	38	51	210	48	208	202	52	219
156	137	124	135	130	125	120	101	220	201	60	199	194	61	56	37
100	118	134	121	128	131	139	157	36	54	198	57	64	195	203	221
158	117	127	132	133	122	140	99	222	53	63	196	197	58	204	35
159	143	129	126	123	136	114	98	223	207	193	62	59	200	50	34
97	141	111	145	113	119	142	160	33	205	47	209	49	55	206	224
153	147	148	108	107	106	105	154	217	211	212	44	43	42	41	218
7	14	13	245	246	247	248	8	71	78	77	181	182	183	184	72
6	19	242	16	240	234	20	251	70	83	178	80	176	170	84	187
252	233	28	231	226	29	24	5	188	169	92	167	162	93	88	69
4	22	230	25	32	227	235	253	68	86	166	89	96	163	171	189
254	21	31	228	229	26	236	3	190	85	95	164	165	90	172	67
255	239	225	30	27	232	18	2	191	175	161	94	91	168	82	66
1	237	15	241	17	23	238	256	65	173	79	177	81	87	174	192
249	243	244	12	11	10	9	250	185	179	180	76	75	74	73	186

Fig. 218

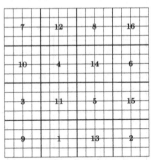

Fig. 219*b*

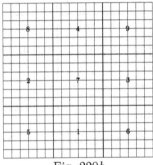

Fig. 220*b*

52	207	202	53	92	167	162	93	60	199	194	61	124	135	130	125
206	49	56	203	166	89	96	163	198	57	64	195	134	121	128	131
55	204	205	50	95	164	165	90	63	196	197	58	127	132	133	122
201	54	51	208	161	94	91	168	193	62	59	200	129	126	123	136
76	183	178	77	28	231	226	29	108	151	146	109	44	215	210	45
182	73	80	179	230	25	32	227	150	105	112	147	214	41	48	211
79	180	181	74	31	228	229	26	111	148	149	106	47	212	213	42
177	78	75	184	225	30	27	232	145	110	107	152	209	46	43	216
20	239	234	21	84	175	170	85	36	223	218	37	116	143	138	117
238	17	24	235	174	81	88	171	222	33	40	219	142	113	120	139
23	236	237	18	87	172	173	82	39	220	221	34	119	140	141	114
233	22	19	240	169	86	83	176	217	38	35	224	137	118	115	144
68	191	186	69	4	255	250	5	100	159	154	101	12	247	242	13
190	65	72	187	254	1	8	251	158	97	104	155	246	9	16	243
71	188	189	66	7	252	253	2	103	156	157	98	15	244	245	10
185	70	67	192	249	6	3	256	153	102	99	160	241	14	11	248

Fig. 219a

131	198	128	196	190	132	59	270	56	268	262	60	149	180	146	178	172	150
189	140	187	182	141	136	261	68	259	254	69	64	171	158	169	164	159	154
134	186	137	144	183	191	62	258	65	72	255	263	152	168	155	162	165	173
133	143	184	185	138	192	61	71	256	257	66	264	151	161	166	167	156	174
195	181	142	139	188	130	267	253	70	67	260	58	177	163	160	157	170	148
193	127	197	129	135	194	265	55	269	57	63	266	175	145	179	147	153	176
23	306	20	304	298	24	113	216	110	214	208	114	41	288	38	286	280	42
297	32	295	290	33	28	207	122	205	200	123	118	279	50	277	272	51	46
26	294	29	36	291	299	116	204	119	126	201	209	44	276	47	54	273	281
25	35	292	293	30	300	115	125	202	203	120	210	43	53	274	275	48	282
303	289	34	31	296	22	213	199	124	121	206	112	285	271	52	49	278	40
301	19	305	21	27	302	211	109	215	111	117	212	283	37	287	39	45	284
77	252	74	250	244	78	5	324	2	322	316	6	95	234	92	232	226	96
243	86	241	236	87	82	315	14	313	308	15	10	225	104	223	218	105	100
80	240	83	90	237	245	8	312	11	18	309	317	98	222	101	108	219	227
79	89	238	239	84	246	7	17	310	311	12	318	97	107	220	221	102	228
249	235	88	85	242	76	321	307	16	13	314	4	231	217	106	103	224	94
247	73	251	75	81	248	319	1	323	3	9	320	229	91	233	93	99	230

Fig. 220a

196	207	202	197	92	311	306	93	188	215	210	189	84	319	314	85	180	223	218	181
206	193	200	203	310	89	96	307	214	185	192	211	318	81	88	315	222	177	184	219
199	204	205	194	95	308	309	90	191	212	213	186	87	316	317	82	183	220	221	178
201	198	195	208	305	94	91	312	209	190	187	216	313	86	83	320	217	182	179	224
76	327	322	77	172	231	226	173	68	335	330	69	164	239	234	165	60	343	338	61
326	73	80	323	230	169	176	227	334	65	72	331	238	161	168	235	342	57	64	339
79	324	325	74	175	228	229	170	71	332	333	66	167	236	237	162	63	340	341	58
321	78	75	328	225	174	171	232	329	70	67	336	233	166	163	240	337	62	59	344
156	247	242	157	52	351	346	53	148	255	250	149	44	359	354	45	140	263	258	141
246	153	160	243	350	49	56	347	254	145	152	251	358	41	48	355	262	137	144	259
159	244	245	154	55	348	349	50	151	252	253	146	47	356	357	42	143	260	261	138
241	158	155	248	345	54	51	352	249	150	147	256	353	46	43	360	257	142	139	264
36	367	362	37	132	271	266	133	28	375	370	29	124	279	274	125	20	383	378	21
366	33	40	363	270	129	136	267	374	25	32	371	278	121	128	275	382	17	24	379
39	364	365	34	135	268	269	130	31	372	373	26	127	276	277	122	23	380	381	18
361	38	35	368	265	134	131	272	369	30	27	376	273	126	123	280	377	22	19	384
116	287	282	117	12	391	386	13	108	295	290	109	4	399	394	5	100	303	298	101
286	113	120	283	390	9	16	387	294	105	112	291	398	1	8	395	302	97	104	299
119	284	285	114	15	388	389	10	111	292	293	106	7	396	397	2	103	300	301	98
281	118	115	288	385	14	11	392	289	110	107	296	393	6	3	400	297	102	99	304

Fig. 221a

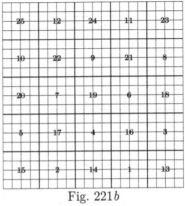

Fig. 221b

C. DIVISION INTO UNEQUAL PARTS

§ 4. Cross in the middle

The method of the cross seen above (II, § 26) extends the possibility of forming composite evenly-odd squares by means of smaller even-order squares. That is what is meant by the author of the *Harmonious arrangement* in a remark made just after explaining this method:[115] *When you know and master the (ways of) filling the square of 4, the square of 6 and the cross as explained, you will be able to break up any square which can be broken up in such a way as to remain in (the domain of) the squares of 4, the squares of 6 and the separation (in form) of a cross according to the model explained.*

In other words, knowing how to fill the cross and the 4×4 and 6×6 squares, we can construct any even-order square. Indeed,

(*a*) if $n = 4k$, the square may be filled with squares of order 4 (II, § 22, and examples above, Fig. 215, 217, 219, 221);

(*b*) if $n = 4k + 2$, we can use a cross as separation and strips of k squares of order 4, one of which will be cut by the cross when k is odd (above, our Fig. 176).

But there are other possibilities in this last case, thus for evenly-odd squares, according to the form of k:

(*c*) if $k = 3t$, thus $n = 12t + 2$, we may consider strips of t squares of order 6 on both sides of the cross;

(*d*) if $k = 3t + 1$, thus $n = 12t + 6$, we may divide the square into strips of order 6 without using the cross;

(*e*) but if $k = 3t + 2$, thus $n = 12t + 10$, we shall consider the cross and squares of order 4.

That author is therefore right. But this was already known in earlier times since Thābit mentions (and has examples of) divisions in keeping with the above possibilities according to the form of k: for case *c*, a square of order 14; for case *d*, a square of order 18 (example above, Fig. 220, and mention of the case $n = 30$); for case *e*, a square of order 10. His examples are thus:

— A 10×10 square with a cross and four 4×4 subsquares (Fig. 181 above).[116]

---

[115] *Un traité médiéval*, p. 83 of the translation, lines 688-690 of the Arabic text; or below, Appendix 17.

[116] *Magic squares in the tenth century*, A.II.46 & 50, and MS. London BL Delhi Arabic 110, fol. 99ᵛ, 1-2 & 100ʳ, 5-6, (mention), and 102ʳ (figure).

— A 14×14 square with a cross and four 6×6 subsquares (Fig. 222, with the remaining rectangular strips in the cross filled with neutral placings; here and below, the numbers underlined are the starting points).[117]

We have already seen an 18×18 square divided into 6×6 subsquares (Fig. 220). Here we consider also (first)

— A 18×18 square with a cross and four 8×8 subsquares in the corners (Fig. 223a-b).[118]

But, as remarked above by the author of the *Harmonious arrangement*, we may also find a composition using only 4×4 squares. Thus the following example in the manuscript reporting Thābit's text:

— A 18×18 square with a cross and sixteen 4×4 subsquares (Fig. 224a-b).[119] In these two examples, the equalization of the cross is made on the same side of a branch (see above, pp. 100-101) and the remaining 2×4 rectangles are filled individually.

41	160	38	158	152	42	88	109	59	142	56	140	134	60
151	50	149	144	51	46	110	87	133	68	131	126	69	64
44	148	47	54	145	153	111	86	62	130	65	72	127	135
43	53	146	147	48	154	<u>85</u>	112	61	71	128	129	66	136
157	143	52	49	150	40	102	95	139	125	70	67	132	58
155	<u>37</u>	159	39	45	156	97	100	137	<u>55</u>	141	57	63	138
<u>77</u>	78	118	117	91	105	98	101	107	<u>89</u>	116	115	83	84
120	119	79	80	106	92	96	99	90	108	81	82	114	113
5	196	2	194	188	6	94	103	23	178	20	176	170	24
187	14	185	180	15	10	104	93	169	32	167	162	33	28
8	184	11	18	181	189	76	121	26	166	29	36	163	171
7	17	182	183	12	190	122	75	25	35	164	165	30	172
193	179	16	13	186	4	123	74	175	161	34	31	168	22
191	<u>1</u>	195	3	9	192	<u>73</u>	124	173	<u>19</u>	177	21	27	174

Fig. 222

---

[117] *Magic squares in the tenth century*, A.II.46 & 50, and MS. London BL Delhi Arabic 110, fol. 99ᵛ, 3-4 & 100ʳ, 6-7 (mention), and 102ᵛ (figure).
[118] MS. London BL Delhi Arabic 110, fol. 104ʳ.
[119] MS. London BL Delhi Arabic 110, fol. 105ʳ.

71	78	77	249	250	251	252	72	148	177	103	110	109	217	218	219	220	104
70	83	246	80	244	238	84	255	178	147	102	115	214	112	212	206	116	223
256	237	92	235	230	93	88	69	179	146	224	205	124	203	198	125	120	101
68	86	234	89	96	231	239	257	145	180	100	118	202	121	128	199	207	225
258	85	95	232	233	90	240	67	168	157	226	117	127	200	201	122	208	99
259	243	229	94	91	236	82	66	158	167	227	211	197	126	123	204	114	98
65	241	79	245	81	87	242	260	166	159	97	209	111	213	113	119	210	228
253	247	248	76	75	74	73	254	161	164	221	215	216	108	107	106	105	222
137	187	186	140	141	183	182	144	162	165	169	155	171	153	152	174	175	149
188	138	139	185	184	142	143	181	160	163	156	170	154	172	173	151	150	176
7	14	13	313	314	315	316	8	136	189	39	46	45	281	282	283	284	40
6	19	310	16	308	302	20	319	190	135	38	51	278	48	276	270	52	287
320	301	28	299	294	29	24	5	191	134	288	269	60	267	262	61	56	37
4	22	298	25	32	295	303	321	133	192	36	54	266	57	64	263	271	289
322	21	31	296	297	26	304	3	132	193	290	53	63	264	265	58	272	35
323	307	293	30	27	300	18	2	194	131	291	275	261	62	59	268	50	34
1	305	15	309	17	23	306	324	195	130	33	273	47	277	49	55	274	292
317	311	312	12	11	10	9	318	129	196	285	279	280	44	43	42	41	286

Fig. 223a

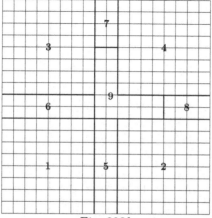

Fig. 223b

64	263	258	65	108	219	214	109	72	253	140	187	182	141	76	251	246	77
262	61	68	259	218	105	112	215	254	71	186	137	144	183	250	73	80	247
67	260	261	62	111	216	217	106	255	70	143	184	185	138	79	248	249	74
257	66	63	264	213	110	107	220	69	256	181	142	139	188	245	78	75	252
100	227	222	101	48	279	274	49	168	157	56	271	266	57	132	195	190	133
226	97	104	223	278	45	52	275	158	167	270	53	60	267	194	129	136	191
103	224	225	98	51	276	277	46	166	159	59	268	269	54	135	192	193	130
221	102	99	228	273	50	47	280	161	164	265	58	55	272	189	134	131	196
40	286	287	37	152	174	175	149	162	165	169	155	171	153	44	282	283	41
285	39	38	288	173	151	150	176	160	163	156	170	154	172	281	43	42	284
92	235	230	93	32	295	290	33	148	177	24	303	298	25	124	203	198	125
234	89	96	231	294	29	36	291	178	147	302	21	28	299	202	121	128	199
95	232	233	90	35	292	293	30	179	146	27	300	301	22	127	200	201	122
229	94	91	236	289	34	31	296	145	180	297	26	23	304	197	126	123	204
16	311	306	17	84	243	238	85	12	313	116	211	206	117	4	323	318	5
310	13	20	307	242	81	88	239	314	11	210	113	120	207	322	1	8	319
19	308	309	14	87	240	241	82	315	10	119	208	209	114	7	320	321	2
305	18	15	312	237	86	83	244	9	316	205	118	115	212	317	6	3	324

Fig. 224a

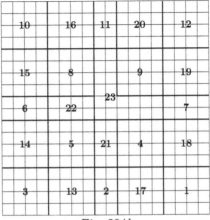

Fig. 224b

§ 5. Central square

Another possibility of having different parts is using the method of the central square (II, § 28). The text of Thābit (thus Anṭākī's) mentions several examples; in the London manuscript, they are actually constructed and said, together with the examples with the central cross, to be from Thābit's translation and to conclude it.[120]

— A 14×14 square with a central 6×6 square surrounded by eight 4×4 squares and four pairs of strips four cells long (Fig. 225a-b).[121]

— A 16×16 square with four 6×6 squares in the corners, a central 4×4 square and four 4×4 squares separated from the central square by four 2×4 rectangles (Fig. 226a-b).[122] Here as elsewhere, we have added a figure showing the order in which the parts are filled (with starting points underlined).

— A 16×16 square with a central 8×8 square surrounded by twelve 4×4 squares (Fig. 227a-b).[123]

— A 18×18 square with a central 10×10 square surrounded by twelve 4×4 squares and four pairs of strips four cells long (Fig. 228a-b).[124]

— A 20×20 square with four 8×8 squares in the corners separated by nine 4×4 squares, one of which in the centre (Fig. 229a-b).[125]

— A 20×20 square with four 6×6 squares in the centre and sixteen 4×4 squares around (Fig. 230a-b).[126]

— A 20×20 square with four 4×4 squares in the centre and, separated by pairs of strips six cells long, eight 6×6 squares (Fig. 231a-b).[127]

— A 20×20 square with a central 8×8 square, four 6×6 squares in the corners, and eight 4×4 squares each completed with a 2×4 rectangle (Fig. 232a-b).[128]

---

[120] MS. London BL Delhi Arabic 110, fol. 107^v.

[121] *Magic squares in the tenth century*, A.II.46, and MS. London BL Delhi Arabic 110, fol. 99^v, 5-6, (mention), and fol. 102^r (figure).

[122] MS. London BL Delhi Arabic 110, fol. 103^r.

[123] MS. London BL Delhi Arabic 110, fol. 103^v.

[124] *Magic squares in the tenth century*, A.II.46, and MS. London BL Delhi Arabic 110, fol. 99^v, 6-8, (mention), and 104^v (figure).

[125] MS. London BL Delhi Arabic 110, fol. 105^v.

[126] MS. London BL Delhi Arabic 110, fol. 106^r. Note the reduction of the central square: we are still with even orders ≤ 10 (above, pp. 121-122).

[127] MS. London BL Delhi Arabic 110, fol. 106^v.

[128] MS. London BL Delhi Arabic 110, fol. 107^r.

78	121	116	79	8	86	113	108	87	189	94	105	100	95
120	75	82	117	190	112	83	90	109	7	104	91	98	101
81	118	119	76	191	89	110	111	84	6	97	102	103	92
115	80	77	122	5	107	88	85	114	192	99	96	93	106
12	186	187	9	21	180	18	178	172	22	16	182	183	13
62	137	132	63	171	30	169	164	31	26	70	129	124	71
136	59	66	133	24	168	27	34	165	173	128	67	74	125
65	134	135	60	23	33	166	167	28	174	73	126	127	68
131	64	61	138	177	163	32	29	170	20	123	72	69	130
185	11	10	188	175	17	179	19	25	176	181	15	14	184
38	161	156	39	4	46	153	148	47	193	54	145	140	55
160	35	42	157	194	152	43	50	149	3	144	51	58	141
41	158	159	36	195	49	150	151	44	2	57	142	143	52
155	40	37	162	1	147	48	45	154	196	139	56	53	146

Fig. 225a

Fig. 225b

97	164	94	162	156	98	36	223	218	37	115	146	112	144	138	116
155	106	153	148	107	102	222	33	40	219	137	124	135	130	125	120
100	152	103	110	149	157	39	220	221	34	118	134	121	128	131	139
99	109	150	151	104	158	217	38	35	224	117	127	132	133	122	140
161	147	108	105	154	96	16	242	243	13	143	129	126	123	136	114
159	93	163	95	101	160	241	15	14	244	141	111	145	113	119	142
28	231	226	29	8	249	88	171	166	89	12	245	44	215	210	45
230	25	32	227	250	7	170	85	92	167	246	11	214	41	48	211
31	228	229	26	251	6	91	168	169	86	247	10	47	212	213	42
225	30	27	232	5	252	165	90	87	172	9	248	209	46	43	216
53	208	50	206	200	54	4	254	255	1	71	190	68	188	182	72
199	62	197	192	63	58	253	3	2	256	181	80	179	174	81	76
56	196	59	66	193	201	20	239	234	21	74	178	77	84	175	183
55	65	194	195	60	202	238	17	24	235	73	83	176	177	78	184
205	191	64	61	198	52	23	236	237	18	187	173	82	79	180	70
203	49	207	51	57	204	233	22	19	240	185	67	189	69	75	186

Fig. 226a

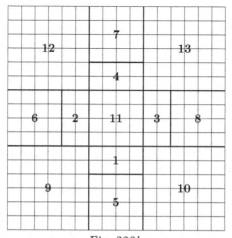

Fig. 226b

28	231	226	29	84	175	170	85	36	223	218	37	92	167	162	93
230	25	32	227	174	81	88	171	222	33	40	219	166	89	96	163
31	228	229	26	87	172	173	82	39	220	221	34	95	164	165	90
225	30	27	232	169	86	83	176	217	38	35	224	161	94	91	168
76	183	178	77	103	110	109	149	150	151	152	104	44	215	210	45
182	73	80	179	102	115	146	112	144	138	116	155	214	41	48	211
79	180	181	74	156	137	124	135	130	125	120	101	47	212	213	42
177	78	75	184	100	118	134	121	128	131	139	157	209	46	43	216
20	239	234	21	158	117	127	132	133	122	140	99	52	207	202	53
238	17	24	235	159	143	129	126	123	136	114	98	206	49	56	203
23	236	237	18	97	141	111	145	113	119	142	160	55	204	205	50
233	22	19	240	153	147	148	108	107	106	105	154	201	54	51	208
68	191	186	69	4	255	250	5	60	199	194	61	12	247	242	13
190	65	72	187	254	1	8	251	198	57	64	195	246	9	16	243
71	188	189	66	7	252	253	2	63	196	197	58	15	244	245	10
185	70	67	192	249	6	3	256	193	62	59	200	241	14	11	248

Fig. 227a

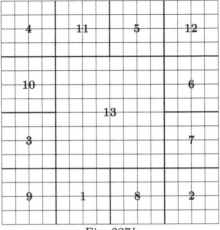

Fig. 227b

92	235	230	93	16	44	283	278	45	100	227	222	101	309	108	219	214	109
234	89	96	231	310	282	41	48	279	226	97	104	223	15	218	105	112	215
95	232	233	90	311	47	280	281	42	103	224	225	98	14	111	216	217	106
229	94	91	236	13	277	46	43	284	221	102	99	228	312	213	110	107	220
12	314	315	9	121	212	114	210	117	206	200	127	196	122	8	318	319	5
84	243	238	85	195	137	144	143	183	184	185	186	138	130	36	291	286	37
242	81	88	239	128	136	149	180	146	178	172	150	189	197	290	33	40	287
87	240	241	82	199	190	171	158	169	164	159	154	135	126	39	288	289	34
237	86	83	244	124	134	152	168	155	162	165	173	191	201	285	38	35	292
28	299	294	29	123	192	151	161	166	167	156	174	133	202	52	275	270	53
298	25	32	295	205	193	177	163	160	157	170	148	132	120	274	49	56	271
31	296	297	26	118	131	175	145	179	147	153	176	194	207	55	272	273	50
293	30	27	300	209	187	181	182	142	141	140	139	188	116	269	54	51	276
313	11	10	316	203	113	211	115	208	119	125	198	129	204	317	7	6	320
76	251	246	77	4	68	259	254	69	20	307	302	21	321	60	267	262	61
250	73	80	247	322	258	65	72	255	306	17	24	303	3	266	57	64	263
79	248	249	74	323	71	256	257	66	23	304	305	18	2	63	264	265	58
245	78	75	252	1	253	70	67	260	301	22	19	308	324	261	62	59	268

Fig. 228a

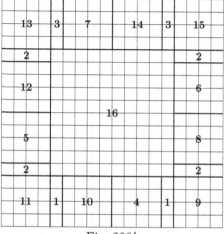

Fig. 228b

143	150	149	253	254	255	256	144	28	375	370	29	175	182	181	221	222	223	224	176
142	155	250	152	248	242	156	259	374	25	32	371	174	187	218	184	216	210	188	227
260	241	164	239	234	165	160	141	31	372	373	26	228	209	196	207	202	197	192	173
140	158	238	161	168	235	243	261	369	30	27	376	172	190	206	193	200	203	211	229
262	157	167	236	237	162	244	139	60	343	338	61	230	189	199	204	205	194	212	171
263	247	233	166	163	240	154	138	342	57	64	339	231	215	201	198	195	208	186	170
137	245	151	249	153	159	246	264	63	340	341	58	169	213	183	217	185	191	214	232
257	251	252	148	147	146	145	258	337	62	59	344	225	219	220	180	179	178	177	226
4	399	394	5	44	359	354	45	132	271	266	133	52	351	346	53	20	383	378	21
398	1	8	395	358	41	48	355	270	129	136	267	350	49	56	347	382	17	24	379
7	396	397	2	47	356	357	42	135	268	269	130	55	348	349	50	23	380	381	18
393	6	3	400	353	46	43	360	265	134	131	272	345	54	51	352	377	22	19	384
71	78	77	325	326	327	328	72	36	367	362	37	103	110	109	293	294	295	296	104
70	83	322	80	320	314	84	331	366	33	40	363	102	115	290	112	288	282	116	299
332	313	92	311	306	93	88	69	39	364	365	34	300	281	124	279	274	125	120	101
68	86	310	89	96	307	315	333	361	38	35	368	100	118	278	121	128	275	283	301
334	85	95	308	309	90	316	67	12	391	386	13	302	117	127	276	277	122	284	99
335	319	305	94	91	312	82	66	390	9	16	387	303	287	273	126	123	280	114	98
65	317	79	321	81	87	318	336	15	388	389	10	97	285	111	289	113	119	286	304
329	323	324	76	75	74	73	330	385	14	11	392	297	291	292	108	107	106	105	298

Fig. 229a

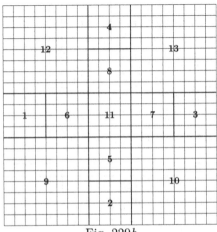

Fig. 229b

144	259	254	145	28	375	370	29	152	251	246	153	96	307	302	97	160	243	238	161
258	141	148	255	374	25	32	371	250	149	156	247	306	93	100	303	242	157	164	239
147	256	257	142	31	372	373	26	155	248	249	150	99	304	305	94	163	240	241	158
253	146	143	260	369	30	27	376	245	154	151	252	301	98	95	308	237	162	159	244
20	383	378	21	71	334	68	332	326	72	187	218	184	216	210	188	88	315	310	89
382	17	24	379	325	80	323	318	81	76	209	196	207	202	197	192	314	85	92	311
23	380	381	18	74	322	77	84	319	327	190	206	193	200	203	211	91	312	313	86
377	22	19	384	73	83	320	321	78	328	189	199	204	205	194	212	309	90	87	316
112	291	286	113	331	317	82	79	324	70	215	201	198	195	208	186	136	267	262	137
290	109	116	287	329	67	333	69	75	330	213	183	217	185	191	214	266	133	140	263
115	288	289	110	169	236	166	234	228	170	53	352	50	350	344	54	139	264	265	134
285	114	111	292	227	178	225	220	179	174	343	62	341	336	63	58	261	138	135	268
44	359	354	45	172	224	175	182	221	229	56	340	59	66	337	345	12	391	386	13
358	41	48	355	171	181	222	223	176	230	55	65	338	339	60	346	390	9	16	387
47	356	357	42	233	219	180	177	226	168	349	335	64	61	342	52	15	388	389	10
353	46	43	360	231	165	235	167	173	232	347	49	351	51	57	348	385	14	11	392
104	299	294	105	36	367	362	37	120	283	278	121	4	399	394	5	128	275	270	129
298	101	108	295	366	33	40	363	282	117	124	279	398	1	8	395	274	125	132	271
107	296	297	102	39	364	365	34	123	280	281	118	7	396	397	2	131	272	273	126
293	106	103	300	361	38	35	368	277	122	119	284	393	6	3	400	269	130	127	276

Fig. 230a

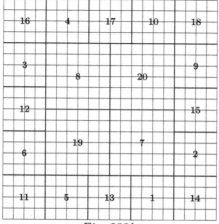

Fig. 230b

151	254	148	252	246	152	12	169	236	166	234	228	170	389	187	218	184	216	210	188
245	160	243	238	161	156	390	227	178	225	220	179	174	11	209	196	207	202	197	192
154	242	157	164	239	247	391	172	224	175	182	221	229	10	190	206	193	200	203	211
153	163	240	241	158	248	9	171	181	222	223	176	230	392	189	199	204	205	194	212
251	237	162	159	244	150	8	233	219	180	177	226	168	393	215	201	198	195	208	186
249	147	253	149	155	250	394	231	165	235	167	173	232	7	213	183	217	185	191	214
24	378	379	21	20	382	36	367	362	37	52	351	346	53	383	17	16	386	387	13
79	326	76	324	318	80	366	33	40	363	350	49	56	347	133	272	130	270	264	134
317	88	315	310	89	84	39	364	365	34	55	348	349	50	263	142	261	256	143	138
82	314	85	92	311	319	361	38	35	368	345	54	51	352	136	260	139	146	257	265
81	91	312	313	86	320	44	359	354	45	28	375	370	29	135	145	258	259	140	266
323	309	90	87	316	78	358	41	48	355	374	25	32	371	269	255	144	141	262	132
321	75	325	77	83	322	47	356	357	42	31	372	373	26	267	129	271	131	137	268
377	23	22	380	381	19	353	46	43	360	369	30	27	376	18	384	385	15	14	388
61	344	58	342	336	62	395	97	308	94	306	300	98	6	115	290	112	288	282	116
335	70	333	328	71	66	5	299	106	297	292	107	102	396	281	124	279	274	125	120
64	332	67	74	329	337	4	100	296	103	110	293	301	397	118	278	121	128	275	283
63	73	330	331	68	338	398	99	109	294	295	104	302	3	117	127	276	277	122	284
341	327	72	69	334	60	399	305	291	108	105	298	96	2	287	273	126	123	280	114
339	57	343	59	65	340	1	303	93	307	95	101	304	400	285	111	289	113	119	286

Fig. 231*a*

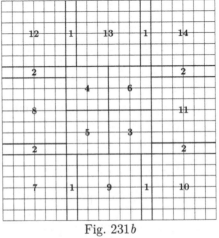

Fig. 231*b*

169	236	166	234	228	170	92	311	306	93	353	47	358	44	187	218	184	216	210	188
227	178	225	220	179	174	310	89	96	307	46	356	41	359	209	196	207	202	197	192
172	224	175	182	221	229	95	308	309	90	43	357	48	354	190	206	193	200	203	211
171	181	222	223	176	230	305	94	91	312	360	42	355	45	189	199	204	205	194	212
233	219	180	177	226	168	25	375	374	28	49	351	350	52	215	201	198	195	208	186
231	165	235	167	173	232	376	26	27	373	352	50	51	349	213	183	217	185	191	214
32	371	366	33	337	64	139	146	145	257	258	259	260	140	361	40	80	323	318	81
370	29	36	367	63	338	138	151	254	148	252	246	152	263	39	362	322	77	84	319
35	368	369	30	62	339	264	245	160	243	238	161	156	137	38	363	83	320	321	78
365	34	31	372	340	61	136	154	242	157	164	239	247	265	364	37	317	82	79	324
341	59	346	56	389	12	266	153	163	240	241	158	248	135	313	88	381	19	386	16
58	344	53	347	11	390	267	251	237	162	159	244	150	134	87	314	18	384	13	387
55	345	60	342	10	391	133	249	147	253	149	155	250	268	86	315	15	385	20	382
348	54	343	57	392	9	261	255	256	144	143	142	141	262	316	85	388	14	383	17
101	304	98	302	296	102	73	327	326	76	21	379	378	24	119	286	116	284	278	120
295	110	293	288	111	106	328	74	75	325	380	22	23	377	277	128	275	270	129	124
104	292	107	114	289	297	4	399	394	5	329	71	334	68	122	274	125	132	271	279
103	113	290	291	108	298	398	1	8	395	70	332	65	335	121	131	272	273	126	280
301	287	112	109	294	100	7	396	397	2	67	333	72	330	283	269	130	127	276	118
299	97	303	99	105	300	393	6	3	400	336	66	331	69	281	115	285	117	123	282

Fig. 232a

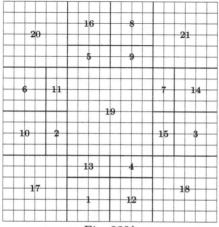

Fig. 232b

Chapter IV

Bordered magic squares

§1. Preliminary observations

The principle for constructing a border, of whatever order, is as follows. Consider we have a magic square of order $n-2$, in which are arranged the $(n-2)^2$ first natural numbers. Its magic sum is therefore

$$M_{n-2} = \tfrac{n-2}{2}\left[(n-2)^2 + 1\right].$$

We now wish to surround it with a border comprising $2n+2(n-2) = 4n-4$ cells, to be filled with the $2n-2$ first numbers and their complements to $n^2 + 1$. So we shall first uniformly increase by $2n-2$ each number of the square already constructed. The sum in any row (line, column, main diagonal) being thus increased by $n-2$ times this quantity becomes

$$M'_{n-2} = \tfrac{n-2}{2}\left[(n-2)^2 + 1 + 2(2n-2)\right] = \tfrac{n-2}{2}\left[n^2 + 1\right],$$

with the smallest number in the square being now $2n-1$ and the largest, $(n-2)^2 + 2n - 2 = n^2 - 2n + 2$. We are now to fill the $4n-4$ cells of the new border with the two sequences

$$1,\ 2,\ \ldots,\ 2n-2$$

$$n^2 - 2n + 3,\ n^2 - 2n + 4,\ \ldots,\ n^2.$$

Since the magic sum must be increased by

$$M_n - M'_{n-2} = n^2 + 1,$$

we shall arrange in pairs of complements the numbers to be placed, that is, we shall superpose the above two sequences with the second taken in reverse order:

$$
\begin{array}{ccccc}
1 & 2 & 3 & \cdots & 2n-2 \\
n^2 & n^2-1 & n^2-2 & \cdots & n^2-2n+3,
\end{array}
$$

and put these pairs at either end of each line, column and main diagonal. We are now left with the problem of obtaining the magic sum in each of the four border rows. But this does not mean that we shall have to do so for each order separately: here again, there exist general methods, valid for the same three types of order we have already encountered.

Indeed, by 'bordered square' is commonly meant a magic square where the removal of *each* border leaves a magic square. For squares of odd orders the smallest square thus reached will be that of order 3, whereas for both evenly-even and evenly-odd orders the smallest inner square will be that of order 4 (which itself cannot be a bordered square since there

© Springer Nature Switzerland AG 2019
J. Sesiano, *Magic Squares*, Sources and Studies in the History of Mathematics and Physical Sciences, https://doi.org/10.1007/978-3-030-17993-9_4

is no magic square of order 2). As in the case of ordinary magic squares, a single general method will suffice to fill squares of odd orders; but not so for squares of even orders since their borders will be alternately of evenly-even and evenly-odd order.

Since in opposite cells of each border, outer or inner, of a bordered magic square of order n there is a pair of complements adding up to n^2+1, and the magic sum of such a square filled with the n^2 first numbers is

$$M_n = \frac{n(n^2+1)}{2},$$

removing repeatedly the outer border will reduce the magic sum each time by $n^2 + 1$, leaving for an inner square of order m the magic sum

$$M_n^{(m)} = \frac{m(n^2+1)}{2},$$

as already mentioned (p. 2). From this follows that for a bordered square of odd order filled with the first natural numbers the central element must be $\frac{n^2+1}{2}$, as was also the case for symmetrical squares (p. 25).

A. SQUARES OF ODD ORDERS

§ 2. Empirical construction

One of the two 10th-century treatises preserved, that by Abū'l-Wafā' Būzjānī, describes precisely the above way to construct successive borders around the central square of order 3 and how from this a general method will appear.[129] As a matter of fact, this general method was known from Thābit ibn Qurra's translation; but Abū'l-Wafā' probably chose, for didactic reasons, to follow what must have been the steps to the discovery of this general method.

As we have seen above, the outer border of a square of order m comprises $4m-4$ cells to be filled with $2m-2$ pairs of complements. Consider then, as does Abū'l-Wafā', a square of order $m = 5$. We first fill the inner square of order 3. To do that, we start by placing in the central cell $\frac{m^2+1}{2} = 13$. We then take the $2m-2 = 4$ numbers on each side of 13 and arrange them in pairs of complements:

$$9 \quad 10 \quad 11 \quad 12$$
$$17 \quad 16 \quad 15 \quad 14,$$

which we write in the border surrounding the central cell in the known way (Fig. 233a & b).[130]

---

[129] *Magic squares in the tenth century*, B.15-17.

[130] Having, in the 3×3 square, 2 in the upper left-hand corner cell and 4 in the right-hand one suits the general method to be taught.

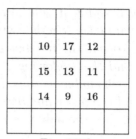

Fig. 233a   Fig. 233b

The sum of opposite cells is thus 26, and this is the sum we are to find in opposite cells of the outer border of order 5, namely using the pairs

$$1 \quad 2 \quad 3 \quad 4 \quad 5 \quad 6 \quad 7 \quad 8$$
$$25 \quad 24 \quad 23 \quad 22 \quad 21 \quad 20 \quad 19 \quad 18.$$

The first placing to be considered is that of the corners. Abū'l-Wafā' chooses to put in the two upper corner cells the smaller numbers 4 and 6, with their complements opposite (Fig. 234). He then chooses among the remaining numbers those which will complete the required sum for this order, 65, first horizontally (Fig. 235), then vertically (Fig. 236), each time with the complements written opposite.

4				6
	10	17	12	
	15	13	11	
	14	9	16	
20				22

Fig. 234

4	24	23	8	6
	10	17	12	
	15	13	11	
	14	9	16	
20	2	3	18	22

Fig. 235

4	24	23	8	6
19	10	17	12	7
21	15	13	11	5
1	14	9	16	25
20	2	3	18	22

Fig. 236

Doing so, he has constructed a bordered magic square. He adds an interesting remark on his choice for the upper corner cells:[131] *One should not think that the two numbers which we have (at first) put in the two consecutive corner cells are there by chance; many pairs of numbers taken in the two conjugate lines[132] and put in these two corner cells would not permit us to complete the magic sum in this square. Thus, putting in these two corners, instead of 4 and 6, 1 and 4, or 1 and 2, or 4 and 5, or 6 and 7, we shall not find in the two conjugate lines numbers the sum of which would, together with the occupants of the two corner cells, make 65. (On the other hand,) putting 1 and 3, or 3 and 7, or 2 and 8, or 4 and*

---

[131] *Magic squares in the tenth century*, B.15.vi; or below, Appendix 18.
[132] The above list of pairs of complements.

6, or 5 and 7, or 6 and 8, it will be possible to find three numbers such that their sum will make, together with the occupants of the corners, 65. So let us decide to put the number preceding the side of the square in the first corner and that following the side of the square in the second corner: this pair will enable us to equalize the row and (therefore) it will be possible to find the other numbers for the other rows. Abū'l-Wafā' is right in observing that the only possible pairs of smaller numbers for the corner cells are 1, 3; 3, 7; 5, 7; 2, 8; 4, 6; 6, 8; they lead to the ten borders given by us in Fig. 237-246.

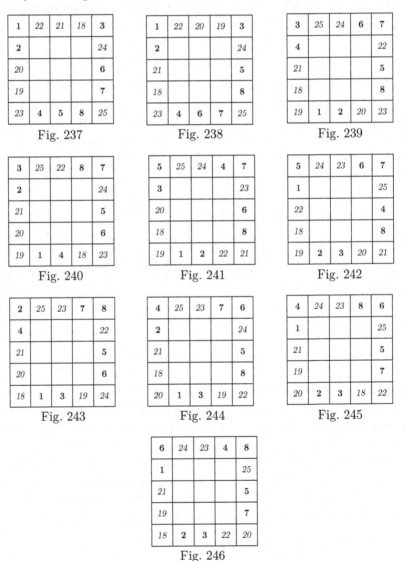

1	22	21	18	3
2				24
20				6
19				7
23	4	5	8	25

Fig. 237

1	22	20	19	3
2				24
21				5
18				8
23	4	6	7	25

Fig. 238

3	25	24	6	7
4				22
21				5
18				8
19	1	2	20	23

Fig. 239

3	25	22	8	7
2				24
21				5
20				6
19	1	4	18	23

Fig. 240

5	25	24	4	7
3				23
20				6
18				8
19	1	2	22	21

Fig. 241

5	24	23	6	7
1				25
22				4
18				8
19	2	3	20	21

Fig. 242

2	25	23	7	8
4				22
21				5
20				6
18	1	3	19	24

Fig. 243

4	25	23	7	6
2				24
21				5
18				8
20	1	3	19	22

Fig. 244

4	24	23	8	6
1				25
21				5
19				7
20	2	3	18	22

Fig. 245

6	24	23	4	8
1				25
21				5
19				7
18	2	3	22	20

Fig. 246

Abū'l-Wafā' certainly knew that the six possibilities he mentions lead to *ten* possible borders. Although all the figures are omitted in the manuscript of his treatise, a faithful copy by a later writer has a list of the numbers to be placed in the border, and only one possibility is missing (our Fig. 240); six are found six centuries later, in the treatise by Shabrāmallisī (our 239, 240, 242, 243, 245, 246).[133]

8	80	78	76	75	12	14	16	10
67	22	64	62	61	26	28	24	15
69	55	32	52	51	36	34	27	13
71	57	47	38	45	40	35	25	11
73	59	49	43	41	39	33	23	9
5	19	29	42	37	44	53	63	77
3	17	48	30	31	46	50	65	79
1	58	18	20	21	56	54	60	81
72	2	4	6	7	70	68	66	74

Fig. 247

To construct the square of order 9, Abū'l-Wafā' proceeds in the same manner. He first takes (Fig. 247) the median number, 41, which he puts in the central cell. Then he arranges the four numbers preceding it and the four numbers following it in the border of order 3. To the border of order 5 he allots the eight numbers from 29 to 36 and from 46 to 53, placed in the same way as were the corresponding ones in the border he has constructed before (Fig. 236). He then considers the two sequences of twelve numbers

17 18 19 20 21 22 23 24 25 26 27 28
65 64 63 62 61 60 59 58 57 56 55 54.

He begins by writing 22 and 24 in the corners, which is in accordance with his previous choices: just as the corner cells in the square of 3 contained the second and the fourth smaller numbers, and those of the square of 5 the fourth and the sixth, he will take here the sixth and the eighth, thus 22 and 24. He is left with finding five numbers making, together with 22 and 24, 287, which is the sum found in the main diagonals (and thus the required magic sum for the 7 × 7 inner square); with the numbers left, he completes the columns. There remain the numbers from 1 to 16 and 66 to 81 for the border of order 9, the eighth and tenth of the first

---

[133] MS. London BL Delhi Arabic 110, fol. 67^v - 68^r; MS. Paris BNF Arabe 2698, fol. 8^r - 8^v.

sequence being attributed to the corner cells, and the others being used to complete, by trial and error, the sums in the border's horizontal and vertical rows.

This being done, Abū'l-Wafā' infers from it *a method to improve the student's skill and (also) intended for those who prefer to save themselves the trouble of working out which numbers to arrange in the square.*[134] It is thus a general way of constructing odd-order bordered squares, which therefore does not involve any computation or trial and error, which is the following.

§ 3. Grouping the numbers by parity

We have seen that, when he constructed the border of order 5, Abū'l-Wafā' mentioned the six possibilities for the corner cells. He also observed that to each of these configurations may be added the various permutations of the elements within one line and within one column, the opposite elements being displaced accordingly. But he chose one particular arrangement, which can be generally applied and is easy to remember. It is this and another way which are mostly encountered in Arabic texts. These two ways have in common that the sequence of natural numbers from 1 on is placed continuously in the borders starting from the outer one. They differ in that the numbers are placed alternately on either side of a corner in one case, while in the other (§ 4) they are placed continuously in each row.

8					12	14	16	10
								15
								13
								11
								9
5								
3								
1								
	2	4	6	7				

Fig. 248

Abū'l-Wafā''s method may be generally described as follows (Fig. 248). Starting from a corner, we put in the rows enclosing it the nat-

---

[134] *Magic squares in the tenth century*, B.17; or below, Appendix 19.

ural sequence, alternately in one and the other, till we reach the middle cells. We write the next number in the cell next to the cell just reached (7 in the figure), then move to the opposite corner (that of the row where we started), then put the next in the middle cell on the other side, the next in the corner of the same row (just one possibility since the other corner is for a complement), from which we resume the alternate placing on either side of this corner, but this time (say) towards the corner cell. Opposite cells are each time left empty for the complements.

This method must have been well known not only in the tenth century, since it occurs in the two earliest treatises, but even earlier since the same principle is used in the highly intricate (Greek) method to be examined in Ch. V (see, in particular, pp. 183-184). (As already said, Abū'l-Wafā''s didactic treatise merely intended to retrace the steps leading to its discovery.) A characteristics of this method is to separate the numbers by parity: in the columns are found, excepting the corner cells, only odd numbers, while the lines contain, except for their middle cells, even numbers. Later variants either change the order of the numbers, ascending numbers moving away from, and not towards, the right-hand (here bottom) corner cell (Fig. 249);[135] or else, the smaller numbers $4k$ and $4k - 2$ are put in the corners (Fig. 250; here both on the bottom).[136]

110	1	3	5	7	111	103	105	107	109	10
2	92	21	23	25	93	87	89	91	28	120
4	22	78	37	39	79	75	77	42	100	118
6	24	38	68	49	69	67	52	84	98	116
8	26	40	50	62	57	64	72	82	96	114
9	27	41	51	63	61	59	71	81	95	113
102	86	74	66	58	65	60	56	48	36	20
104	88	76	70	73	53	55	54	46	34	18
106	90	80	85	83	43	47	45	44	32	16
108	94	101	99	97	29	35	33	31	30	14
112	121	119	117	115	11	19	17	15	13	12

Fig. 249

---

[135] *Un traité médiéval*, p. 29; *Herstellungsverfahren III*, pp. 194-195.

[136] Shabrāmallisī, MS. Paris BNF Arabe 2698, fol. 9ʳ-10ʳ; *Quelques méthodes*, p. 63; *An Arabic treatise*, pp. 15, 23, and lines 36-44 of the Arabic text.

102	106	108	110	112	9	8	6	4	2	104
1	86	90	92	94	27	26	24	22	88	121
3	21	74	78	80	41	40	38	76	101	119
5	23	37	66	70	51	50	68	85	99	117
7	25	39	49	60	65	58	73	83	97	115
111	93	79	69	59	61	63	53	43	29	11
109	91	77	67	64	57	62	55	45	31	13
107	89	75	54	52	71	72	56	47	33	15
105	87	46	44	42	81	82	84	48	35	17
103	34	32	30	28	95	96	98	100	36	19
18	16	14	12	10	113	114	116	118	120	20

Fig. 250

§ 4. Placing together consecutive numbers

Whereas Abū'l-Wafā''s didactic treatise uses some kind of induction for reaching his general method, most later treatises just describe how to place the numbers, without any justification or proof. Thus the method to be considered here (Fig. 251) is explained as follows in the 11th century:[137] *You put 1, thus the smaller extreme, in the median cell of the right-hand row, the following one in the following (cell), going down (in such a way) to the lower right-hand corner, which you do not fill. (You put) the following (number) in the left-hand bottom corner, the following in the following (cell) of the bottom row, (and so on) until (you reach) its median cell. (You put) the following (number) in the middle of the top row, the following above the median (cell) of the left-hand row, the following in the following (cell), and going then upwards to the left-hand top corner, which you fill. (You put) the next in the (cell) following the median of the top row, (and so on) as far as the (cell) next to the upper right-hand corner. Then you put their complements opposite, in the known way. You do the same with each border, the smaller number reached taking the place of the smaller extreme, till you reach the square of 3, which you construct by the same rule.*

This method is then regularly met with up until modern times.[138]

---

[137] *Un traité médiéval*, pp. 30-31, lines 122-131 of the Arabic text; or below, Appendix 20. On the vocabulary used, see above, p. 67n. Our figures read left to right.

[138] See *L'Abrégé*, pp. 108-110; *Herstellungsverfahren II & II'*, p. 253 and lines 47-69;

163	24	23	22	21	20	13	158	159	160	161	162	19
152	140	44	43	42	41	35	136	137	138	139	40	18
153	131	121	60	59	58	53	118	119	120	57	39	17
154	132	114	106	72	71	67	104	105	70	56	38	16
155	133	115	101	95	80	77	94	79	69	55	37	15
156	134	116	102	92	88	83	84	78	68	54	36	14
1	25	45	61	73	81	85	89	97	109	125	145	169
2	26	46	62	74	86	87	82	96	108	124	144	168
3	27	47	63	91	90	93	76	75	107	123	143	167
4	28	48	100	98	99	103	66	65	64	122	142	166
5	29	113	110	111	112	117	52	51	50	49	141	165
6	130	126	127	128	129	135	34	33	32	31	30	164
151	146	147	148	149	150	157	12	11	10	9	8	7

Fig. 251

§5. Method of Stifel

16	81	79	77	75	11	13	15	2
78	28	65	63	61	25	27	18	4
76	62	36	53	51	35	30	20	6
74	60	50	40	45	38	32	22	8
9	23	33	39	41	43	49	59	73
10	24	34	44	37	42	48	58	72
12	26	52	29	31	47	46	56	70
14	64	17	19	21	57	55	54	68
80	1	3	5	7	71	69	67	66

Fig. 252

Bordered squares are sometimes called 'Stifel's squares', for they were first known in Europe from a passage in Michael Stifel's *Arithmetica integra* (1544; see above, p. 15). His explanations for placing the numbers

MS. Paris BNF Arabe 2698, fol. 10ʳ-11ʳ; *Quelques méthodes*, pp. 64-65.

in the outer border (and thus also in the inner ones) are as follows (Fig. 252).

First border. *You are to place half of the smaller odd terms one after the other in the lower row. For their other half, you are to put its first term in the middle cell of the left-hand row and the remainder in the upper row. Next you are to place half of the smaller even terms (going) downwards in the right-hand row, and likewise their other half on the left, and putting the largest in the first cell (on the top).* Second border. *Clearly, there is just one rule to fill any border if its sides have an odd number of cells.*[139] Observe that Stifel too arranges the numbers according to parity, as in the first method above.

§ 6. Zigzag placing

112	19	104	17	106	15	108	13	110	11	56
9	95	34	89	32	91	30	93	28	57	113
114	26	82	45	78	43	80	41	58	96	8
7	97	39	73	52	71	50	59	83	25	115
116	24	84	48	68	55	60	74	38	98	6
5	99	37	75	53	61	69	47	85	23	117
118	22	86	46	62	67	54	76	36	100	4
3	101	35	63	70	51	72	49	87	21	119
120	20	64	77	44	79	42	81	40	102	2
1	65	88	33	90	31	92	29	94	27	121
66	103	18	105	16	107	14	109	12	111	10

Fig. 253

Starting from the cell next to a corner, we fill (Fig. 253) alternately the cells of two opposite rows. When we reach the cell underneath the corner in the initial row, we put the next number in the diagonally opposite corner, and resume the zigzag move in the other pair of rows, starting with the row opposite. Having reached the cell next to the corner, we proceed with the subsequent border, do as seen, and continue, border by border, until we reach the central square of order 3, which, we may observe, is treated in the same way. The still empty diagonal is then filled continuously with the subsequent numbers, which leads us to the corner

---

[139] *Arithmetica integra*, fol. 25^r - 25^v; or below, Appendix 21.

next to the cell first filled. The remaining empty cells are then filled with
the complements.

This nice method, particularly easy to remember, is explained in the
13th century by Būnī; which is surprising since he is mainly known for
his putting magic squares to esoteric use.[140]

§ 7. Mathematical basis of these general methods

Although these methods differ in appearance, they all nevertheless
obey the same principle. Since the outer border of a square of order
$n = 2k + 1$ contains $2(2k + 1) + 2(2k - 1) = 8k$ cells, it will be filled with
four sequences of k smaller numbers

$$\sum_1^k a_g, \quad \sum_1^k b_h, \quad \sum_1^k c_i, \quad \sum_1^k d_j,$$

and four sequences of larger numbers, namely their complements. We are
now to choose from these sequences numbers which will produce in each
border row, after placing the complements, the required sum. Or, which
should be simpler, we may consider putting at first each of the above
sequences in each of the border rows, and the complements opposite, and
then modify one or two of their elements so as to reach the required
sum. Note too that, since two of the smaller numbers are to occupy two
consecutive corner cells, and therefore their complements the two opposite
corners, each of two consecutive rows will contain $k + 1$ smaller numbers
and k complements, whereas the other two will accordingly contain k
smaller numbers and $k + 1$ complements.

Application to the methods seen

1. Variation in the corner cells

First case.

Suppose first (Fig. 254) that the a_g are within a row, say the first
column, the b_h in the next row, with one of them, say b_s, in the corner,
then the c_i in the subsequent row with one of them, say c_t, in the still
empty corner, and finally the d_j in the last row, within it since the corners
are already (or will be) occupied. Thus the k terms a_g will be, together
with $k + 1$ complements, in column I, while line II will contain the b_h,
again with $k + 1$ complements. We shall therefore take for the a_g and

---

[140] Carra de Vaux, p. 211, filling with the first numbers from the inside; in the MS.
Istanbul Esat Efendi 3558 (fol. 67^r-73^v, taken, we are told, from Būnī) the filling
proceeds from the outside, as in the above figure.

the b_h the two sequences making the lesser sums. Proceeding now with the equalization, we shall consider keeping b_s unchanged and modifying c_t, which will then give its place to some other number, say $c_t + \delta$; if this latter element exceeds the values of the c_i, and becomes one of the d_j, two terms of the last two original sequences will be exchanged.

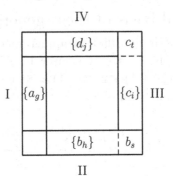

IV

I   III

II

Fig. 254

In order to now determine δ, we shall consider the sum in the first column after inserting the complements. It equals

$$\sum_1^k a_g + [(n^2 + 1) - b_s] + [(n^2 + 1) - (c_t + \delta)] + \sum_{i \neq t}[(n^2 + 1) - c_i]$$

$$= \sum_1^k a_g - \sum_1^k c_i - b_s - \delta + (k+1)(n^2 + 1)$$

and this must equal $M_n = \frac{n}{2}(n^2 + 1) = (k + \frac{1}{2})(n^2 + 1)$; so we obtain the condition

$$\sum_1^k c_i - \sum_1^k a_g + b_s + \delta = \frac{1}{2}(n^2 + 1) = 2k^2 + 2k + 1. \qquad (\star)$$

(α) Let us first take for the sequences of smaller numbers the four sequences of k consecutive numbers, namely

$\{a_g\}$ $1, 2, ..., k$ with sum $\frac{k(k+1)}{2}$

$\{b_h\}$ $k+1, k+2, ..., 2k$ with sum $k^2 + \frac{k(k+1)}{2}$

$\{c_i\}$ $2k+1, 2k+2, ..., 3k$ with sum $2k^2 + \frac{k(k+1)}{2}$

$\{d_j\}$ $3k+1, 3k+2, ..., 4k$ with sum $3k^2 + \frac{k(k+1)}{2}$.

Let us take $b_s = b_1$ for the lower corner cell. Relation $(\star)$ becomes $2k^2 + (k+1) + \delta = 2k^2 + 2k + 1$, whereby $\delta = k$; so the corner term c_t, whatever it is, will be replaced by the corresponding term in the d_j sequence. Thus, if $c_t = c_1 = 2k+1$, the first term of the third sequence will become $c_1 + \delta =$

$3k + 1$, and the first term of the fourth sequence will be $2k + 1$. Putting now (in order to better memorize the placing) the terms in increasing order, the four sequences to be inserted will be the following ones (with the corner elements in bold and the terms exchanged underlined)

$\{a_g\}$ 1, 2, ..., k

$\{b_h\}$ $\mathbf{k+1}$, $k + 2$, ..., $2k$

$\{c_i'\}$ $2k + 2$, ..., $3k$, $\mathbf{\underline{3k+1}}$ with sum $2k^2 + \frac{k(k+1)}{2} + k$

$\{d_j'\}$ $\underline{2k + 1}$, $3k + 2$, ..., $4k$ with sum $3k^2 + \frac{k(k+1)}{2} - k$.

Adding the complements will equalize all the border rows.

Such is the origin of the placing together of consecutive numbers (§ 4). The above exchange also enables us to understand the instructions given by the 11th-century *Brief treatise* (above, p. 14), which are as follows (Fig. 255, $k = 5$):[141]

6	117	118	119	120	121	12	13	14	15	16
7	25	98	99	100	101	30	31	32	33	115
8	26	40	83	84	85	44	45	46	96	114
9	27	41	51	72	73	54	55	81	95	113
10	28	42	52	58	65	60	70	80	94	112
111	93	79	69	63	61	59	53	43	29	11
105	88	75	66	62	57	64	56	47	34	17
104	87	74	67	50	49	68	71	48	35	18
103	86	76	39	38	37	78	77	82	36	19
102	89	24	23	22	21	92	91	90	97	20
106	5	4	3	2	1	110	109	108	107	116

Fig. 255

You divide the small numbers to be placed in the border into four groups (of numbers taken) consecutively, except that you put the first of the third group as the first of the fourth group and the first of the fourth group as the last of the third group. Next you put the first group below, excluding the corner, the second group on the right (here left) starting from the upper corner, the third group above with its end in the left-hand

---

[141] *L'Abrégé*, pp. 108 & (Arabic text) 146, 1–10; or below, Appendix 22. As usual, the orientation of the figure has been modified by us.

corner, and the fourth group on the left, excluding the corner. Then you put their complements opposite. You then proceed likewise with the border of the inner square, (and so on) until you reach the square of 3, which you fill in the known way.[142] Each time (in these inner squares) you will place the remaining small numbers as if they were the beginning of the small numbers, and divide the numbers to be placed in the border into four groups in the manner explained (above). (In each case) the quantity of the small numbers to be placed in the (whole) border will (equal) half the cells of the border.

This is indeed the construction we have seen in §4, except that here we begin at the bottom.

(β) Let us now divide the $4k$ smaller numbers according to parity. The four sequences will be

$\{a_g\}$	1, 3, ..., $2k-1$	with sum k^2
$\{b_h\}$	2, 4, ..., $2k$	with sum $k^2 + k$
$\{c_i\}$	$2k+1$, $2k+3$, ..., $4k-1$	with sum $3k^2$
$\{d_j\}$	$2k+2$, $2k+4$, ..., $4k$	with sum $3k^2 + k$.

Let us consider once again that b_1 will occupy the lower corner. Relation ($\star$) becomes $3k^2 - k^2 + 2 + \delta = 2k^2 + 2k + 1$, whereby $\delta = 2k - 1$. Taking here too $c_t = c_1 = 2k + 1$, the modified term will again be one of the d_j, namely $c_1 + \delta = 4k$, which will join the third sequence while $2k+1$ will replace $4k$ in the fourth sequence. Putting the terms in ascending order, we obtain the sequences

$\{a_g\}$	1, 3, ..., $2k-1$	
$\{b_h\}$	**2**, 4, ..., $2k$	
$\{c_i'\}$	$2k+3$, ..., $4k-1$, **$4k$**	
$\{d_j'\}$	$2k+1$, $2k+2$, ..., $4k-2$.	

Placing these sequences successively as seen before (Fig. 254), thus inside a row, in the next from the corner, in the next ending in the corner, inside the fourth, we obtain Stifel's configuration, which starts at the bottom (Fig. 252, $k = 4$).

Second case.

For filling the corners we have considered thus far the two middle sequences. We could, however, use the two extreme ones, with d_t the fixed element and a_s the provisional element, to be increased by δ (Fig. 256). The sum in column I will become, after placing of the complements,

---

[142] As a matter of fact, it obeys —as a particular case— the same instructions.

$$\sum_1^k a_g + \delta + (n^2 + 1) - d_t + k(n^2 + 1) - \sum_1^k c_i = M_n$$

and thus

$$\sum_1^k c_i - \sum_1^k a_g + d_t - \delta = 2k^2 + 2k + 1. \qquad (\star\star)$$

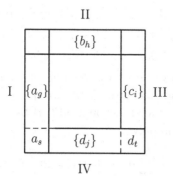

II

I {a_g} {b_h} {c_i} III

a_s {d_j} d_t

IV

Fig. 256

(α) Let us once again take sequences of consecutive numbers, namely

{a_g}	1, 2, ..., k	with sum	$\frac{k(k+1)}{2}$
{b_h}	$k+1, k+2, ..., 2k$	with sum	$k^2 + \frac{k(k+1)}{2}$
{c_i}	$2k+1, 2k+2, ..., 3k$	with sum	$2k^2 + \frac{k(k+1)}{2}$
{d_j}	$3k+1, 3k+2, ..., 4k$	with sum	$3k^2 + \frac{k(k+1)}{2}$.

40	1	5	6	38	39	46
43	31	13	16	30	35	7
42	33	28	21	26	17	8
41	32	23	25	27	18	9
2	14	24	29	22	36	48
3	15	37	34	20	19	47
4	49	45	44	12	11	10

Fig. 257

Let us take as the fixed element the first of the fourth sequence, thus $d_t = d_1 = 3k+1$ and as the variable element $a_s = a_1$, to become $a_1 + \delta = 1 + \delta$. After placing the complements, we shall have in column I, according to relation ($\star\star$), the sum $2k^2 + 3k + 1 - \delta = 2k^2 + 2k + 1$, whence $\delta = k$. Therefore we shall adopt, as the modified term of the first sequence,

$a_1 + \delta = k + 1$; since this is the first term of the second sequence, we are to replace b_1 by 1, the value of a_1. Fig. 257 represents the corresponding square of order 7. This is also an attested placing.[143]

(β) We may also once again choose to consider the four sequences grouping the numbers by parity, that is

$\{a_g\}$	1, 3, ..., $2k-1$	with sum k^2
$\{b_h\}$	2, 4, ..., $2k$	with sum $k^2 + k$
$\{c_i\}$	$2k+1, 2k+3, ..., 4k-1$	with sum $3k^2$
$\{d_j\}$	$2k+2, 2k+4, ..., 4k$	with sum $3k^2 + k$.

Suppose we take for the two lower corner cells the last of the $\{a_g\}$, thus $a_s = 2k - 1$, and the last of the $\{d_j\}$, thus $d_t = 4k$, the latter unchanged while the former is to be modified by the additive quantity δ. Now we find by ($\star\star$), namely

$$\sum_1^k c_i - \sum_1^k a_g + d_t - \delta = 2k^2 + 2k + 1,$$

that $\delta + 2k^2 + 2k + 1 = d_t + 2k^2$, thus $\delta = 2k - 1$ and $a_s + \delta = 4k - 2$. Therefore the quantity $4k - 2$ will replace $a_s = 2k - 1$ as the last term of the first group, while a_s will join the second group, the last term of which is to replace $4k - 2$ in the last set. Putting the terms in ascending order, the four sequences become (corner elements in bold)

$\{a'_g\}$	1, 3, ..., $2k-3$, **$4k-2$**
$\{b'_h\}$	2, 4, ..., $2k-2$, $2k-1$
$\{c_i\}$	$2k+1, 2k+3, ..., 4k-1$
$\{d'_j\}$	$2k, 2k+2, ..., 4k-4$, **$4k$**.

On this relies the configuration seen in Fig. 250 ($k = 5$) —though simply a rearrangement of that of Fig. 249.

Third case.

We have thus far considered modifications involving no other terms than the first $4k$ numbers. This condition is not a necessary one, and we may well just leave one element undetermined. Thus in order to establish the basis for the zigzag placing, we shall consider the sequences

$\{a_g\}$	1, 3, ..., $2k-1$	with sum k^2
$\{b_h\}$	$2k, 2k+2, ..., 4k-2$	with sum $3k^2 - k$
$\{c_i\}$	2, 4, ..., $2k-2$, C	with sum $k^2 - k + C$
$\{d_j\}$	$2k+1, 2k+3, ..., 4k-1$	with sum $3k^2$

---

[143] Example taken from the MS. Mashhad Āstān-i Quds 12235, p. 407; also 12167, fol. 168^{v}.

where, according to the arrangement shown in Fig. 258, the undetermined element C and $b_1 = 2k$ occupy the corners. The sum in column I will then equal

$$\sum a_g + (n^2 + 1) - b_1 + k(n^2 + 1) - \sum c_i = M_n,$$

that is,

$$\sum c_i - \sum a_g + b_1 = 2k^2 + 2k + 1.$$

Since the left-hand side equals $k^2 - k + C - k^2 + 2k$, that is, $C + k$, we find that $C = 2k^2 + k + 1$, that is, $C = \frac{1}{2}(n^2 + 1) - k$. In the example of Fig. 225 ($n = 11$, $k = 5$), this gives $C = 56$. As for $4k$, which is the only one of the first $4k$ smaller numbers to remain unplaced, it will be put in the next border, as the first of its elements (this will change the parity of the numbers placed in the rows).

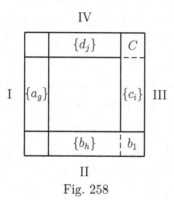

IV

Fig. 258

2. Variation in the corner and middle cells

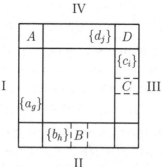

IV

Fig. 259

Suppose we once again separate the $4k$ first numbers according to parity, but take the $k-1$ first terms of the two lesser sequences and the $k-1$ last terms of the other two. We shall write them continuously inside the rows, with opposite rows thus containing terms of same parity. As to the four remaining terms, they will serve to equalize the rows and will be placed in two corners and two median cells (Fig. 259).

Let then the incomplete sequences be

$$\{a_g\} \quad 1, 3, ..., 2k - 3 \qquad \text{with sum} \ \ k^2 - 2k + 1$$
$$\{b_h\} \quad 2, 4, ..., 2k - 2 \qquad \text{with sum} \ \ k^2 \ \ - \ \ k$$
$$\{c_i\} \quad 2k + 3, 2k + 5, ..., 4k - 1 \quad \text{with sum} \ \ 3k^2 - 2k - 1$$
$$\{d_j\} \quad 2k + 4, 2k + 6, ..., 4k \qquad \text{with sum} \ \ 3k^2 - k - 2,$$

the omitted numbers thus being $2k - 1, 2k, 2k + 1, 2k + 2$, respectively. The sum in column I will be

$$\sum a_g + A + (k-1)(n^2 + 1) - \sum c_i + (n^2 + 1) - C + (n^2 + 1) - D$$
$$= k^2 - 2k + 1 + A + (k+1)(n^2 + 1) - 3k^2 + 2k + 1 - C - D$$
$$= A - C - D - 2k^2 + 2 + (k+1)(n^2 + 1).$$

Since this must equal $M_n = (k + \frac{1}{2})(n^2 + 1)$, we shall have

$$C + D - A + 2k^2 - 2 = 2k^2 + 2k + 1$$

and therefore

$$C + D - A = 2k + 3. \qquad (*)$$

On the other hand, the sum in line IV will be

$$\sum d_j + A + D + (k-1)(n^2 + 1) - \sum b_h + (n^2 + 1) - B$$
$$= A + D - B + 3k^2 - k - 2 - k^2 + k + k(n^2 + 1)$$
$$= A + D - B + 2k^2 - 2 + k(n^2 + 1),$$

and we shall therefore put

$$A + D - B + 2k^2 - 2 = 2k^2 + 2k + 1,$$

whence

$$A + D - B = 2k + 3. \qquad (**)$$

Now using the four still unplaced numbers, $2k - 1, 2k, 2k + 1, 2k + 2$, we have the two combinations

$$(2k + 2) + (2k + 1) - 2k = 2k + 3$$
$$(2k + 2) + 2k - (2k - 1) = 2k + 3.$$

Comparing the terms in these expressions with our two relations $(*)$ and $(**)$, we shall take $A = 2k$, $B = 2k - 1$, $C = 2k + 1$, $D = 2k + 2$. This gives the early placing according to parity (Fig. 247-249).

B. SQUARES OF EVENLY-EVEN ORDERS

Since the construction methods differ for evenly-even and evenly-odd orders, our illustrations will only be of outer borders. In the case of evenly-even orders, the smallest possible order is 8.

§8. Equalization by means of the first numbers

The *Brief treatise* teaches how to construct a bordered square of evenly-even order as follows (Fig. 260).[144] *You take, starting with the first of the small numbers which you wish to place in the border, four consecutive numbers, of which you place the two extremes below, but not in the corners, and the two middle in the corners above. Next you take two numbers following those four, of which you place the lesser on the side of the larger corner number, be it on the right or the left, and the larger on the side of the lesser (corner number). Then you put their complements opposite to them. The rows are thus equalized. Then you take, for each (group of) four cells remaining empty, four consecutive numbers of which you place the extremes in one row and the middle in the opposite row, (and proceed so) until, with the placing of the complements, the rows are filled.*

The same method is taught in various other sources.[145]

2	64	61	51	13	12	54	3
60							5
6							59
58							7
8							57
9							56
55							10
62	1	4	14	52	53	11	63

Fig. 260

A variant for completing the rows when the order is larger than 8 is mentioned by (e.g.) Zanjānī. We take of the subsequent consecutive smaller numbers as many as there are empty cells remaining in a row, of which we put a fourth in a row, half in the opposite one, and the last fourth in the initial row (leaving each time the opposite cells empty for the complements). We proceed in the same way for the remaining pair of opposite rows (Fig. 261). These are just the two forms of neutral placings already seen in the method of the cross (p. 98).

---

[144] *L'Abrégé*, p. 112, Arabic text p. 147, 4-11; or below, Appendix 23.

[145] *Un traité médiéval*, p. 55, lines 360–365 of the Arabic text; *Herstellungsverfahren II & II'*, p. 257 and lines 113-124 of the Arabic text; *An Arabic treatise*, pp. 20, 24-25, lines 113-124 of the Arabic text.

2	144	141	7	8	136	135	134	133	13	14	3
140											5
6											139
15											130
16											129
128											17
127											18
126											19
125											20
21											124
22											123
142	1	4	138	137	9	10	11	12	132	131	143

Fig. 261

§9. Alternate placing

15	1	255	254	4	5	251	250	8	9	10	246	245	244	243	16
240															17
18															239
19															238
237															20
236															21
22															235
23															234
233															24
232															25
26															231
27															230
229															28
228															29
30															227
241	256	2	3	253	252	6	7	249	248	247	11	12	13	14	242

Fig. 262

Consider an empty square of order $n = 4k$, supposing here $k \geq 3$. Starting from the upper left-hand corner, we put (Fig. 262) 1 in the cell next to it, then alternately below and above the subsequent pairs of numbers until there remain in the upper row (without the corners) seven empty cells; we then write the next three numbers above and the four subsequent below. We then fill the upper left-hand corner and the right-hand one. We put in the cell below it the next number, then alternately the subsequent pairs, and end by placing a single number on the left, namely the last small number to be written.

This method is applicable from order 12 on. It will also be applicable to order 8 if we fill the two lines by placing initially two numbers above and then four below (Fig. 263a).

7	1	2	62	61	60	59	8
56							9
10							55
11							54
53							12
52							13
14							51
57	64	63	3	4	5	6	58

Fig. 263a

7	14	13	53	54	55	56	8
6							59
60							5
4							61
62							3
63							2
1							64
57	51	52	12	11	10	9	58

Fig. 263b

This arrangement was known in the 10th century, for it is fully described by Anṭākī, though in a slightly different arrangement (see, for order 8, Fig. 263b, or Fig. 214c —thus Greek in origin). We find it then in the *Harmonious arrangement* (with various modifications in the choice of numbers placed initially), then again later on.[146]

§ 10. Method of Stifel

The *Arithmetica integra* already mentioned (p. 149) gives the following instructions (Fig. 264, $n = 16$) —it is supposed that the first two numbers have already been placed in the corners of the right-hand column, and, opposite, their complements.[147] *When (the side of) the border has a*

---

[146] *Magic squares in the tenth century*, $\mathcal{A}$.II.41-43, $\mathcal{B}$.19.iv ; *Un traité médiéval*, p. 51 seqq.; *Herstellungsverfahren III*, pp. 196 & 199; *An Arabic treatise*, pp. 20-21, 24-25, lines 125-135 of the Arabic text.

[147] *Arithmetica integra*, fol. 27^r & 28^r; or below, Appendix 24.

number of cells divisible by 8, the terms are placed going downwards in the left-hand and right-hand sides alternately, until the number of cells filled equals half the number of cells of the side.[148] Then, interrupting this descent, we move to the upper side, and proceed through the upper side and the lower one just as for the descent through the right-hand and left-hand side; that is, always placing two terms, one even and one odd, next to one another on one side. Exception is made for four cells: the bottom one on the right, which is that of the first term; the top one of the same side; the second cell of the upper side; the penultimate of the lower side. When this progression through the top and bottom side is finished, the previous descent is resumed —it always stops on the left side, and is there resumed, whereby it appears that four cells are filled continuously, and four left empty on the other side (. . .).

256	9	247	246	12	13	243	242	16	17	239	238	20	21	235	2
3															254
4															253
252															5
251															6
7															250
8															249
23															234
24															233
232															25
231															26
27															230
28															229
228															29
227															30
255	248	10	11	245	244	14	15	241	240	18	19	237	236	22	1

Fig. 264

When (the side of) the border has a number of cells divisible by 4

---

[148] Including the complements of the two numbers placed, and those of the two corner cells already filled.

but not by 8 (Fig. 265, $n = 12$), one proceeds exactly as with the border having the side divisible by 8 —as were the first and the fifth—[149] with the sole exception that the descent is not interrupted by putting two terms in one and the same side, but first a single one in the right-hand side, and then the next one, which ends this part of the descent, isolated in the left-hand side. The same holds for resuming the descent: one (term) is first put in the left side, then the next in the right side. In this way eight (numbers) are placed singly; it is clearly seen which are the eight cells containing smaller numbers in isolation: the bottom and the top ones on the right side; the second on the upper side; the penultimate on the lower side; then the aforementioned four.[150]

144	7	137	136	10	11	133	132	14	15	129	2
3											142
4											141
140											5
6											139
17											128
127											18
19											126
20											125
124											21
123											22
143	138	8	9	135	134	12	13	131	130	16	1

Fig. 265

§ 11. General principles of placing for even orders

It will appear that the aim when constructing borders of even orders, whether or not divisible by 4, is the same: to reach a situation where only 'neutral placings' are needed, that is, adding tetrads of consecutive numbers of which the extremes are put in one row and the medians in the opposite row (above, p. 98): indeed, with the complements of these tetrads in place, not only will the sum due be obtained in cells facing one another, but also in all groups of four cells in each row. For such

---

[149] The first and the fifth border in the previous example.

[150] Namely those near the middle of the columns; 5, 6, 17, 18 in the figure.

a situation to be attained in the two types of even order, a preliminary placing is required which must obey the following conditions.

(i) The preliminary placing must be such that, after addition of the complements, each row displays its sum due, thus $m \cdot \frac{n^2+1}{2}$ if m cells are filled.

(ii) The preliminary placing must be uniform to be applicable to any order of the same type, excepting at most small orders.

(iii) The number of cells filled thereby must be as small as possible in order for the method to be applicable, as far as possible, to the smallest orders.

(iv) The preliminary placing must settle the question of the corner cells, each common to two rows.

(v) The number of cells left empty thereafter must be divisible by 4 for subsequent neutral placings.

(vi) The remaining numbers must form groups of four consecutive numbers in order to facilitate neutral placings.

(vii) Finally, considering here the outer border, the numbers placed preliminarily should, ideally, be the first natural numbers, for that will make the neutral placings straightforward. But this condition may be disregarded if the preliminary placing made with other numbers is particularly easy to remember.

Let us apply these general instructions to the present case, that of evenly-even orders. Consider thus a square of order $n = 4k$ with $k \geq 2$. There are various possibilities for the preliminary placing.

We shall have a quantity of empty cells divisible by 4 by placing two smaller numbers in each row, two of which must occupy consecutive corner cells (Fig. 266). Indeed, since with the complements in place four cells in each row will be occupied, $4(k-1)$ will still be empty. Thus only six smaller numbers are needed for this preliminary placing.

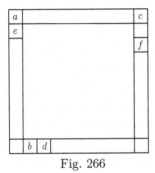

Fig. 266

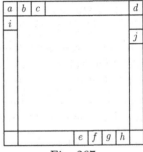

Fig. 267

We might also place two more numbers in one row and two more in its opposite; this will leave $4(k-2)$ empty cells in these two rows and $4(k-1)$ in the other two, as before. Here we shall have placed ten smaller numbers (Fig. 267). This method will also be general since it is applicable to the smallest square, of order 8.

This remains true if we place another pair of smaller numbers in the columns, thus altogether 14 numbers. We could of course continue placing further pairs, but that would be at the cost of simplicity and no longer be applicable to smaller orders.

Application to the methods seen

1. Equalization by means of the first numbers (§ 8)

Suppose we place six numbers as in Fig. 266. Let us designate by a prime their complements, that of a being thus $a' = (n^2 + 1) - a$. After writing in the complements, we must have the sum due for four cells filled, thus, for the first line and the first column, respectively,

$$a + b' + d' + c = 2(n^2 + 1),$$
$$a + c' + f' + e = 2(n^2 + 1).$$

Since $b' + d' = 2(n^2 + 1) - b - d$ and $c' + f' = 2(n^2 + 1) - c - f$, the two previous relations become

$$\begin{cases} a + c = b + d \\ a + e = c + f, \end{cases}$$

a and c being common to the two equalities since they occupy the corner cells. We shall have attained our aim if we can solve this pair of equations by means of the first six natural numbers. The remainder of the border, each row of which contains $4(k-1)$ empty cells, will be filled either by means of single tetrads, with pairs of extremes and pairs of middle numbers on opposite sides, or by considering groups of $k-1$ consecutive numbers. But the possibilities offered by these placings of the first six numbers are limited: we may either place 2 and 3 in the corners and 1 and 4 in the opposite row, or 4 and 5 in the corners and 3 and 6 in the facing row, the two remaining numbers being used in both cases to cancel the difference of 1 left in the columns. Such was the method of § 8. At least this preliminary placing is particularly easy to remember.

The placing of ten numbers offers notably more possibilities. In that case we have the relations (Fig. 267)

$$\begin{cases} a + b + c + d = e + f + g + h \\ a + i = d + j, \end{cases}$$

which gives, with the first ten numbers, 128 solutions.[151]

2. Alternate placing (§ 8)

This too relies on the preliminary placing of ten numbers, but which are just not the first ones, namely, in Fig. 262,

$$15 + 9 + 10 + 16 \ = \ 11 + 12 + 13 + 14,$$
$$15 + 30 \ = \ 16 + 29.$$

Generally, we shall write in the top line $n - 1$, $n - 7$, $n - 6$, n and in the bottom $n - 5$, $n - 4$, $n - 3$, $n - 2$, which make equal sums; in the columns, we shall put, along with $n - 1$, $2n - 2$, and, along with n, $2n - 3$, which make equal sums as well. The remaining smaller numbers, from 1 to $n-8$ and from $n + 1$ to $2n - 4$, are then just written as tetrads of consecutive numbers. As to the square of order 8 (Fig. 263a), which differed in the description of the placing (it did not show the initial alternance in the lines), it simply obeys the initial placing of ten consecutive numbers ($1, \ldots, 10$ in Fig. 263a, and $5, \ldots, 14$ in Fig. 263b). By the way, these two arrangements could just be kept for any evenly-even order and completed with neutral placings.

3. Method of Stifel (§ 10)

This is once again the placing of ten numbers, namely, in Fig. 264 (which is the same as that of the *Arithmetica integra*, with $n = 16$),

$$21 + 2 \ = \ 22 + 1$$
$$3 + 8 + 23 + 24 \ = \ 2 + 25 + 30 + 1,$$

generally

$$\left(\tfrac{3n}{2} - 3\right) + 2 \ = \ \left(\tfrac{3n}{2} - 2\right) + 1$$

in the horizontal rows, and

$$3 + \tfrac{n}{2} + \left(\tfrac{3n}{2} - 1\right) + \tfrac{3n}{2} \ = \ 2 + \left(\tfrac{3n}{2} + 1\right) + (2n - 2) + 1$$

in the vertical rows. The other numbers to be placed in the border indeed form tetrads of consecutive numbers: since n has the form $8t$, there remain intervals of numbers from 4 to $\tfrac{n}{2} - 1 = 4t - 1$, from $\tfrac{n}{2} + 1 = 4t + 1$ to $\tfrac{3n}{2} - 4 = 12t - 4$, from $\tfrac{3n}{2} + 2 = 12t + 2$ to $2n - 3 = 16t - 3$, thus intervals with, respectively, $4t - 4$, $8t - 4$, $4t - 4$ numbers ($t \geq 2$).

For the case where n has the form $8t + 4$, we may observe in figure 265 ($n = 12$) the presence of the following equalities

$$15 + 2 \ = \ 16 + 1$$
$$3 + 4 + 6 + 17 \ = \ 2 + 5 + 22 + 1,$$

or, generally,

---

[151] Listed in *Magic squares in the tenth century*, pp. 57-58.

$$\left(\tfrac{3n}{2} - 3\right) + 2 \;=\; \left(\tfrac{3n}{2} - 2\right) + 1$$

in the horizontal rows and, slightly different from the above,

$$3 + \left(\tfrac{n}{2} - 2\right) + \tfrac{n}{2} + \left(\tfrac{3n}{2} - 1\right) \;=\; 2 + \left(\tfrac{n}{2} - 1\right) + (2n - 2) + 1$$

which leaves intervals of tetrads from 4 to $\tfrac{n}{2} - 3 = 4t - 1$ (provided that $t \geq 2$), from $\tfrac{n}{2} + 1 = 4t + 3$ to $\tfrac{3n}{2} - 4 = 12t + 2$, from $\tfrac{3n}{2} = 12t + 6$ to $2n - 3 = 16t + 5$, thus intervals with, respectively, $4t - 4$, $8t$, $4t$ numbers.

C. SQUARES OF EVENLY-ODD ORDERS

§ 12. Equalization by means of the first numbers

1	194	192	190	19	20	176	175	174	173	25	26	10	4
195													2
6													191
8													189
188													9
11													186
12													185
184													13
183													14
182													15
181													16
17													180
18													179
193	3	5	7	178	177	21	22	23	24	172	171	187	196

Fig. 268

Since we are again to place initially a small quantity of numbers, this is once more the ideal situation for instructions enumerating individual placings of numbers, as often practised in treatises within the reach of general readers (see p. 15). Such is for instance the description by Zanjānī:[152] *You place (Fig. 268; here inverted, but the instructions remain valid) 1 in the first diagonal, 2 in the last column, in a cell outside the diagonal, 3 in a cell outside the diagonal in the last line, 4 in the second diagonal*

---

[152] *Herstellungsverfahren II & II'*, pp. 254-256, lines 78-86 and 98-108 of the Arabic text; or below, Appendix 25. Same construction in the MSS. Istanbul Ayasofya 2794, fol. 21ʳ, Mashhad Āstān-i Quds 12235, p. 443.

in the first line, 5 in the row of 3, in a cell outside the diagonal, 6 in the
first column, in a cell outside the diagonal, 7 in the row of 3, in a cell
outside the diagonal, 8 in the row of 6, in a cell outside the diagonal, 9
in the row of 2, in a cell outside the diagonal, 10 in the first line. But a
number must not be placed opposite to a number.[153]

After describing another placing for the first ten numbers (with 1, 8,
6 in the top row —1 and 6 in the corners— 2, 4, 9 below, 1, 5, 10 and
6, 3, 7 in the columns), the author goes on to explain how to place the
remaining numbers (as seen above, pp. 98, 159, 165). If the square is that
of 6, its border is completed: you place in each empty cell of the border
the number completing, together with that of the opposite cell, the square
of 6 plus 1, thus 37. (...) If the order of the square is other than 6, such
as 10 and other similar ones, the method is the following. You place
the ten numbers and complete the opposite cells as we have explained.
There then remains in each row a number of cells always divisible by
4. Fill then a fourth of the (empty) cells of one row, beginning with the
(smaller) number reached, thus 11, (by taking the numbers) consecutively,
then half the cells of the opposite row, next a fourth of the initial row;
next fill a fourth of the cells of the third row, half the cells of the fourth,
then a fourth of the cells of the third, (all that) on condition that no
number be (placed) opposite to another. Then complete the filling of the
(border of the) square with their complements. Other authors prefer, as
does that of the Harmonious arrangement, to complete with tetrads.[154]
When the four rows are equalized by means of the ten numbers, either
your treatment was for the square of 6 and it is finished after the placing
of the complements opposite, or it is for another square and there will
remain, of each row, four cells, or eight, and so on by adding repeatedly
four; (in such a case) you will take (repeatedly) four consecutive numbers
of which you will place the extremes in one row and the middle in the
opposite row.

§ 13. Cyclical placing

This method is, or should be, less tedious to describe and memorize
than the previous one, and seems very different; it obeys, though, the
same principle, as we shall see. Kharaqī describes it as follows (Fig. 269,

---

[153] We have already encountered this filling (Fig. 185, p. 103). Some of the indications
'outside the diagonal' are superfluous, since all the corner cells are already allotted to
a number.

[154] Un traité médiéval, p. 70 of the translation, lines 521-524 of the Arabic text; or
below, Appendix 26.

$n = 14$):[155] *We put 1, or what takes its place,*[156] *next to the right-hand corner of the bottom row, the next in the third cell of the top row, the next in the third cell (after) 1 in the bottom row, the next in the left-hand row, the next in the top row, the next on the right, the next on the bottom, the next on the left, (and so on) until the number placed equals the side of the next inner square. Then we write the next in the right-hand corner of the upper row, the next in the left-hand corner, and this (number) will equal the side of the square considered. Then we write the next two numbers on the right, a single number below, one on the left, (and so on) until, with putting one number on the left, we have filled half the (cells of the) border's rows, the quantity of (the numbers thus placed) being equal to the sum of the side of the square considered and that of the next inner square. Then we complete them by means of the equalizing number for the square considered.*[157]

13	196	2	194	5	190	9	186	180	19	176	23	172	14
6													191
10													187
15													182
16													181
20													177
24													173
171													26
175													22
179													18
185													12
189													8
193													4
183	1	195	3	192	7	188	11	17	178	21	174	25	184

Fig. 269

---

[155] *Herstellungsverfahren III*, pp. 196-198, lines 105-115 of the Arabic text; or below, Appendix 27; the instructions are from right to left.

[156] Case of inner borders of evenly-odd orders.

[157] The 'equalizing number' or 'balancing number' (*'adad mu'tadil*) is the quantity $n^2 + 1$. We also find with the same meaning *'idl* (below, p. 267).

To sum up, we write 1 next to the lower left-hand corner, 2 above, 3 below, then the next ones cyclically until we reach $4k = n - 2$, which will be in a cell of the right-hand column if we have adopted the above rotating movement. We then write the next two numbers, $n - 1$ and n, in the upper left and right corners. Next, we put the two following numbers on the left, then the subsequent ones around the border, in the same rotary motion as before, until we reach on the right side, after having thus put $4k$ more numbers, $8k + 2 = 2n - 2$. Half the border cells are thus filled, and putting their complements opposite will fill the border completely. This method is also applicable to the order 6 (Fig. 270; see Fig. 214b).

5	36	2	34	28	6
7					30
8					29
27					10
33					4
31	1	35	3	9	32

Fig. 270

This is the oldest method we know since it already appears in the (Greek) source of Anṭākī's 10th-century treatise. It is often encountered later on.[158]

Remark. To place the smaller numbers in this way, Anṭākī does not name them in succession, but by parity, and thus fills alternately the two columns with sequences of (mainly) even numbers, then the two lines with sequences of (smaller) odd numbers.

§ 14. Method of Stifel

After completing the border of the square of order 16 (above, § 10), Stifel proceeds with the next one, the filling of which he describes as follows (Fig. 271, $n = 14$).[159] *When the border has an evenly-odd number of cells, the descent on either side as described previously proceeds until*

---

[158] *Magic squares in the tenth century*, A.II.39; *Un traité médiéval*, p. 70 & lines 527-532 of the Arabic text; *An Arabic treatise*, pp. 17-18, 23, lines 66-86 of the Arabic text.

[159] *Arithmetica integra*, fol. 27ᵛ; or below, Appendix 28. He does not say how to place the first two numbers which, as before (p. 161), will occupy the end cells of the right-hand column.

there remains, of the cells allotted to the smaller numbers, a single cell on the right and on the left; the descent then ends on the left with three terms placed next to one another. This being thus concluded, we pass on to the bottom side, whence we proceed alternately to the top and to the bottom side as for the previous (evenly-even) border and as you see here. Having attained the end of this movement, thus the penultimate cell on the bottom, you do not put in it the term reached with the progression, but you put it in the cell omitted in the descent, thus in the penultimate cell on the right; as for the next term, you put it in the penultimate cell on the bottom. You may observe here that there are five cells receiving single numbers; namely, on the right, the bottom, penultimate, top ones; on the bottom, the second and the penultimate.

196	183	15	16	180	179	19	20	176	175	23	24	171	2
3													194
4													193
192													5
191													6
7													190
8													189
188													9
187													10
11													186
12													185
13													184
172													25
195	14	182	181	17	18	178	177	21	22	174	173	26	1

Fig. 271

§ 15. Principles of these methods

The basic principle is the same as for evenly-even squares: filling a small number of cells, including the corner ones, with the first natural numbers or some small group of numbers in such a way that after placing the complements we shall have the sum due and be left in each row with a number of empty cells divisible by 4.

Application to the general methods seen above

1. Equalization by means of the first numbers (§ 12)

A minimal placing, also applicable to the smallest order, 6, is to put ten smaller numbers, three in each row (including two consecutive corners); with the complements in place, this will leave, if the order in question is $n = 4k + 2$ with $k > 1$, $4(k - 1)$ empty cells in each row.

Thus, with the ten quantities represented in Fig. 272, we are to find a solution to

$$\begin{cases} a + b + c = d + e + f \\ a + g + h = c + i + j. \end{cases}$$

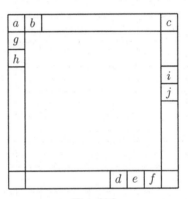

Fig. 272

Using the first ten numbers, we have 140 possibilities.[160] While Zanjānī's treatise has two (above, § 12), the *Harmonious arrangement* gives eleven, Khāzinī in the early 12th century nine (among the 'many') and Shabrāmallisī's later treatise twenty-four.[161]

2. Cyclical placing (§ 13)

This (oldest) method relies on the preliminary placing of ten numbers as well but, except for order 6, discontinuously. Those of the upper row are 2, $n - 1$ and n, those of the lower row 1, $n - 3$ and $n + 3$, thus 2, 13, 14 and 1, 11, 17, respectively, in Fig. 269; since their sums are equal, we shall have the sum due after adding the complements. The numbers in the columns are $n - 1$, $n + 1$, $n + 2$ on the left and n, $n - 2$, $n + 4$ on the right, thus 13, 15, 16 and 14, 12, 18 in the figure, both sets making equal sums.

---

[160] Listed in our *Magic squares in the tenth century*, pp. 63-64.

[161] *Un traité médiéval*, pp. 70-75; MS. London BL Delhi Arabic 110, fol. 78ᵛ; MS. Paris BNF Arabe 2698, fol. 41ᵛ - 43ʳ (items 1-8, 12, 21, 25, 30-32, 39, 43, 48, 59, 60, 98, 118, 127, 134, 140 in our list).

The cyclical placing distributes the remaining numbers appropriately: the upper row associates 5 and $2n - 5$, 9 and $2n - 9$, and so on to $n - 5$ and $2n - (n - 5)$, whereby the row is increased by $k - 1$ times $2n$; the lower row associates the other odd numbers, thus 3 and $2n - 3$, 7 and $2n - 7$, and so on to $n - 7$ and $2n - (n - 7)$, whereby here too the sum in the row is increased by $(k - 1)2n$. The columns will likewise comprise, together with the initial numbers, the pairs 6 and $2n - 4$, 10 and $2n - 8$, ..., $n - 4$ and $2n - (n - 6)$ for one of them, and for the other the remaining even numbers, thus 4 and $2n - 2$, 8 and $2n - 6$, ..., $n - 6$ and $2n - (n - 8)$; this adds to each column two sequences of $k - 1$ numbers making the same sum, namely $(k - 1)(2n + 2)$. Since the sums in these two opposite rows are equal, the complements of one will produce in the other the sum due.

3. Method of Stifel (§ 14)

This too relies on the placing of ten numbers. Fig. 271 $(n = 14)$ shows the equalities

$$\begin{cases} 23 + 24 + 2 = 22 + 26 + 1 \\ 3 + 12 + 13 = 2 + 25 + 1 \end{cases}$$

thus, generally,

$$\begin{cases} (2n - 5) + (2n - 4) + 2 = (2n - 6) + (2n - 2) + 1 \\ 3 + (n - 2) + (n - 1) = 2 + (2n - 3) + 1, \end{cases}$$

and the intervals from 4 to $n - 3$ and from n to $2n - 7$, each with $4(k - 1)$ elements, are indeed formed by tetrads of consecutive numbers.

Remark. In all the above cases the equality of the sums in opposite rows was obtained with filling an equal number of cells on each side. A Persian example shows why this is a necessary condition: the complements will not display an equal sum.[162]

1	3	5	33	30	2
31					6
27					10
8					29
9					28
35	34	32	4	7	36

Fig. 273

---

[162] MS. Mashhad Āstān-i Quds 12235, p. 372.

To close this chapter, we have included two larger squares. That of Fig. 274 is of order 20, filled by means of the methods taught in §§ 9 et 13 (except for the inner 4 × 4 square, the arrangement of which follows the pattern of Fig. 76 & 79). Fig. 275a reproduces an 18 × 18 composite square, taken from Persian manuscripts;[163] the lateral 3 × 3 subsquares follow the arrangement of a border of order 6 (corresponding to the placing of ten numbers, with $1 + 10 + 2 = 3 + 4 + 6$ in the lines and $1 + 7 + 8 = 2 + 9 + 5$ in the columns), while those within follow the well-known pandiagonal structure of a 4 × 4 square (Fig. 275b). The orientation of the 3 × 3 squares is uniform, except for those of the upper corners.

19	1	399	398	4	5	395	394	8	9	391	390	12	13	14	386	385	384	383	20
380	55	362	40	360	43	356	47	352	51	348	342	61	338	65	334	69	330	56	21
22	359	87	73	327	326	76	77	323	322	80	81	82	318	317	316	315	88	42	379
23	44	312	115	298	104	296	107	292	111	288	282	121	278	125	274	116	89	357	378
377	355	90	295	139	129	271	270	132	133	134	266	265	264	263	140	106	311	46	24
376	48	91	108	260	159	250	152	248	155	244	238	165	234	160	141	293	310	353	25
26	351	309	291	142	247	175	169	170	230	229	228	227	176	154	259	110	92	50	375
27	52	308	112	143	156	224	187	218	184	216	210	188	177	245	258	289	93	349	374
373	347	94	287	257	243	178	215	193	207	206	196	186	223	158	144	114	307	54	28
372	57	95	117	256	161	179	189	204	198	199	201	212	222	240	145	284	306	344	29
30	58	305	118	146	162	221	190	200	202	203	197	211	180	239	255	283	96	343	371
31	341	304	281	147	237	220	209	205	195	194	208	192	181	164	254	120	97	60	370
369	62	98	122	253	166	182	213	183	217	185	191	214	219	235	148	279	303	339	32
368	337	99	277	252	233	225	232	231	171	172	173	174	226	168	149	124	302	64	33
34	66	301	126	150	241	151	249	153	246	157	163	236	167	242	251	275	100	335	367
35	333	300	273	261	272	130	131	269	268	267	135	136	137	138	262	128	101	68	366
365	70	102	285	103	297	105	294	109	290	113	119	280	123	276	127	286	299	331	36
364	329	313	328	74	75	325	324	78	79	321	320	319	83	84	85	86	314	72	37
38	345	39	361	41	358	45	354	49	350	53	59	340	63	336	67	332	71	346	363
381	400	2	3	397	396	6	7	393	392	10	11	389	388	387	15	16	17	18	382

Fig. 274

---

[163] MSS. Mashhad Āstān-i Quds 12235, p. 416, and Āstān-i Quds 12167, fol. 177ʳ.

8	1	6	305	300	301	296	291	292	278	273	274	89	84	85	17	10	15
3	5	7	298	302	306	289	293	297	271	275	279	82	86	90	12	14	16
4	9	2	303	304	299	294	295	290	276	277	272	87	88	83	13	18	11
62	57	58	98	93	94	215	210	211	188	183	184	161	156	157	269	264	265
55	59	63	91	95	99	208	212	216	181	185	189	154	158	162	262	266	270
60	61	56	96	97	92	213	214	209	186	187	182	159	160	155	267	268	263
71	66	67	197	192	193	152	147	148	107	102	103	206	201	202	260	255	256
64	68	72	190	194	198	145	149	153	100	104	108	199	203	207	253	257	261
69	70	65	195	196	191	150	151	146	105	106	101	204	205	200	258	259	254
251	246	247	143	138	139	170	165	166	233	228	229	116	111	112	80	75	76
244	248	252	136	140	144	163	167	171	226	230	234	109	113	117	73	77	81
249	250	245	141	142	137	168	169	164	231	232	227	114	115	110	78	79	74
287	282	283	224	219	220	125	120	121	134	129	130	179	174	175	44	39	40
280	284	288	217	221	225	118	122	126	127	131	135	172	176	180	37	41	45
285	286	281	222	223	218	123	124	119	132	133	128	177	178	173	42	43	38
314	309	310	26	21	22	35	30	31	53	48	49	242	237	238	323	318	319
307	311	315	19	23	27	28	32	36	46	50	54	235	239	243	316	320	324
312	313	308	24	25	20	33	34	29	51	52	47	240	241	236	321	322	317

Fig. 275a

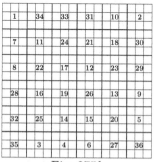

Fig. 275b

Chapter V

Bordered squares with separation by parity

This admirable construction also involves bordered squares, but where the odd numbers are all placed in an oblique square with its corners at the middle of each side of the main square and the even numbers around it (Fig. 276). This configuration naturally occurs in the square of order 3 (Fig. 277), which certainly led to the search for an analogous configuration in odd squares of higher orders, which is by no means an easy matter.

Fig. 276   Fig. 277

The construction of such squares of larger size may certainly be obtained by trial and error, as witnessed in the tenth century by Abū'l-Wafā' Būzjānī's construction (see below), who also tells us that this type of arrangement is *used by mathematicians*.[164] But this is a step backwards: the Greek text translated by Thābit ibn Qurra, textually reproduced in the tenth century by Anṭākī, gives a true *method of placing* since, once the order is known, the construction can be performed right away. The difficulty of understanding the Greek method may account for this return to empirical constructions.

§ 1. The main square and its parts

As we see in the particular case of order 3, just as the main square contains an oblique square, the latter includes a smaller square (in this case the central cell). Now this situation occurs in all larger squares, and Abū'l-Wafā' gives a precise description of the size of each such inner part when the order n of the main square is known.[165]

The squares which appear within the (main) square are of two kinds. One is the square with its corners placed in the middle of the sides of the large square, with an odd number of cells, and in the cells of which will be found the odd numbers. When we consider the rows of such (squares),

---

[164] *Magic squares in the tenth century*, B.20.i.

[165] *Magic squares in the tenth century*, B.21.i-ii, and MS. London BL Delhi Arabic 110, fol. 87ʳ, 13-25; or below, Appendix 29.

which are (therefore) parallel to the sides having their extremities placed
in the middle of the sides of the large square, we find that the lateral ones
equal the consecutive numbers taken in natural order beginning with 2.
Indeed, one sees that the side of the (oblique) square within the square of
3 is 2, that the side of the square within the square of 5 is 3, that the side
of the square within the square of 7 is 4, that the side of the square within
the square of 9 is 5, and so on following the succession of the consecutive
numbers. Therefore, when we wish to know the side of such an oblique
square, we add 1 to the side of the (main) square and take half the result;
this will give the side of the oblique square. As for their (inner) rows,
they are unequal: the (number of) cells in a row may be equal to the side
and less than it by one cell.   The second kind (of square included in the
large square) is the square in its centre (...). But each square (of this
kind) appears in two squares according to the succession of consecutive
odd numbers. Indeed, in the square of 5 and in the square of 7 appears
one and the same square, which is the square of 3; in the square of 9 and
in the square of 11 appears one and the same square, which is the square
of 5; likewise, in the square of 13 and the square of 15 appears inside the
square of 7.

Let us explain this situation in our terms. The main square, of odd
order $n = 2k+1$, contains $n^2 = 4k^2+4k+1$ cells. The inner oblique square
will contain as many cells as there are odd numbers, thus $2k^2 + 2k + 1$,
and the quantity of $2k^2 + 2k$ even numbers will be equally divided among
the outer triangles, thus $\frac{1}{2}k(k+1)$ elements in each —an integer, since
the product of k by $k+1$ is even. Now, as observed in the text, each
side of the oblique square will comprise $\frac{n+1}{2}$, thus $k+1$, cells. As stated
further, the inner rows of the oblique square do not all have the same
number of cells: whereas the lateral ones have $k+1$ cells, the subsequent
ones contain alternately k and $k+1$ cells. The second square, thus the
largest square within the oblique square, will be of the same size for
two successive orders, since it has the side $2t+1$ for both $n = 4t+1$
and $n = 4t+3$, that is, when in $n = 2k+1$ the quantity k takes two
consecutive values, first even then odd.

This enables us to better conceive the structure of the oblique square
(Fig. 278-280). Of three consecutive odd orders $4t+1$, $4t+3$, $4t+5$, the
oblique squares of the first two contain on the one hand (largest) inner
squares of equal size, but on the other hand triangular parts, surrounding these equal inner squares, which are unequal since these squares are
surrounded by, respectively, t and $t+1$ borders; for the last two orders,
the largest inner squares will be of unequal size, with sides $2t+1$ and

$2(t+1)+1$ respectively, while the triangular parts of the oblique square will be equal since both inner squares are surrounded by $t+1$ borders.

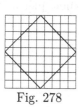
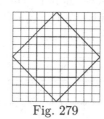
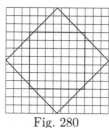

Fig. 278	Fig. 279	Fig. 280

Remarks.

(1) This structure is clearly expounded by Abū'l-Wafā', but is for him of little use since he has no precise placing method (except for filling the inner square, which is banal). There is nothing about it in Thābit ibn Qurra's translation, at least in the part preserved.

(2) As we shall see, arranging the odd numbers supposes knowledge of the older method for filling odd-order bordered squares (IV, §3), both for filling the inner squares and the triangular parts within the oblique square. Furthermore, for the even numbers, a precisely determined preliminary placing is followed by groups of neutral placings, just as in the construction of even-order bordered squares (IV, §11) —except that here this preliminary placing is by far more complex than that for usual even-order bordered squares. This indicates not just knowledge but familiarity with the construction of common bordered squares in Greek antiquity.

§2. Filling the inner square

Placing the numbers in the inner square is easy, and both texts explain it in the same way: we shall put $\frac{n^2+1}{2}$ (odd) in the central cell, then fill the borders successively with the subsequent smaller and larger numbers, just as we did with bordered squares, but here taking only odd numbers. The largest inner square will thus be occupied by two sequences of consecutive odd numbers on either side of $\frac{n^2+1}{2}$, and it will display its sum due and the characteristics of a bordered square.

This is exemplified in Fig. 281, of order $n = 11$ (see above, Fig. 236 or 248). The smallest term placed in the outer border will be 37.

Remark. If we wish, as with other bordered squares, to start in the outer border of this inner square and place ascending sequences of smaller numbers (that border being at least of order 5), we shall put the smallest number above the lower left-hand corner cell; this number will be $4t^2 + 1$ if n (≥ 9) has the form $4t + 1$, but $4(t+1)^2 + 1$ for the form $4t + 3$. Next we shall put the subsequent smaller odd numbers alternately around the corner, then in the middle and corner cells, finally alternately on both sides of the opposite corner, just as we did for usual bordered odd-order squares.

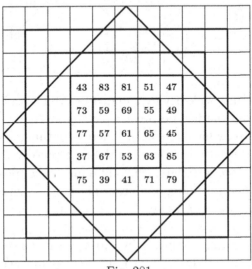

Fig. 281

§3. Filling the remainder of the square by trial and error

The inner square being thus filled with two continuous sequences and displaying the characteristics of a bordered square, Abū'l-Wafā' proceeds with filling the *whole* part surrounding the inner square. After choosing its corner elements, he fills each border by trial and error, using the remaining odd numbers to complete the oblique square and putting the even ones around it.[166]

Abū'l-Wafā' constructs thus the squares of order 5 and 7, and then —but this time in a slightly simplified way— that of order 9 (see below). Before considering said way, we shall report the trial-and-error construction for a square of order 11, as described by a later follower of Abū'l-Wafā'.[167]

We first put 61 in the central cell and fill the inner square of order five as known for bordered squares, except that we use here only odd numbers

---

[166] *Magic squares in the tenth century*, B.22, and MS. London BL Delhi Arabic 110, fol. 87^v, 15 - 89^r.

[167] MS. London BL Delhi Arabic 110, fol. 89^r - 90^r. Inner square here rotated to match the previous figure.

(see Fig. 282). We are then left with placing the pairs of odd numbers

1	3	5	7	9	11	13	15	17	19	21	23	25	27	29	31	33	35
121	119	117	115	113	111	109	107	105	103	101	99	97	95	93	91	89	87

and those of even numbers

2	4	6	8	10	12	14	16	18	20	22	24	26	28	30
120	118	116	114	112	110	108	106	104	102	100	98	96	94	92

32	34	36	38	40	42	44	46	48	50	52	54	56	58	60
90	88	86	84	82	80	78	76	74	72	70	68	66	64	62.

86	120	114	112	106	9	4	6	12	14	88
90	46	18	24	97	107	7	100	102	48	32
84	66	60	3	91	103	89	23	58	56	38
72	68	17	43	83	81	51	47	105	54	50
70	95	29	73	59	69	55	49	93	27	52
111	5	35	77	57	61	65	45	87	117	11
40	109	101	37	67	53	63	85	21	13	82
30	42	121	75	39	41	71	79	1	80	92
28	44	64	119	31	19	33	99	62	78	94
26	74	104	98	25	15	115	22	20	76	96
34	2	8	10	16	113	118	116	110	108	36

Fig. 282

For the 7 × 7 inner square, the text places in two consecutive corner cells, here the upper ones, the largest pair of the smaller (even) numbers, 58, 60, and their complements opposite. Since the sum due is (7·61 =) 427, we are to complete the row between 60 and 58 with five odd numbers adding up to 309, and to 303 for the column of 60 and 64. Choosing for the first set 3, 91, 103, 89, 23, and for the second 17, 29, 35, 101, 121, the border will be filled after putting their complements opposite. For the 9 × 9 border, with the sum due, 549, the text takes 46, 48 for the corner cells and completes the row of 46, 48 by choosing among the still available numbers four even ones and three odd ones completing the sum due, the same being done for the column, with the complements then being written opposite. We shall proceed in the same way with the 11 × 11 border, its sum due being 671.[168]

What Abū'l-Wafā' suggests as a simplification for filling the cells

---

[168] The choice of the corner elements will turn out to be the same as with the truly mathematical method described by Thābit (and Antākī).

around the inner square is as follows.[169] We are to consider the usual bordered square of the same order (Fig. 283) and transfer to correspondingly situated empty cells of the target square the numbers meeting the condition of parity, provided that the odd numbers have not been already used for the inner square (Fig. 284). The remaining cells are filled by making up, as above, the remainder of the sum due (Fig. 285, numbers underlined). But this is in fact not much of a simplification: the numbers in the bordered square initially considered being distributed by parity, it merely enables us to fill part of the cells.

8	80	78	76	75	12	14	16	10
67	22	64	62	61	26	28	24	15
69	55	32	52	51	36	34	27	13
71	57	47	38	45	40	35	25	11
73	59	49	43	41	39	33	23	9
5	19	29	42	37	44	53	63	77
3	17	48	30	31	46	50	65	79
1	58	18	20	21	56	54	60	81
72	2	4	6	7	70	68	66	74

Fig. 283

8	80	78	76	75	12	14	16	10
	22	64				28	24	
		23	63	61	31	27		
		53	35	49	39	29		
73		57	45	41	37	25		9
		17	43	33	47	65		
		55	19	21	51	59		
	58	18				54	60	
72	2	4	6	7	70	68	66	74

Fig. 284

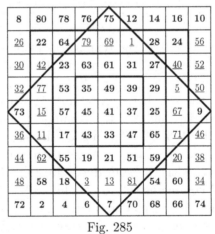

8	80	78	76	75	12	14	16	10
26	22	64	79	69	1	28	24	56
30	42	23	63	61	31	27	40	52
32	77	53	35	49	39	29	5	50
73	15	57	45	41	37	25	67	9
36	11	17	43	33	47	65	71	46
44	62	55	19	21	51	59	20	38
48	58	18	3	13	81	54	60	34
72	2	4	6	7	70	68	66	74

Fig. 285

A. METHODICAL FILLING OF THE OBLIQUE SQUARE

As we have just seen, once the inner square is filled, Abū'l-Wafā''s construction gives way to placing by trial and error. That is not the case

---

[169] *Magic squares in the tenth century*, B.23, and MS. London BL Delhi Arabic 110, fol. 90ᵛ - 91ʳ, 20.

in Thābit's text, which systematically (but tediously) explains where to put the odd numbers in the borders surrounding the central square, starting from the smallest border. But we shall use the more convenient way of filling the borders of the oblique square, starting from its outer border. The connection with the method of filling usual bordered squares of odd order will then be evident.

§4. Completing the placing of odd numbers

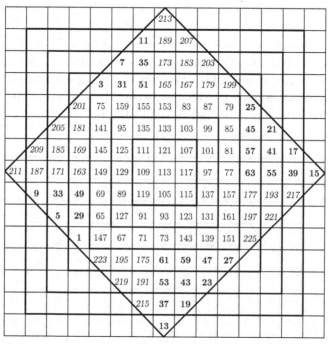

Fig. 286

Consider thus the oblique square as an entity, with its lateral rows as borders, and the starting point, 1, being the cell following the middle of its lower side, or else, if preferred, the cell next to the lower corner of the inner square. We shall place the odd numbers alternately in the upper halves of the two contiguous left-hand lateral rows until we arrive at the two corner cells, which we leave empty whereas we fill the other two, starting with the one below. Then we resume the alternate placing from the cell next to the one last reached, this time in the right-hand lower halves, until we reach the sides of the inner square. We do the same for each inner border of the oblique square. The placing will end with the two cells on the diagonals of the oblique square and next to the inner square, which we shall fill, following the same movement, first the one at the bottom then the one on the right. Finally, we shall place the complements of the numbers already put in —considering here, of course, the borders of the main square. From Fig. 286 & Fig. 287 ($n = 15$, thus order $4t + 3$ with $t = 3$, and $n = 17$, thus order $4(t + 1) + 1$) the perfect regularity of the procedure becomes apparent. As we have seen (pp. 178-179), the triangular parts of the oblique squares are the same for these two consecutive orders, and thus will contain just the same smaller odd numbers.

Seen that way, this placing indeed recalls the method taught earlier for odd-order bordered squares, with its alternate movement placing the successive numbers in two contiguous rows (Fig. 247-249), and is obviously inferred from it. This, though, does not appear from the extant text, which merely states which number will occupy which border, these borders being taken in turn. Whoever wrote these instructions must have just described the figure he had in front of him.

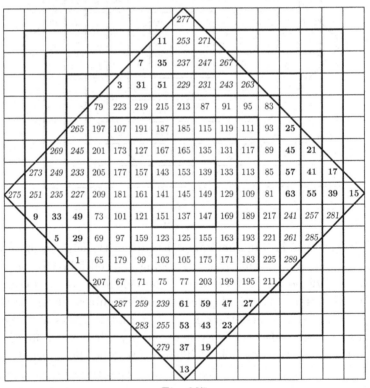

Fig. 287

B. METHODICAL PLACING OF THE EVEN NUMBERS

The treatment for orders $n = 4t + 1$ and $n = 4t + 3$ has so far been the same. But it now becomes necessary to differentiate between these two orders. As Thābit writes, *at this point you will find that the squares are divided into classes requiring each a treatment distinct from that of its predecessor for the arrangement of the even numbers in the corners. There are (those of order 5,) 9, 13, 17, and so on by adding successively 4; and (those of order) 7, 11, 15, 19, and so on by adding successively 4.*[170]

§5. Situation after filling the oblique square

Before placing the even numbers, we are to consider the present situation. That is, we must determine, for each border, both the number of cells still empty in each row and the sum this row already contains. We shall then know its excess or deficit relative to the sum due for the cells already filled, and how many cells are at our disposal to eliminate the difference. The text of Thābit (and Anṭākī) teaches all that, in a very precise though sometimes too concise manner. This being one of the finest pieces of ancient Greek mathematics, we shall present ample quotations from the text.

Fig. 288   Fig. 289

1. Determining the number of cells remaining empty

(1) Case of order $n = 4t + 1$ (Fig. 288).[171]

For the figure of 5 and those of the same kind: There remain as empty cells in the first border, which surrounds the inner square, the four cells at the corners and eight adjacent to them, two on each side; in the second, the cells remaining empty are those at the corners and twenty-four cells adjacent to them; in the third, the empty cells are those at the corners and forty cells adjacent to them. It will always be like that: each border has, excepting the corner cells, 16 more than the preceding border.

(2) Case of order $n = 4t + 3$ (Fig. 289).[172]

For the figure of seven and those of the same kind: There remain as

---

[170] *Magic squares in the tenth century*, A.II.12, and MS. London BL Delhi Arabic 110, fol. 82ʳ, 9-11; or below, Appendix 30.

[171] *Magic squares in the tenth century*, A.II.13, and MS. London BL Delhi Arabic 110, fol. 82ʳ, 12-15; or below, Appendix 31.

[172] *Magic squares in the tenth century*, A.II.16, and MS. London BL Delhi Arabic 110, fol. 82ᵛ, 7-10; or below, Appendix 32.

empty cells in the first border, which surrounds the inner square, only the four cells at the corners; in the second border, the cells at the corners and 16 cells adjacent to them, 4 on each side; in the third border, the empty cells are those at the corners and 32 cells adjacent to them. It will always be like that: each border has, excepting the corner cells, 16 more than the preceding border.

The number of empty cells in each border p cut by the sides of the oblique square, starting with the innermost $(p = 1)$, is thus as indicated in Fig. 290.

	$p = 1$	$p = 2$	$p = 3$	$p = 4$	$p = 5$
$n = 4t + 1$	$4 + 8$	$4 + 24$	$4 + 40$	$4 + 56$	$4 + 72$
$n = 4t + 3$	4	$4 + 16$	$4 + 32$	$4 + 48$	$4 + 64$

Fig. 290

Expressed in more mathematical terms, this means that:

In a square of order $4t + 1$, with an inner square of order $2t + 1$ completely filled and thus with t borders partly filled, the pth border starting from the inner square contains $16p - 4$ empty cells, $4p$ in each of its rows $(p = 1, \ldots, t)$; in a square of order $4t + 3$, with again an inner square of order $2t + 1$ and thus $t + 1$ borders partly filled, the pth border contains $16p - 12$ empty cells, $4p - 2$ in each of its rows $(p = 1, \ldots, t + 1)$.

2. Determining the sum required

Since the magic sum for a square of order n is $M_n = n \cdot \frac{n^2+1}{2}$, the sum due for m cells in one row is $m \cdot \frac{n^2+1}{2}$. Let us examine the situation after the previous placing of odd numbers —disregarding the inner square since each of its rows makes the sum due. Relative to the cells already filled, the upper rows in each incomplete border display an excess over the sum due, as do also the (for us) left-hand columns, while the opposite rows show a deficit of equal amount since they are filled with complements. Thābit's (and Anṭākī's) text gives all necessary information about the excesses or deficits for the orders $n = 4t + 1$ and $n = 4t + 3$, starting from the border surrounding the inner square —in fact, in the absence of symbolism, it merely gives the values for the first orders, but this is sufficient to set out the general rule.

(1) Case of order $n = 4t + 1$.[173] *After placing the odd numbers as indicated, the first upper row of the square of 9 is in excess over its sum due,*

---

[173] *Magic squares in the tenth century*, A.II.14, and MS. London BL Delhi Arabic 110, fol. 82ʳ, 16 - 82ᵛ, 1; or below, Appendix 33. Remember that in the Arabic text the columns in excess are on the right.

required for it, by 20, the lower one is in deficit by 20; the first right-hand row is in excess over its sum due by 18 and the left-hand one is in deficit by 18. The second upper (row) is in excess by 36 and the lower one is in deficit by the same amount; the second right-hand (row) is in excess by 34 and the left-hand one is in deficit by the same amount. For the square of 13, the first upper (row) is in excess by 28 and the lower one is in deficit by the same amount; the first right-hand (row) is in excess by 26 and the left-hand one is in deficit by the same amount. The second upper (row) is in excess by 52 and the lower one is in deficit by the same amount; the second right-hand (row) is in excess by 50 and the left-hand one is in deficit by the same amount. The third upper (row) is in excess by 76 and the lower one is in deficit by the same amount; the third right-hand (row) is in excess by 74 and the left-hand one is in deficit by the same amount. Likewise for the others. The text then indicates (again, rather too concisely) the increments for (horizontal or vertical) rows of successive borders within the same square, and for rows of corresponding borders in successive squares: *Nine: the second (row) has 16 more than that before. Thirteen: the first has 8 more than the first of nine, then each row 24 more than that before. Seventeen: the first has 8 more than the first of thirteen, then each row 32 more than that before. Then each row always more than that before in the same manner.*

This enables us to set out a table of the differences displayed by the lines and columns according to order $n = 4t + 1$ (we include $n = 5$) and number p of the border (Fig. 291):

	$p = 1$		$p = 2$		$p = 3$		$p = 4$		$p = 5$	
$n = 5$ $(t = 1)$	12	10								
$n = 9$ $(t = 2)$	20	18	36	34						
$n = 13$ $(t = 3)$	28	26	52	50	76	74				
$n = 17$ $(t = 4)$	36	34	68	66	100	98	132	130		
$n = 21$ $(t = 5)$	44	42	84	82	124	122	164	162	204	202

Fig. 291

We may now express these differences in a formula. Let us designate by $\Delta_p^{L,\,4t+1}$ the excess of the upper line belonging to the pth border if the order of the main square is $n = 4t+1$, and by $\Delta_p^{C,\,4t+1}$ that of the left-hand column of the same border. We infer from the table that $\Delta_1^{L,\,4t+1} = 8t+4$, $\Delta_2^{L,\,4t+1} = 16t+4$, $\Delta_3^{L,\,4t+1} = 24t+4$, $\Delta_4^{L,\,4t+1} = 32t+4$, thus, generally,

$$\Delta_p^{L,\,4t+1} = 8pt + 4 = 2p\,(n - 1) + 4.$$

As for the columns of these borders, their differences happen to be less than those of the lines by 2, and we shall thus have

$$\Delta_p^{C,\,4t+1} = 8pt + 2 = 2p\,(n - 1) + 2.$$

(2) Case of order $n = 4t + 3$.[174] *In this kind of square, the first upper row has (always) an excess of 4 over its sum due, and the lower one has a deficit of the same amount; the first right-hand (row) has an excess of 2 over its sum due and the left-hand (row) has a deficit of the same amount. The second upper row of the (square of) seven has an excess of 20 over its sum due, and the second right-hand (row) 18. In the square of 11, the second upper (row) has an excess of 28 over its sum due, the second right-hand (row) 26, the third upper (row) 52, the third right-hand (row) 50, and all the opposite have a deficit equal to the excess. Fifteen: the second upper (row) exceeds the first by 32, the third the second by 32. And so on for the others: the excess increases each time by 8.* Presented in tabular form, this gives Fig. 292.

	$p = 1$		$p = 2$		$p = 3$		$p = 4$		$p = 5$	
$n = 7$ $(t = 1)$	4	2	20	18						
$n = 11$ $(t = 2)$	4	2	28	26	52	50				
$n = 15$ $(t = 3)$	4	2	36	34	68	66	100	98		
$n = 19$ $(t = 4)$	4	2	44	42	84	82	124	122	164	162

Fig. 292

From this we may infer, as before, what these differences are. Let us designate by $\Delta_p^{L,\,4t+3}$ and $\Delta_p^{C,\,4t+3}$ the excesses of the upper line and of the left-hand column in the pth border when the main square has the order $n = 4t + 3$. Then, according to the table, $\Delta_1^{L,\,4t+3} = 4$, $\Delta_2^{L,\,4t+3} = 8(t+1) + 4$, $\Delta_3^{L,\,4t+3} = 16(t+1) + 4$, $\Delta_4^{L,\,4t+3} = 24(t+1) + 4$, whence, generally,

$$\Delta_p^{L,\,4t+3} = 8(p - 1)(t + 1) + 4 = 2\,(p - 1)(n + 1) + 4.$$

As for the columns of these borders, their differences are again smaller by 2 than those of the lines, so

$$\Delta_p^{C,\,4t+3} = 8(p - 1)(t + 1) + 2 = 2\,(p - 1)(n + 1) + 2.$$

§6. Rules for placing the even numbers

We now know, for each incomplete row of a border, the (even) number of empty cells it contains and its difference to the sum due. We are then to proceed as follows.

A preliminary placing must first eliminate the difference in each incomplete border's row ('equalization'); that is, the cells already occupied

---

[174] *Magic squares in the tenth century*, A.II.17, and MS. London BL Delhi Arabic 110, fol. 82^{v}, 10-16; or below, Appendix 34. In the Arabic text the columns in excess are on the right.

by odd numbers and those about to receive even numbers during this preliminary placing must then display their sum due, thus, for the order n, as many times $\frac{n^2+1}{2}$ as the number of occupied cells. But this preliminary placing must obey certain conditions:

— it must be uniform so that it may be applied to all squares of the same kind of order, either $n = 4t + 1$ or $n = 4t + 3$;

— it must involve the least possible number of cells in order to be applicable to the smallest orders, namely $n = 9$ and $n = 7$, respectively (order 5 has a treatment of its own);

— it must fill the corner cells, each common to two rows;

— it must leave a number of empty cells divisible by 4;

— the remaining even numbers must form groups of four, or an even number of pairs of, consecutive numbers.

If all these conditions have been met in the preliminary placing, completing the treatment is an easy matter since it will involve only neutral placings. (This procedure reminds us of the one for filling bordered squares of even orders; placing even numbers only makes no difference.)

Let us first consider the set of even numbers to be put in the square considered, as in Thābit's text.[175] For the order $n = 2k + 1$, there are $2k^2 + 2k$ even numbers, namely $k(k + 1)$ smaller ones $\left(\text{less than } \frac{n^2+1}{2}\right)$ and $k(k + 1)$ larger ones, their complements; furthermore, since $k(k + 1)$ is even, all these numbers, smaller and larger, may be arranged in pairs of consecutive even numbers. See Fig. 293; here these pairs are numbered starting from the highest smaller numbers, the jth pair of smaller numbers having thus the form $\frac{n^2-8j+3}{2}$, $\frac{n^2-8j+7}{2}$ $(j = 1, 2, \ldots, \frac{1}{2}k(k + 1))$.

2	4	$\ldots$	$\frac{n^2-8j+3}{2}$	$\frac{n^2-8j+7}{2}$	$\ldots$	$\frac{n^2-5}{2}$	$\frac{n^2-1}{2}$
$n^2 - 1$	$n^2 - 3$	$\ldots$	$\frac{n^2+8j-1}{2}$	$\frac{n^2+8j-5}{2}$	$\ldots$	$\frac{n^2+7}{2}$	$\frac{n^2+3}{2}$
$\frac{1}{2}k(k+1)$		$\ldots$	j		$\ldots$		1

Fig. 293

We are, as far as possible, to equalize the rows without breaking these pairs of numbers, for that will facilitate later neutral placings. Furthermore, since the difference to the sum due is the same (but with opposite

[175] *Magic squares in the tenth century*, A.II.18, and MS. London BL Delhi Arabic 110, fol. 82^{v}, 16-19.

signs) in two opposite rows, we may carry out the equalization for just one of two opposite rows.

1. First main equalization rule: excess or deficit of the form $\pm 2s$, $s \geq 1$

When you write the first small number, or any small number, on one side and the subsequent number on the other, opposite side, and you write the two large numbers which are their complements opposite to them, the side where you have written the first small number will be less than its sum due by 2, and the other will be in excess by 2. If you write some small number on one side and you write on the other, opposite side the third small number counted from this one, the side containing the first small number will be less than its sum due by 4. And so on: whenever you increase the distance between these two by one number, the deficit is always increased by 2.[176] That is, with a number α written in one row then $\alpha+2s$ in the opposite row, and next, facing them, their complements, the row of α will display a deficit of $2s$ and the opposite row an excess of $2s$ (Fig. 294).

	α	$n^2 + 1 - (\alpha + 2s)$		$\Longrightarrow n^2 + 1 - 2s$
	$n^2 + 1 - \alpha$	$\alpha + 2s$		$\Longrightarrow n^2 + 1 + 2s$

Fig. 294

With this rule, we may eliminate any even difference, thus, in fact, any difference which may occur here. But its use with $s \neq 1$ should be avoided as far as possible since we must try to keep all pairs of consecutive even numbers unbroken in order to complete the placing.

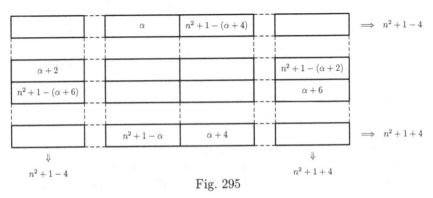

Fig. 295

---

[176] *Magic squares in the tenth century*, A.II.21, and MS. London BL Delhi Arabic 110, fol. 83^r, 9-14; or below, Appendix 35.

Particular case

Four consecutive even numbers placed around the border will, together with the complements, produce a difference of 4, which will be a deficit in the line and the column containing the lesser pair (Fig. 295).[177]

2. Second main equalization rule: excess or deficit of the form $\pm (8j - 4)$, $j \geq 1$

Placing the jth pair on the same side will, after placing the complements, produce a deficit of $8j - 4$ on the side of that pair and an excess of the same amount opposite (Fig. 296). This is indeed what the text asserts, of course in a less general way since it considers one pair at a time, beginning with $j = 1$: *You will find that if the last small number and the preceding one are placed on one side, and opposite to them their complements, the side containing the two small numbers will be less than its sum due by 4, and the other more by 4; placing the next two numbers, the differences will be 12 and 12, and with the next two numbers 20 and 20, then 28, 36, 44, 52, 60 and so on always till (placing) 4 and 2.*[178]

$\dfrac{n^2 - 8j + 3}{2}$	$\dfrac{n^2 - 8j + 7}{2}$
$\dfrac{n^2 + 8j - 1}{2}$	$\dfrac{n^2 + 8j - 5}{2}$

$\implies n^2 + 1 - (8j - 4)$

$\implies n^2 + 1 + (8j - 4)$

Fig. 296

Particular case

If these pairs are written in the corners (Fig. 297), they will in addition produce a constant difference of 2 in the other two rows, which will be a deficit in the row containing the least element, as stated in the text after a repetition of the previous rule: *If four numbers are put in the corners, each pair in consecutive (corners) and their complements diagonally opposite, and if you put the two small consecutive numbers on the upper side, this side will be less than its sum due by 4 if they are the last two, by 12 if they are the two previous (numbers), and so on always, with regular additions of 8, until you reach 2 and 4. The right-hand side will always have, relative to its sum due, an excess of 2, or a deficit of 2, without any*

---

[177] *Magic squares in the tenth century,* A.II.29, and MS. London BL Delhi Arabic 110, fol. 84^r, 5-7.

[178] *Magic squares in the tenth century,* A.II.19, and MS. London BL Delhi Arabic 110, fol. 82^v, 19-83^r, 1; or below, Appendix 36.

augmentation and diminution.[179]

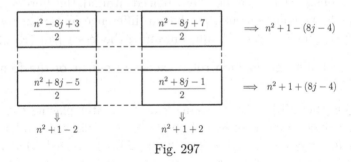

Fig. 297

Having thus given the equalization rules, the text concludes: *You must understand all this: it belongs to what you need for writing the even numbers in (squares of) this kind.*

Remark. As seen above, the differences to be eliminated are always even. Now by applying once, twice or at most three times the above rules using pairs of consecutive numbers, we can eliminate differences of the form $8u$, $8u \pm 2$, $8u \pm 4$, $8u \pm 6$, that is, any difference which might occur. To do this, however, the number of empty cells available must be sufficient. If that is not the case, we shall be obliged to eliminate the difference by means of two even, non-consecutive numbers (α, $\alpha + 2s$, with $s \neq 1$; first main equalization rule); but then we shall have to apply it a second time in order to use the other two terms of the two broken pairs. See example below, in the equalization of the first (smallest) border for the order $n = 4t + 1$ (pp. 193-195). It may also happen that the difference can be eliminated with fewer cells than the quantity required to leave a number of empty cells divisible by 4; then we shall just use more cells than necessary (one application may only partly eliminate the difference, or even increase it). See example below, in the equalization of the other borders' lines for the order $n = 4t + 1$ (p. 195 *seqq.*). In all other cases we shall be able to apply these rules to the least possible number of cells, and so as to leave a number of empty cells divisible by 4, thus paving the way to the neutral placings.

3. Neutral placings

Once the equalization has been effected, completing the arrangement is easy: we are to place groups of four —or groups of two pairs of— consecutive numbers, with the extremes on one side and the means opposite,

---

[179] *Magic squares in the tenth century*, $\mathcal{A}$.II.22, and MS. London BL Delhi Arabic 110, fol. 83^r, 17-21; or below, Appendix 37.

as we have seen for the method of the cross (Fig. 174, p. 98). But since we are dealing here with even numbers, the four numbers will have the form α, $\alpha + 2$, $\alpha + 4$, $\alpha + 6$ if they are consecutive (Fig. 298) or α, $\alpha + 2$, β, $\beta + 2$ if they form two pairs. This is just what the text says after giving the equalization rules: *If there are remaining empty cells, it can be only four facing four, eight facing eight, twelve facing twelve (and so on); you will equalize them four by four using the still available sequences of either four numbers in progression (...) or four numbers of which each pair is in (the same) progression: you put the first of the first pair and the second of the second pair on one side, the second of the first pair and the first of the second pair on the facing side, and opposite to each its complement.*[180]

α	$n^2+1-(\alpha+2)$	$n^2+1-(\alpha+4)$	$\alpha+6$		$\implies 2(n^2+1)$
$n^2+1-\alpha$	$\alpha+2$	$\alpha+4$	$n^2+1-(\alpha+6)$		$\implies 2(n^2+1)$

Fig. 298

§7. Case of the order $n = 4t + 1$ (with $t \geq 2$)[181]

Since the bottom lines are to receive the complements, we shall consider only the top ones, thus those in excess; likewise, we shall consider the left-hand columns, also in excess. As we have seen (p. 186), the lines and columns of the pth border ($p = 1,\ldots,t$, counting from the inner square already filled) comprise $4p$ empty cells. Making up for the differences must thus be effected with four cells for the rows of all first borders, whereas we may use eight cells for the rows of the other borders, this being applicable to the outer border of the lowest order $n = 9$ as well.

1. First border

We are thus to eliminate the excess in the upper line, $\Delta_1^{L,\,4t+1} = 8t+4$, by means of two pairs of numbers. Let us take for the first pair the largest dyad ($j = 1$), namely $\frac{n^2-5}{2}$ and $\frac{n^2-1}{2}$, to be put in the corners (say in the left-hand and right-hand corners, respectively).[182] Since their sum is $n^2 - 3$, they display a deficit of 4 relative to their sum due $n^2 + 1$, and this

---

[180] *Magic squares in the tenth century*, A.II.31, and MS. London BL Delhi Arabic 110, fol. 84^r, 11-14; or below, Appendix 38.

[181] We shall consider the particular case of order 5 at the end.

[182] This choice will be in keeping with that for the other borders.

leaves us with eliminating $8t$ by means of two numbers. The only way is to use two numbers from different dyads, say the smallest number 2 and thus the larger number $n^2 + 1 - (2 + 8t)$, the sum of which eliminates the excess completely and fills the last two available cells.

This is just what the text says, except that it is $2 + 8t$ (that is, $2n$) which is said to be put in the opposite row: *Put the last small even term in the upper left-hand* (for us right-hand) *corner of the first border* —that is, the border following the inner square which you have completely filled with odd numbers— *and opposite to it diagonally, in the lower right-hand corner of the first border, its complement. Put the preceding small term in the upper right-hand corner of the first border, and opposite to it diagonally, in the lower left-hand corner, the term which is its complement.* (...) *Put then 2 in the upper row; consider the excess of the upper row, take its half, and count after 2 as many small even numbers as this half;*[183] *put then the number reached in the lower row. Put opposite to each of these two numbers its complementary term.*[184] See Fig. 301-303, and, for the values, Fig. 299, I-IV.

	$\frac{n^2-5}{2}$	$\frac{n^2-1}{2}$	2	$8t+2$	4	$8t+4$
$n = 9$ $(t = 2)$	38	40	2	18	4	20
$n = 13$ $(t = 3)$	82	84	2	26	4	28
$n = 17$ $(t = 4)$	142	144	2	34	4	36
	I	II	III	IV	V	VI

Fig. 299

Consider now the columns. The left-hand corner cells are now occupied by the smaller of the numbers placed first and the complement of the other, thus by $\frac{n^2-5}{2}$, $\frac{n^2+3}{2}$, the sum of which is $n^2 - 1$; the initial excess on the left-hand side, $\Delta_1^{C, 4t+1} = 8t + 2$, is thus reduced to $8t$, just as before. Since there are only two empty cells available, we must again eliminate this excess by two numbers from different dyads. Then, as we are told in the same place, *put 4 on the right side; then count after 4 as many small even numbers as half of the excess, and put the number reached in the left-hand row. Put opposite to each of these two numbers its complementary term.* Indeed, putting 4 in the (for us) left-hand column and, in the right-hand one, $4 + 8t$, that is, the $4t$th even number after 4, not

---

[183] The excess being now $8t$, we shall arrive at $8t + 2$ by counting $4t$ even numbers after 2.

[184] *Magic squares in the tenth century*, A.II.24, and MS. London BL Delhi Arabic 110, fol. 83^v, 3-10; or below, Appendix 39.

only shall we have eliminated the excess of $8t$ since the left-hand column now contains 4 and $n^2 + 1 - (4 + 8t)$, but we shall also have placed the remaining terms of the two pairs previously broken, thereby avoiding the problem of using subsequently two single numbers. See Fig. 301-303 and 299, V-VI.

2. Other borders

The excesses in the upper rows are $\Delta_p^{L,\,4t+1} = 8pt+4$, and there are $4p$ empty cells. A direct elimination by one dyad would be possible (putting $j = pt+1$), but unsuitable since the number of cells left empty would not be divisible by 4. Since elimination by two dyads is not possible, we shall eliminate these excesses by means of eight cells, this being applicable to the smallest order $n = 9$. The same will be done with the excesses in the left-hand columns, originally $\Delta_p^{C,\,4t+1} = 8pt + 2$. All that will be reached in five steps (i-v below). In the first, we shall fill the two corner cells of each column in excess (for us the left-hand ones). In the second, two numbers put within these columns will reduce their excesses to a uniform quantity. In the third step, this same situation will be attained for the lines, using two of the six cells still available. The last two equalization steps, applied together to lines and columns, will use the remaining four cells to remove the constant differences left.

(i) We first look for two (consecutive) even numbers, α and $\alpha+2$, adding up to $(n^2 + 1) - (8p - 4)$. The two numbers are thus

$$\frac{n^2 - 8p + 3}{2}, \qquad \frac{n^2 - 8p + 7}{2},$$

the lesser of which is put in the upper corner. For these placings, the text explains that we are to put, at either end of each of these columns, a pair of consecutive small numbers the sum of which is, respectively, less than their sum due by 12, 20, 28, thus by our $8p - 4$ (this subtractive quantity being thus the same for same borders).[185] See the resulting values for the first orders in Fig. 300, I and II, and Fig. 301-303. Note that the above formulas are valid for the first border as well ($p = 1$).

(ii) With the placing of these two numbers, adding up to $(n^2+1)-(8p-4)$, the initial excess of the pth left-hand column $\Delta_p^{C,\,4t+1} = 8pt + 2$ has become $8p(t - 1) + 6$. Now we know from the second main equalization rule how to reduce an excess of such a form: putting $8p(t-1)+6 = 8j-4$, we have $8j = 8p(t - 1) + 10$. Taking the closest value, $j = p(t - 1) + 1$, we shall put in the pth left-hand column ($p \geq 2$) the pair

---

[185] *Magic squares in the tenth century*, A.II.25i, 26i, 27i, and MS. London BL Delhi Arabic 110, fol. 83^{v}, 11-13, 17-18, 22-23.

$$\frac{n^2 - 8[p(t-1)+1]+3}{2} = \frac{n^2 - 2pn + 10p - 5}{2}$$

$$\frac{n^2 - 8[p(t-1)+1]+7}{2} = \frac{n^2 - 2pn + 10p - 1}{2},$$

the values of which for the first orders are seen in Fig. 300, III and IV, or Fig. 301-303. This then leaves an excess of 2 in each left-hand column and four cells to complete the equalization.

For the above placings, the text explains that we are to put in the (for us left-hand) columns pairs of consecutive small numbers adding up to their sum due less an amount equal to the (initial) excess of the column minus, respectively, 14, 22, 30, thus minus the quantity $8p-2$; that is, the sum of these two numbers α, $\alpha+2$ must be $(n^2+1)-[8pt+2-(8p-2)] = n^2 - 8[p(t-1)+1]+5$, whence the above pair of values.[186]

$\frac{n^2-8p+3}{2}$	$\frac{n^2-8p+7}{2}$	$\frac{n^2-2pn+10p-5}{2}$	$\frac{n^2-2pn+10p-1}{2}$	$\frac{n^2-2pn+2p+3}{2}$	$\frac{n^2-2pn+2p+7}{2}$
34	36	30	32	26	28
78	80	66	68	62	64
74	76	58	60	50	52
138	140	118	120	114	116
134	136	106	108	98	100
130	132	94	96	82	84
I	II	III	IV	V	VI

with row labels: $n=9,\ p=2$; $n=13,\ p=2$; $p=3$; $n=17,\ p=2$; $p=3$; $p=4$.

Fig. 300

(*iii*) Consider now the upper rows. Their excesses have now changed since one number has been placed in each pth left-hand corner while the right-hand one is occupied by a complement. Since these numbers, namely

$$\frac{n^2 - 8p + 3}{2}, \qquad \frac{n^2 + 8p - 5}{2},$$

add up to $n^2 - 1 = (n^2+1) - 2$, the excesses in the upper rows have been reduced from $8pt+4$ to $8pt+2$, and there are $4p-2$ empty cells. Putting

$$8pt + 2 = 8j - 4,$$

we shall take $j = pt$ and thus place the ptth pair of smaller numbers

$$\frac{n^2 - 8pt + 3}{2} = \frac{n^2 - 2pn + 2p + 3}{2}$$

$$\frac{n^2 - 8pt + 7}{2} = \frac{n^2 - 2pn + 2p + 7}{2}$$

within the pth upper row, the values of which for the first orders are seen in Fig. 300, V and VI, or Fig. 301-303. This then leaves an excess of 6

---

[186] *Magic squares in the tenth century*, A.II.25*ii*, 26*ii*, 27*ii*, and MS. London BL Delhi Arabic 110, fol. 83^v, 13-15, 18-20, 23-25.

in each upper row and four cells to complete the equalization. The text expresses the choice of these numbers in the same way as before: we are to put in the upper rows two consecutive even smaller numbers, say again α and $\alpha+2$, the sum of which is less than their sum due by the (initial) excess of the upper row minus 8 (thus minus a quantity independent of border and order); that is, their sum must be $2\alpha + 2 = (n^2 + 1) - [(8pt + 4) - 8]$, which gives the above two values.[187]

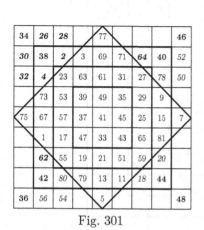

Fig. 301

34	26	28		77				46
30	38	2	3	69	71	64	40	52
32	4	23	63	61	31	27	78	50
	73	53	39	49	35	29	9	
75	67	57	37	41	45	25	15	7
	1	17	47	33	43	65	81	
	62	55	19	21	51	59	20	
	42	80	79	13	11	18	44	
36	56	54		5				48

Fig. 302

74	50	52				161						94
58	78	62	64		7	145	155				90	112
60	66	82	2	3	23	137	139	151	144	84	104	110
	68	4	47	131	127	125	55	59	51	166	102	
		153	113	67	107	105	75	71	57	17		
	157	141	117	97	83	93	79	73	53	29	13	
159	143	135	121	101	81	85	89	69	49	35	27	11
	5	21	41	61	91	77	87	109	129	149	165	
		1	37	99	63	65	95	103	133	169		
	142	119	39	43	45	115	111	123	28			
	86	168	167	147	33	31	19	26	88			
80	108	106				163	25	15			92	
76	120	118				9						96

With this and what we have found for the first border, we can now set out the table in Fig. 304 (the asterisk marks cells to receive complements). Although it gives the required quantities for only five borders, their forms clearly show how the terms within the same oblique line can be found using arithmetical progressions —the only exceptions being in the first border. That is therefore sufficient for constructing any square of order $n = 4t + 1$ with separation by parity; indeed, what remains to complete its rows is straightforward.

Indeed, as stated in the text,[188] there remains —except for the rows of the first border, which are completely filled (and equalized)— a number of cells divisible by 4 in each row and uniform excesses of 6 in the upper lines and of 2 in the left-hand columns (Fig. 305). In order to have an equalization valid from the smallest order (here $n = 9$), we are now to fill

[187] *Magic squares in the tenth century*, A.II.25*iii*, 26*iii*, 27*iii*, and MS. London BL Delhi Arabic 110, fol. 83ᵛ, 15-16, 20-22, 25 - 84ʳ, 1.

[188] *Magic squares in the tenth century*, A.II.28, and MS. London BL Delhi Arabic 110, fol. 84ʳ, 2-5.

four cells in each row. We further know that the numbers still available form groups of four, or pairs of dyads of, consecutive even numbers —since we have already placed the two pairs which had been broken.

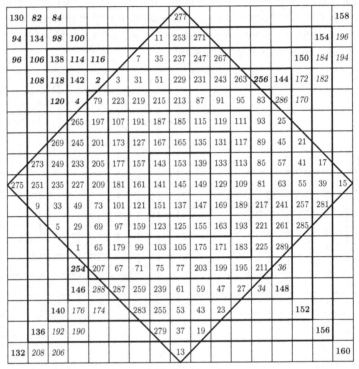

Fig. 303

Eliminating the remaining excesses will be carried out in the last two steps.[189]

Fig. 305   Fig. 306

(*iv*) First, we put four consecutive (even) numbers around each border, beginning in its upper line and turning (for us) towards the left. As we know from the particular case of the first main equalization rule (p. 191), with the complements in place, that will change the excesses and deficits of ±6 to ±2 and those of ±2 to ∓2 (Fig. 305-306). In Fig. 307 (*n* = 9) we

---

[189] *Magic squares in the tenth century*, $\mathcal{A}$.II.29-30, and MS. London BL Delhi Arabic 110, fol. 84^r, 5-11.

have thus placed, starting in the upper row, the groups of four numbers 6, 8, 10, 12 ($p = 2$); in Fig. 308 ($n = 13$), the groups 14, 16, 18, 20 ($p = 3$) and 6, 8, 10, 12 ($p = 2$); in Fig. 309 ($n = 17$), the groups 22, 24, 26, 28 ($p = 4$), 14, 16, 18, 20 ($p = 3$) and 6, 8, 10, 12 ($p = 2$).

C1	C2	C3	C4	C5	C6	C7	C8	C9	C10	C11	C12	C13
$\frac{n^2+35}{2}$	*	*										*
	$\frac{n^2+27}{2}$	*	*								*	
		$\frac{n^2+19}{2}$	*	*						*		
			$\frac{n^2+11}{2}$	*	*				*			
				$\frac{n^2-1}{2}$	*		$2n+2$	*				
				*			$2n$					
⋯	⋯	⋯	⋯	⋯	⋯	⋯	⋯	⋯	⋯	⋯	⋯	⋯
			$\frac{n^2-4n+11}{2}$	2				*	*			
		$\frac{n^2-6n+13}{2}$	$\frac{n^2-4n+7}{2}$	$\frac{n^2-5}{2}$	4		*	*	*	*		
	$\frac{n^2-8n+15}{2}$	$\frac{n^2-6n+9}{2}$	$\frac{n^2-13}{2}$	$\frac{n^2-4n+15}{2}$	$\frac{n^2-4n+19}{2}$				*	*	*	
$\frac{n^2-10n+17}{2}$	$\frac{n^2-8n+11}{2}$	$\frac{n^2-21}{2}$	$\frac{n^2-6n+25}{2}$	$\frac{n^2-6n+29}{2}$						*	*	*
$\frac{n^2-10n+13}{2}$	$\frac{n^2-29}{2}$	$\frac{n^2-8n+35}{2}$	$\frac{n^2-8n+39}{2}$								*	*
$\frac{n^2-37}{2}$	$\frac{n^2-10n+45}{2}$	$\frac{n^2-10n+49}{2}$										*

Fig. 304

(v) Second, using the simplest case ($s = 1$) of the first equalization rule, take any two pairs of available consecutive numbers and place each in a pair of opposite rows, with the lesser terms on the side of the excess. With that done for all borders, and the complements written in, all remaining differences will be eliminated. In Fig. 307 ($n = 9$), we have thus placed 14, 16 and 22, 24 ($p = 2$); in Fig. 308 ($n = 13$), 34, 36 and 38, 40 ($p = 3$), then 22, 24 and 30, 32 ($p = 2$); in Fig. 309 ($n = 17$), 42, 44 and 54, 56 ($p = 4$), then 38, 40 and 50, 52 ($p = 3$), finally 30, 32 and 46, 48 ($p = 2$). Anṭākī has reproduced from his source the simplest example, with $n = 9$; Thābit's text has also the squares of orders 13 and 17.[190]

Remark. In all these examples the numbers chosen just fill up progressively the gaps left by the former placings in the continuous number sequence.

34	26	28	6	77	72	14	66	46
30	38	2	3	69	71	64	40	52
32	4	23	63	61	31	27	78	50
8	73	53	39	49	35	29	9	74
75	67	57	37	41	45	25	15	7
70	1	17	47	33	43	65	81	12
60	62	55	19	21	51	59	20	22
24	42	80	79	13	11	18	44	58
36	56	54	76	5	10	68	16	48

Fig. 307

74	50	52	14	152	34	161	134	42	126	124	48	94
58	78	62	64	6	7	145	155	160	22	146	90	112
60	66	82	2	3	23	137	139	151	144	84	104	110
16	68	4	47	131	127	125	55	59	51	166	102	154
150	8	153	113	67	107	105	75	71	57	17	162	20
132	157	141	117	97	83	93	79	73	53	29	13	38
159	143	135	121	101	81	85	89	69	49	35	27	11
40	5	21	41	61	91	77	87	109	129	149	165	130
54	158	1	37	99	63	65	95	103	133	169	12	116
114	140	142	119	39	43	45	115	111	123	28	30	56
100	32	86	168	167	147	33	31	19	26	88	138	70
72	80	108	106	164	163	25	15	10	148	24	92	98
76	120	118	156	18	136	9	36	128	44	46	122	96

Fig. 308

We have now completed the first two borders around the central square and are left in the subsequent borders ($p \geq 3$) with numbers of empty cells all divisible by 4. Since the rows of these borders display neither excesses nor deficits, we may fill them repeatedly with neutral placings, thus two pairs of even numbers of which we write the extreme terms on one side and the middle terms on the other, the complements providing then the sum due for four cells.[191] In Fig. 308 ($n = 13$), 42,...,48 and

---

[190] *Magic squares in the tenth century*, pp. 160 & 310, and MS. London BL Delhi Arabic 110, fol. 84ᵛ, 85ʳ, 86ʳ; in all these examples the 3 × 3 inner square is rotated, as in the last example here (with the smallest element placed on the lateral side).

[191] *Magic squares in the tenth century*, A.II.31, and MS. London BL Delhi Arabic

the pairs 54, 56 & 70, 72 complete the outer border (50, 52 and 58, ..., 68 have been used for the preliminary placing); in Fig. 309 ($n = 17$), the two upper lines and the ones opposite have been filled with the tetrads 58, ..., 72, and 74, ..., 80, respectively, the columns with 86, ..., 92 and the pairs 102, 104 & 110, 112, and 122, ..., 128, respectively.

130	82	84	22	264	42	246	58	277	230	228	64	66	222	220	72	158
94	134	98	100	14	272	38	11	253	271	250	74	214	212	80	154	196
96	106	138	114	116	6	7	35	237	247	267	280	30	258	150	184	194
24	108	118	142	2	3	31	51	229	231	243	263	256	144	172	182	266
262	16	120	4	79	223	219	215	213	87	91	95	83	286	170	274	28
236	270	8	265	197	107	191	187	185	115	119	111	93	25	282	20	54
56	240	269	245	201	173	127	167	165	135	131	117	89	45	21	50	234
86	273	249	233	205	177	157	151	141	143	133	113	85	57	41	17	204
275	251	235	227	209	181	161	137	145	153	129	109	81	63	55	39	15
202	9	33	49	73	101	121	147	149	139	169	189	217	241	257	281	88
200	52	5	29	69	97	159	123	125	155	163	193	221	261	285	238	90
92	122	278	1	65	179	99	103	105	175	171	183	225	289	12	168	198
102	166	244	254	207	67	71	75	77	203	199	195	211	36	46	124	188
186	164	48	146	288	287	259	239	61	59	47	27	34	148	242	126	104
180	128	140	176	174	284	283	255	53	43	23	10	260	32	152	162	110
112	136	192	190	276	18	252	279	37	19	40	216	76	78	210	156	178
132	208	206	268	26	248	44	232	13	60	62	226	224	68	70	218	160

Fig. 309

§8. Case of the order $n = 4t + 3$ (with $t \geq 1$)

We shall again consider only the rows in excess, thus the upper lines and the (for us) left-hand columns.

As already seen (pp. 186-188), the number of empty cells in each row of the pth border is $4p - 2$ and the excesses are $\Delta_p^{L, 4t+3} = 8(p - 1)(t + 1) + 4$ and $\Delta_p^{C, 4t+3} = 8(p - 1)(t + 1) + 2$. Thus, for all first borders the equalization *must* be effected with two cells. A single pair will turn out to be sufficient for the lines of larger borders ($p \geq 2$) as well, while for the columns, of which two corner cells are now occupied, at least four cells will be available for equalization.

110, fol. 84^{r}, 11-14.

1. First border

Since the excesses are $\Delta_1^{L,\,4t+3} = 4$ for the lines and $\Delta_1^{C,\,4t+3} = 2$ for the columns, we shall apply the particular case of the second main equalization rule (pp. 191-192) with $j = 1$. That is, putting in the upper corners of the first row the largest pair of smaller numbers, thus

$$\frac{n^2 - 5}{2}, \qquad \frac{n^2 - 1}{2},$$

with the larger in our right-hand corner, where the column displays a deficit, we shall have equalized all first borders: *once you have done that, you will have equalized the first border for all squares of this kind*, as Thābit's (and Anṭākī's) text has.[192] (Note that the same pair had been used for the first borders in the case $n = 4t + 1$ —though without then completing the equalization.) In Fig. 311 ($n = 7$) we have thus placed in the upper corners of the first border 22, 24; in Fig. 312 ($n = 11$), 58, 60; in Fig. 313 ($n = 15$), 110, 112.

2. Other borders

Equalization of the other borders will be carried out in three steps. As said above, the choice of a single pair will suffice for all lines, but two will be needed for the columns —of which the differences have now changed with occupation of the corner cells.

(i) Since the excess in the upper rows is $\Delta_p^{L,\,4t+3} = 8(p - 1)(t + 1) + 4$, we shall choose, as said in the text, a pair of (consecutive) small numbers less than their sum due $n^2 + 1$ by that amount.[193] In our terms, putting $8(p-1)(t+1)+4 = 8j - 4$ we shall choose $j = (p-1)(t+1)+1$ and write in the end cells of the pth upper row ($p \geq 2$) the corresponding pair, that is,

$$\frac{n^2 - 8[(p - 1)(t + 1) + 1] + 3}{2} = \frac{n^2 - 2(p - 1)n - 2p - 3}{2},$$

$$\frac{n^2 - 8[(p - 1)(t + 1) + 1] + 7}{2} = \frac{n^2 - 2(p - 1)n - 2p + 1}{2},$$

with the lesser term in the column with an excess. (Note that the pair for $p = 1$ just obeyed the same formula.) See Fig. 310, I and II, and Fig. 311-313. As then stated in the text, we are left with filling the incomplete lines with neutral placings.[194]

---

[192] *Magic squares in the tenth century*, A.II.32, and MS. London BL Delhi Arabic 110, fol. 84^r, 18-19; or below, Appendix 40.

[193] *Magic squares in the tenth century*, A.II.33, and MS. London BL Delhi Arabic 110, fol. 84^r, 19-20.

[194] *Magic squares in the tenth century*, A.II.33 *in fine*, and MS. London BL Delhi Arabic 110, fol. 84^r, 21-23.

	$\frac{n^2-2(p-1)n-2p-3}{2}$	$\frac{n^2-2(p-1)n-2p+1}{2}$	$\frac{n^2+8p-5}{2}$	$\frac{n^2+8p-1}{2}$	$\frac{n^2-2(p-1)n-10p+5}{2}$	$\frac{n^2-2(p-1)n-10p+9}{2}$
$n=7,\ \ p=2$	14	16	30	32	10	12
$n=11,\ p=2$	46	48	66	68	42	44
$p=3$	34	36	70	72	26	28
$n=15,\ p=2$	94	96	118	120	90	92
$p=3$	78	80	122	124	70	72
$p=4$	62	64	126	128	50	52
	I	II	III	IV	V	VI

Fig. 310

(*ii*) With the above placing, the situation in the left-hand columns has changed: the number of empty cells is now 4 ($p=2$) or a multiple of 4 ($p \geq 3$), while, since the corners are now occupied by the first number of the pair above and the complement of the other, the sum of which is $(n^2 + 1) - 2$, there remains on the pth left-hand side an excess of $8(p-1)(t+1)$, to be eliminated now with four numbers.

Here the text is imprecise:[195] it gives a single condition for determining the two pairs of consecutive even numbers to be put in the (for us) left-hand columns, two larger and two smaller, leaving thus one number (or pair) optional. As a matter of fact, since the corners opposite to those filled in the first border are occupied by the two larger complements $\frac{n^2+7}{2}$ and $\frac{n^2+3}{2}$, we shall just take, as the pairs of larger numbers, the subsequent ones, namely $\frac{n^2+15}{2}$, $\frac{n^2+11}{2}$; $\frac{n^2+23}{2}$, $\frac{n^2+19}{2}$; ... . That is, generally, the pairs of larger numbers to be put in the left-hand columns of the pth border will be

$$\frac{n^2 + 8p - 1}{2}, \qquad \frac{n^2 + 8p - 5}{2},$$

the sum of which is $n^2 + 8p - 3 = (n^2 + 1) + 8p - 4$.

(*iii*) The two larger numbers being thus determined, the pair of smaller numbers to be put with them is easy to find. Since the excess in the pth left-hand column has now increased to $8(p - 1)(t + 1) + 8p - 4 = 8[(p-1)(t+1)+p] - 4$, we shall put $j = (p-1)(t+1)+p$, and thus take as the pair of smaller numbers

$$\frac{n^2 - 8[(p - 1)(t + 1) + p] + 3}{2} = \frac{n^2 - 2(p - 1)n - 10p + 5}{2},$$

$$\frac{n^2 - 8[(p - 1)(t + 1) + p] + 7}{2} = \frac{n^2 - 2(p - 1)n - 10p + 9}{2}.$$

---

[195] *Magic squares in the tenth century*, A.II.34, and MS. London BL Delhi Arabic 110, fol. 84^r, 23 - 84^v, 5.

Once these four numbers —two larger and two smaller— have been placed, all excesses in the left-hand columns are eliminated (so also in the right-hand ones after placing of the complements). See the values in Fig. 310, III-VI and Fig. 311 ($n = 7$), Fig. 312 ($n = 11$), Fig. 313 ($n = 15$).

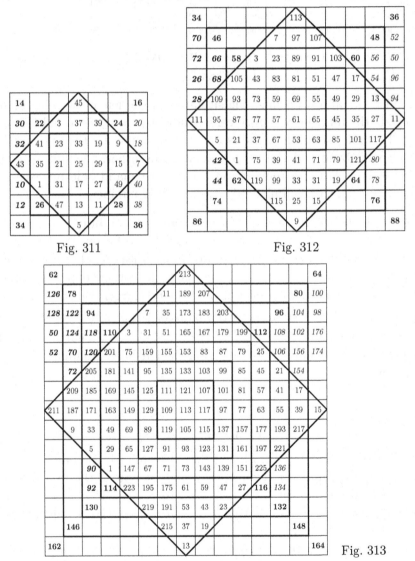

Fig. 311

Fig. 312

Fig. 313

With all these relations, we may set out Fig. 314, which gives the placings required for the first five borders. As before (Fig. 304), that enables us to fill the squares for any order of this type since determining the terms of the preliminary placing becomes evident.

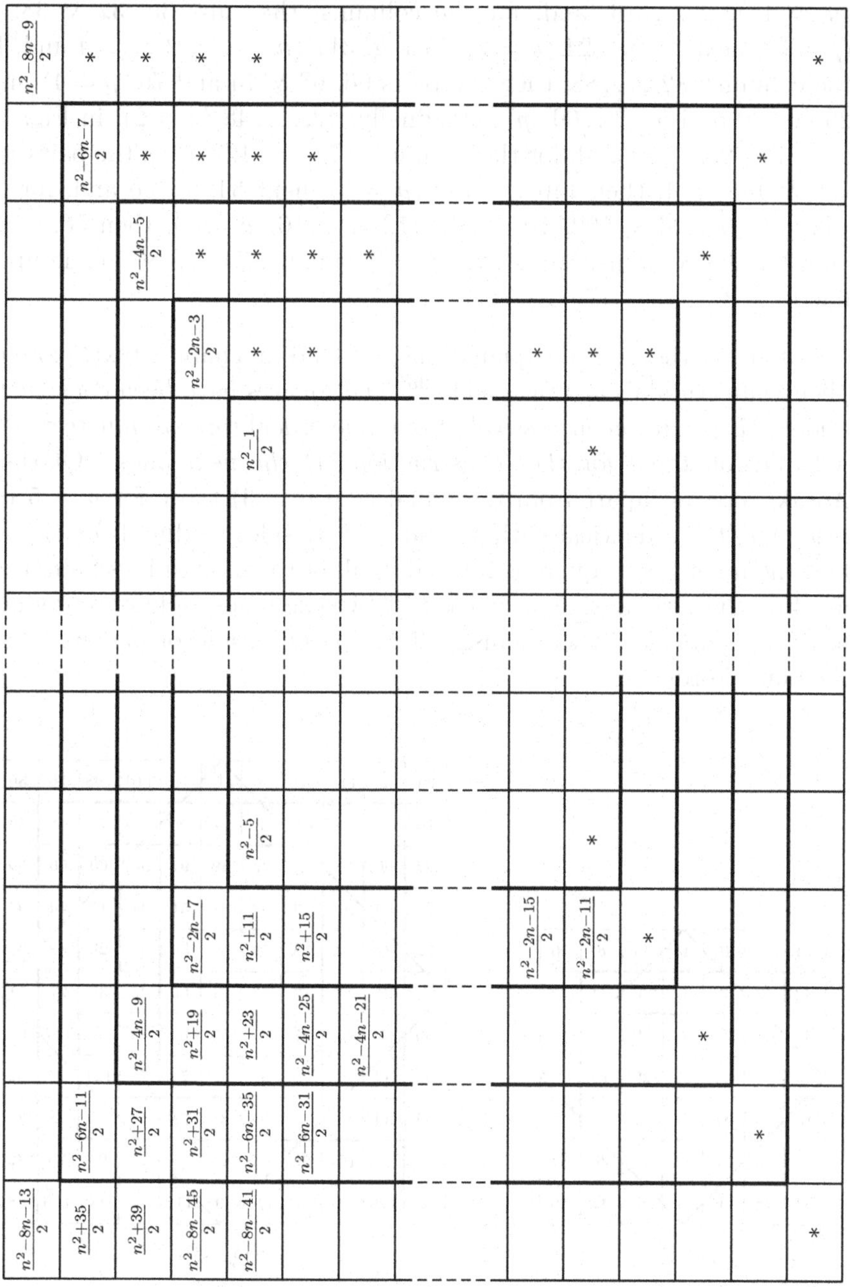

$\frac{n^2-8n-9}{2}$	*	*	*	*								*
	$\frac{n^2-6n-7}{2}$	*	*	*	*						*	
		$\frac{n^2-4n-5}{2}$	*	*	*	*				*		
			$\frac{n^2-2n-3}{2}$	*	*		*	*	*			
				$\frac{n^2-1}{2}$				*				
				$\frac{n^2-5}{2}$				*				
			$\frac{n^2-2n-7}{2}$	$\frac{n^2+11}{2}$	$\frac{n^2+15}{2}$			$\frac{n^2-2n-15}{2}$	$\frac{n^2-2n-11}{2}$	*		
		$\frac{n^2-4n-9}{2}$	$\frac{n^2+19}{2}$	$\frac{n^2+23}{2}$	$\frac{n^2-4n-25}{2}$	$\frac{n^2-4n-21}{2}$					*	
	$\frac{n^2-6n-11}{2}$	$\frac{n^2+27}{2}$	$\frac{n^2+31}{2}$	$\frac{n^2-6n-35}{2}$	$\frac{n^2-6n-31}{2}$						*	
$\frac{n^2-8n-13}{2}$	$\frac{n^2+35}{2}$	$\frac{n^2+39}{2}$	$\frac{n^2-8n-45}{2}$	$\frac{n^2-8n-41}{2}$								*

Fig. 314

Since in that case each row displays the sum due for the number of cells filled, it can be completed with neutral placings. In Fig. 315 ($n = 7$), a single group of four numbers $2, \ldots, 8$ completes the square; in Fig. 316

($n = 11$), $2, \ldots, 16$, and, for the columns, the pairs 30, 32 & 38, 40 ($p = 3$), next $18, \ldots, 24$ ($p = 2$); in Fig. 317 ($n = 15$), $2, \ldots, 24$ and, for the columns, $82, \ldots, 88$, then the pairs 66, 68 & 74 and 76 ($p = 4$), next $26, \ldots, 40$ and $54, \ldots, 60$ ($p = 3$), finally, $42, \ldots, 48$ ($p = 2$); in Fig. 318 ($n = 19$), $2, \ldots, 32$ and, for the columns, $122, \ldots, 128$, then the pairs 134, 136 & 142, 144, then $146, \ldots, 152$ ($p = 5$), next $34, \ldots, 56$ and, for the columns, 94, 96 & 102, 104, then $110, \ldots, 116$ ($p = 4$), then $58, \ldots, 72$ and, for the columns, $86, \ldots, 92$ ($p = 3$), finally, $74, \ldots, 80$ in the upper line ($p = 2$).

After the theoretical explanations, we read in Thābit's text, more explicit than Anṭākī's at that point:[196] *The squares have been constructed under this condition successively from (the order) three to nineteen. You will rely on them for the (construction of) the next ones.* Of these, Anṭākī reports (apart from the banal case $n = 3$) those for $n = 5$ (see below), 7, 9, 11, identical with Thābit's.[197] Besides reading from right to left, the inner 3×3 square is rotated in all of them, with its smallest element on the vertical side, as in the last two examples we have reproduced here; there are also at times insignificant permutations of the numbers in the outer border.

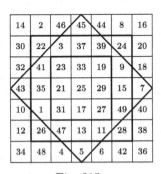

Fig. 315

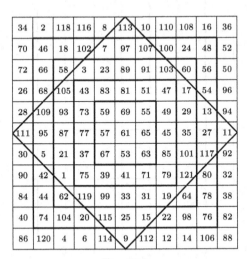

Fig. 316

---

[196] MS. London BL Delhi Arabic 110, fol. 84ᵛ, 11-12 (comp. *Magic squares in the tenth century*, A.II.35 *in fine*, lines 801-803 of the Arabic text).

[197] *Magic squares in the tenth century*, pp. 160 & 163-164 and 309-311; MS. London BL Delhi Arabic 110, fol. 84ᵛ - 86ᵛ.

62	2	222	220	8	10	214	213	212	16	18	206	204	24	64
66	78	26	198	196	32	11	189	207	34	190	188	40	80	160
158	54	94	42	182	7	35	173	183	203	180	48	96	172	68
152	170	120	110	3	31	51	165	167	179	199	112	106	56	74
76	168	118	201	75	159	155	153	83	87	79	25	108	58	150
82	60	205	181	141	95	135	133	103	99	85	45	21	166	144
142	209	185	169	145	125	119	109	111	101	81	57	41	17	84
211	187	171	163	149	129	105	113	121	97	77	63	55	39	15
140	9	33	49	69	89	115	117	107	137	157	177	193	217	86
88	124	5	29	65	127	91	93	123	131	161	197	221	102	138
128	122	90	1	147	67	71	73	143	139	151	225	136	104	98
126	70	92	114	223	195	175	61	59	47	27	116	134	156	100
50	72	130	184	44	219	191	53	43	23	46	178	132	154	176
52	146	200	28	30	194	215	37	19	192	36	38	186	148	174
162	224	4	6	218	216	12	13	14	210	208	20	22	202	164

Fig. 317

Remark. Towards the end of the 11th century, the algebrist Khayyām
worked with Asfizārī (above, p. 14) as an astronomer, and it may have
been the latter who aroused his interest in magic squares. In any
event one particular construction is attributed to *al-Imām Abū Ḥafṣ
'Umar ibn Ibrāhīm al-Khayyāmī*, namely a 28 × 28 square comprising
sixteen 7 × 7 subsquares with separation by parity, each filled with
consecutive numbers on the model of Fig. 315, all placed according
to a pandiagonal arrangement for a 4 × 4 square.[198] The notable
feature about it is that, since the first subsquare contains the sequence
$1, \ldots, 49$, and therefore the second the sequence $50, \ldots, 98$, and so on,
odd and even numbers will alternately change their places in these
subsquares. As to the choice of constructing a 28 × 28 square, it is said
to have a threefold justification: 28 is a 'perfect' number (*'adad tāmm*
= τέλειος ἀριθμός, thus a number equal to the sum of its divisors:
$28 = 1 + 2 + 4 + 7 + 14$ —see Euclid's *Elements* VII, def. 23 and
proposition IX, 36); next, it is the number of lunar mansions; finally,
it is the number of letters in the Arabic alphabet.

---

[198] MS. London BL Delhi Arabic 110, fol. 107^v - 109^r.

98	2	358	356	8	10	350	348	16	345	18	342	340	24	26	334	332	32	100
200	118	34	326	324	40	42	318	15	313	339	316	48	50	310	308	56	120	162
198	196	138	58	302	300	64	11	47	289	307	335	66	294	292	72	140	166	164
82	194	192	158	74	286	7	43	71	273	283	303	331	284	80	160	170	168	280
84	106	190	188	178	3	39	67	87	265	267	279	299	327	180	174	172	256	278
122	108	130	186	329	115	259	255	251	249	123	127	131	119	33	176	232	254	240
238	94	132	333	301	233	143	227	223	221	151	155	147	129	61	29	230	268	124
236	266	337	305	281	237	209	163	203	201	171	167	153	125	81	57	25	96	126
128	341	309	285	269	241	213	193	187	177	179	169	149	121	93	77	53	21	234
343	311	287	271	263	245	217	197	173	181	189	165	145	117	99	91	75	51	19
134	13	45	69	85	109	137	157	183	185	175	205	225	253	277	293	317	349	228
226	260	9	41	65	105	133	195	159	161	191	199	229	257	297	321	353	102	136
220	104	86	5	37	101	215	135	139	141	211	207	219	261	325	357	276	258	142
144	110	274	154	1	243	103	107	111	113	239	235	231	247	361	208	88	252	218
146	250	272	156	182	359	323	295	275	97	95	83	63	35	184	206	90	112	216
214	248	92	202	288	76	355	319	291	89	79	59	31	78	282	204	270	114	148
212	116	222	304	60	62	298	351	315	73	55	27	296	68	70	290	224	246	150
152	242	328	36	38	322	320	44	347	49	23	46	314	312	52	54	306	244	210
262	360	4	6	354	352	12	14	346	17	344	20	22	338	336	28	30	330	264

Fig. 318

C. PARTICULAR CASE OF THE ORDER 5

As observed in the text, the general construction for orders of the type $n = 4t + 1$ (represented in our figure 304) does not apply to the smallest order, with $t = 1$.[199] Indeed, since in this case $\frac{n^2-5}{2}$ and $\frac{n^2-1}{2}$ happen to be numerically equal to, respectively, $2n$ and $2n + 2$, the even numbers in the corners will also occupy cells within the border. Ṯhābit's (and Anṭākī's) text therefore gives a particular construction for this square, with the result as in our Fig. 319 (orientation changed and inner square rotated).[200]

---

[199] *Magic squares in the tenth century*, A.II.13 *in fine*, and MS. London BL Delhi Arabic 110, fol. 82ʳ, 15-16.

[200] *Magic squares in the tenth century*, A.II.23, and MS. London BL Delhi Arabic 110, fol. 83ʳ, 21 - 83ᵛ, 2.

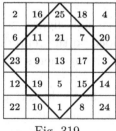

Fig. 319

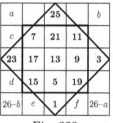

Fig. 320

Whereas this text merely indicates where to put each number, without any general consideration, Abū'l-Wafā' observes that if we place 2 in a corner, the next corner (horizontally or vertically, if we keep 1 below) must be occupied by the (smaller) number 4, 6, 8 or 12, which leads to *numerous figures*.[201] As a matter of fact, this gives eleven figures. But if we do not keep 2 in a corner, there are altogether twenty-one possibilities with the first twenty-five numbers (Fig. 320 & 321).[202]

	a	b	$26 - b$	c	d	e	f
1	2	4	22	6	12	8	10
2	2	4	22	8	10	6	12
3	2	6	20	4	16	8	12
4	2	6	20	8	12	4	16
5	2	14	12	6	22	10	18
6	2	14	12	10	18	6	22
7	2	18	8	10	22	12	20
8	2	18	8	12	20	10	22
9	2	20	6	12	22	16	18
10	2	20	6	16	18	12	22
11	4	6	20	2	16	8	14
12	6	10	16	8	12	4	24
13	6	22	4	14	18	16	24
14	8	2	24	4	6	10	12
15	8	10	16	4	14	6	24
16	8	16	10	4	20	12	24
17	12	4	22	2	6	10	18
18	14	8	18	4	6	10	24
19	14	18	8	4	16	20	24
20	16	8	18	2	6	14	22
21	16	18	8	6	12	22	24

Fig. 321

---

[201] *Magic squares in the tenth century*, B.22.ii, and MS. London BL Delhi Arabic 110, fol. 87ᵛ, 16-88ʳ, 5.

[202] Then, if we consider permutations of c, d, e, f within their rows together with the eight aspects resulting from inverting and rotating the inner 3×3 square, we arrive at $21 \cdot 4 \cdot 8 = 672$ configurations for this 5×5 square.

Remarks.

(1) Whereas Abū'l-Wafā' refers here to a single figure, seven different ones, all with 2 in a corner, are appended to the reproduction of his text in the London manuscript (above arrangements 1, 3, 4, 6, 8, 9, 10).[203]

(2) In all these arrangements, the two smallest odd terms occupy two (consecutive) corners of the oblique square, for then the 3 × 3 square may be filled with a progression displaying the same common difference. But other possibilities are also found at that time, such as that of Fig. 322.[204]

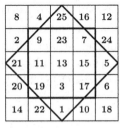

Fig. 322

[203] MS. London BL Delhi Arabic 110, fol. 88ʳ.
[204] MS. London BL Delhi Arabic 110, fol. 91ʳ (3 × 3 square rotated).

Chapter VI

Magic squares with
non-consecutive numbers

Particular case: The given numbers
form arithmetical progressions

The study of such cases is old: it is the subject of the last section of Abū'l-Wafā' Būzjānī's 10th-century treatise. He first considers the simple case in which all the numbers to be arranged form one arithmetical progression, then that in which they form n progressions (with n the order of the square) all with the same constant difference.

1. The numbers form a single progression

As Abū'l-Wafā' expresses it, *the magic quantity to appear in the square will still equal the product of its equalizing number by half the side of the square, (just) as has been taught before.*[205] Indeed, just as we calculated for placing the numbers from 1 to n^2

$$M_n = \tfrac{n}{2}\left(n^2 + 1\right),$$

we shall calculate for the case in which the numbers $a_1, a_2, \ldots, a_{n^2}$ have the constant difference d, thus with $a_{n^2} = a_1 + (n^2 - 1)\, d$,

$$M = \tfrac{n}{2}\left(a_{n^2} + a_1\right) = \tfrac{n}{2}\left[(n^2 - 1)\, d + 2a_1\right].$$

Abū'l-Wafā' gives the example with $n = 4$, $a_1 = 7$, $d = 5$, thus $M = 178$ and $a_{n^2} + a_1 = 89$ for the sum of the complements; we shall arrange the square with one of the known methods (Fig. 323).[206]

7	72	57	42
62	37	12	67
32	47	82	17
77	22	27	52

Fig. 323

---

[205] *Le traité d'Abū'l-Wafā'*, p. 189, lines 1025-1026 of the Arabic text; or below, Appendix 41.

[206] Since there are no figures in the manuscript preserving his work, we have chosen one method of placing.

© Springer Nature Switzerland AG 2019
J. Sesiano, *Magic Squares*, Sources and Studies in the History of Mathematics and Physical Sciences, https://doi.org/10.1007/978-3-030-17993-9_6

2. The numbers form n progressions

If the numbers to be placed in the square of order n are divided into n sequences, it will be easy to construct a magic square provided these sequences all have the same constant difference r and their first terms (or, which comes to the same, their terms having the same rank) form arithmetical progressions with constant difference s. Then, if the first term is a_1, the last term of its sequence will be $a_n = a_1 + (n-1)r$, the next one will be $a_{n+1} = a_1 + s$, and the last term of the square will be $a_{n^2} = a_1 + (n-1)r + (n-1)s$; the magic sum will then once again be the product of half the order by the sum of the two extreme terms, thus

$$M = \tfrac{n}{2}\left(a_{n^2} + a_1\right) = \tfrac{n}{2}\left[(n-1)(r+s) + 2a_1\right].$$

Here again, this is clearly expressed by Abū'l-Wafā':[207] *If the numbers to be arranged in the square are in progression with equal quantities in the horizontal rows,*[208] *and the first number of the vertical rows does not exceed the last number of the previous row by the same quantity as that of the horizontal rows, we shall say that these numbers are 'in progression according to a discontinuous progression'. (...) If we wish to know its magic quantity, we multiply the sum of the constant differences of the vertical and horizontal rows by the side of the square less 1, then we add to the result twice the first number; the result will be the equalizing number; its multiplication by half the side of the square will produce the magic quantity.*

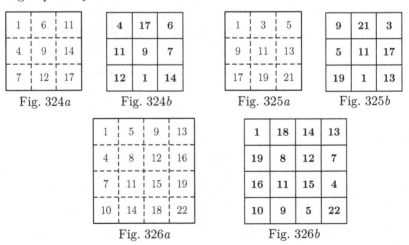

Fig. 324a   Fig. 324b   Fig. 325a   Fig. 325b

Fig. 326a   Fig. 326b

---

[207] *Le traité d'Abū'l-Wafā'*, pp. 190-191 in the translation, lines 1053-1056 and 1061-1064 of the Arabic text; or below, Appendix 42.

[208] Here one has to imagine these n progressions to n terms in an $n \times n$ table, just as with the natural square for the consecutive natural numbers.

Abū'l-Wafā''s examples are the following (arrangement ours): $n = 3$ with $r = 5$ and $s = 3$ (Fig. 288a-b), or with $r = 2$ and $s = 8$ (Fig. 325a-b); $n = 4$ with $r = 4$ and $s = 3$ (Fig. 326a-b, see Fig. 76).

3. Magic square with a set sum

A magic square remains magic if all its elements are increased or multiplied by the same quantity. If s, natural, is the quantity added, the magic sum will become $M = M_n + s \cdot n$ since each row is increased by n times said quantity. Likewise, if t is the multiplying factor, the magic sum will become $M = t \cdot M_n$. Applying this last operation and then the previous one, we shall obtain for a square of order n a magic sum of the form

$$M = t \cdot M_n + s \cdot n.$$

These two transformations were already in use at the time of Abū'l-Wafā' since he tells us:[209] *Some users seemed to think that in the squares could appear only the basic magic quantity or what arises from its multiplication or increment by the side of the square. Such is not the case.* He thus means that we may also construct squares with magic sums of the form

$$M = t \cdot M_n + s \cdot n + q.$$

Abū'l-Wafā' shows that by means of examples, but limited to order 4 since, with it, it is possible to arrive at any magic sum $M > M_4 = 34$ (whereas the 3×3 square does not allow a continuous increment). Thus, increasing by a unit the last sequence in a pandiagonal square we shall have $M = 35$ (Fig. 327, not pandiagonal), increasing by a further unit the last but one, or by two the last, we shall have $M = 36$ (Fig. 328 and 329; first one pandiagonal).[210] Generally, if in a pandiagonal square with sum $M_4 = 34$ the terms of the second sequence are increased by x, those of the third by y more, those of the last by z more, the sums in the lines, columns and main diagonals will become $M_4 + q = 34 + 3x + 2y + z$; thus we shall choose the increments in function of the required magic sum.

The same holds for other orders as long as the sequence increased has an element in each row (line, column, main diagonal) —which therefore excludes squares constructed by methods such as the diagonal placing seen in II, §2. See Fig. 330 (remains pandiagonal);[211] or (based on the

---

[209] *Le traité d'Abū'l-Wafā'*, p. 191 in the translation, lines 1065-1067 of the Arabic text; or below, Appendix 43.

[210] Various examples in Abū'l-Wafā''s treatise, pp. 191-192.

[211] MS. Mashhad Āstān-i Quds 12235, p. 445 (cf. *ibid.* p. 429). Also MS. Istanbul Ayasofya 2794, fol. 8^r and 23^v.

square of Fig. 331, which —since it is of evenly-odd order— cannot be pandiagonal) Fig. 332.[212]

1	15	11	8
12	7	2	14
6	9	17	3
16	4	5	10

Fig. 327

1	15	12	8
13	7	2	14
6	10	17	3
16	4	5	11

Fig. 328

1	16	11	8
12	7	2	15
6	9	18	3
17	4	5	10

Fig. 329

1	26	19	13	7
14	8	2	22	20
23	16	15	9	3
10	4	24	17	11
18	12	6	5	25

Fig. 330

1	33	28	12	16	21
17	11	2	20	29	32
30	22	15	3	31	10
19	26	9	35	4	18
8	13	34	27	24	5
36	6	23	14	7	25

Fig. 331

1	34	28	12	16	21
17	11	2	20	29	33
30	22	15	3	32	10
19	26	9	36	4	18
8	13	35	27	24	5
37	6	23	14	7	25

Fig. 332

4. Magic products

It would seem that Stifel, who introduced bordered squares in Europe (p. 149), was the first to consider magic squares obtained by multiplication, the product remaining the same horizontally, vertically and diagonally. To construct such a square, we shall choose a quantity which we shall raise to the successive powers, here with successive exponents from 0 to $n^2 - 1$. These exponents (plus 1) will determine the magic arrangement. As Stifel himself says, his two examples suffice to show how such squares are constructed (Fig. 333-334, the latter from Fig. 335, with the exponents of 2 being these numbers minus 1).[213]

8	256	2
4	16	64
128	1	32

Fig. 333

32768	4	2	4096
256	32	64	2048
16	512	1024	128
8	16384	8192	1

Fig. 334

16	3	2	13
9	6	7	12
5	10	11	8
4	15	14	1

Fig. 335

---

[212] The latter in MS. Mashhad Āstān-i Quds 12235, p. 445.

[213] *Arithmetica integra*, fol. 29ᵛ-30ʳ. Fig. 335 is Fig. 27 with inverted median lines —which is possible since the square is symmetrical, see p. 8.

General case: Squares with arbitrary given numbers

The construction of such squares appears early, in any case we find it at an advanced stage in the 11th century. As mentioned in the introduction (p. 10), the alphabetical numeration system made it possible to give each letter of a word or each word of a sentence a numerical value. Thus one could place a word or sentence in (say) the top row of a square, which thus determined both order and magic sum. It then remained to complete the square so as to obtain that sum in each of the other lines, and in the columns and main diagonals. (The production of talismans ensued.) But since constructing a magic square with n arbitrary numbers is not always possible, conditions of feasibility arise.

Our main source will be the 11th-century *Harmonious arrangement*, which studies this problem in depth for the first orders, beginning with $n = 3$.

A. SQUARES OF ODD ORDERS

§ 1. Square of order 3

1. General observations

Let M be the magic sum to be found in the square of order 3. As we have already seen (p. 5), the sum of all quantities in the square is on the one hand equal to $3M$ (the sum of all three rows) and, on the other, to $4(M - r) + r$ (the sum of the four pairs of opposite cells forming the border plus the content of the central cell). Setting the equality of these two quantities leads to the relation $M = 3r$. Thus M must in any case be divisible by 3.

The quantity r is just one of the three quantities allowing us to determine the square completely. Indeed, the necessary and sufficient condition for obtaining a square of order 3 with nine numbers is that these nine numbers form three arithmetical progressions with the same common difference, say y, and that their first terms also form an arithmetical progression, say with common difference x (see above, p. 212). These three progressions, here set in function of the central element r, are:

$$r - (x + y), \qquad r - x, \qquad r - (x - y);$$
$$r - y, \qquad r, \qquad r + y;$$
$$r + (x - y), \qquad r + x, \qquad r + (x + y).$$

Placing them according to Fig. 336 will then produce a magic square.

$r-y$	$r+(x+y)$	$r-x$
$r-(x-y)$	r	$r+(x-y)$
$r+x$	$r-(x+y)$	$r+y$

Fig. 336

Now, in our case, the three given numbers do not necessarily have to be in the first row. Indeed, various configurations are found, mainly for reasons of esthetics.

2. The given numbers are in the first row

Let s_1, s_2, s_3 be the three given numbers and M, their sum. The three conditions to be fulfilled are clearly set out by the author of the *Harmonious arrangement*, as follows. *For these numbers, there are conditions. The first is that their sum be divisible by three in order to infer the part due for the central cell. The second is that the two numbers at the extremities be such that the sum of each and the complement of the other be less than the sum of the three numbers in order that there remain, after the placement of their complements, enough to fill the median right and left cells. The third is that these numbers not be repeated, either during the treatment or to begin with.*[214]

s_1	s_2	s_3
$r+s_3-s_1$	r	$r+s_1-s_3$
$2r-s_3$	$2r-s_2$	$2r-s_1$

Fig. 337

92	52	54
28	66	104
78	80	40

Fig. 338a

حمو	بن	محمد
١٠٤	٦٦	٢٨
٤٠	٨٠	٧٨

Fig. 338b

The third condition is evident, and inherent in the definition of a magic square. It thus excludes three-letter words or three-word sentences where elements recur, and also cases in which equal (or negative) quantities appear during the construction. The first condition is clear from what we have seen in the previous section. The second condition appears from the construction. Having the first row and the central element, thus the content of four cells, we can fill all remaining cells: the elements of the last horizontal row are immediately determined, and from these the numbers in the two median lateral cells (Fig. 337). In order for those two numbers to be positive, we must have $r > |\, s_1 - s_3\, |$, or —which amounts to the same but is in keeping with the text— $s_1 + (2r - s_3) < M = 3r$ and $s_3 + (2r - s_1) < 3r$; in addition, $s_i < 2r$. The author constructs the

---

[214] *Un traité médiéval*, p. 85, lines 710-714 of the Arabic text; or below, Appendix 44.

example of Fig. 338*a-b* (the repetition of two letters in the word *mḥmd* is irrelevant since it is the sum of its elements which is considered).

A contemporary text treats another example (Fig. 339).[215] It gives the first two conditions, but omits the third, which indeed is self-evident. Its example is 66, 129, 84 adding up to 279 (*Allāh laṭīf bi-'ibādihi*). Another common example is that of Fig. 340 (*lā-ilaha illā Allah*, there is no other god than Allah), with 67, 32, 66 in the top line (1A-Alh AlA Allh = 165) and consecutive numbers in the middle row.[216] In all these examples, the three arithmetical progressions, each with one term in the first line, are evident.

66	129	84
111	93	75
102	57	120

Fig. 339

67	32	66
54	55	56
44	78	43

Fig. 340

3. The middle number is in the median lower cell

The *Harmonious arrangement* gives the following instructions. *You take a number which, when added to the content of the two extreme cells makes the sum divisible by three. You then consider the number taken to be for the middle cell of the upper row while a third of the sum will be for the central cell. Then, if the two rows are equalized, it is appropriate. Otherwise, we shall consider their difference.*[217] Here the author proceeds using trial and error. With the chosen number, say t, we have in the upper row the sum M_0, a third of which is to occupy the central cell. All the elements of the middle column being then determined, their sum should equal M_0; if such is not the case, we shall change the initial choice until we reach equality in the upper row and middle column. In his example (same numbers as before, thus 92, 52, 54) the author first chooses $t = 4$, giving for the top row $M_0 = 150$ and thus 50 for the middle cell, inappropriate since there is then a deficit of $150 - (52 + 50 + 4) = 44$ in the median column. Adding 30 to the first choice, thus taking $t = 34$, will give $M_0 = 180$, thus 60 in the middle cell and a deficit of $180 - (34 + 60 + 52) = 34$. Since, he says, adding 30 to the first choice reduces the deficit by 10, respectively adding 3 reduces it by 1, while the initial deficit is 44, we shall add to the initial choice $3 \cdot 44$, thus 132. This allows the author to complete the

---

[215] *L'Abrégé*, pp. 128-129 & (Arabic) 151-152.

[216] This example (together with the previous one) in MS. Mashhad Āstān-i Quds 12235, p. 447.

[217] *Un traité médiéval*, p. 86, lines 725-728 of the Arabic text; or below, Appendix 45.

square (Fig. 341).

92	136	54
56	94	132
134	52	96

Fig. 341

As a matter of fact, that can be obtained directly from the data. With s_1, s_2, s_3 the given numbers and α in the median upper cell, we have $M = s_1 + \alpha + s_3$ and $r = \frac{1}{3} M$ in the central cell. Thus $s_1 + \alpha + s_3 = \alpha + r + s_2$, whence $r = s_1 + s_3 - s_2$ and $M = 3s_1 + 3s_3 - 3s_2$, and so $\alpha = M - (s_1 + s_3) = 2s_1 + 2s_3 - 3s_2$.

Therefore, here too, with three given numbers satisfying the conditions, the square is fully determined (Fig. 342). By the way, r could also be determined using Fig. 336, which gives immediately its link with the three given quantities.

s_1	$2s_1+2s_3-3s_2$	s_3
$2s_3-s_2$	r	$2s_1-s_2$
$2s_1+s_3-2s_2$	s_2	$s_1+2s_3-2s_2$

Fig. 342

4. The given numbers are in the diagonal

As the text puts it, *the condition is that the middle number be a third of the sum*.[218] Now since $M = s_1 + s_2 + s_3$ while $M = 3s_2$, we must have $s_1 + s_3 = 2s_2$, a very restrictive condition for such placings. If this is fulfilled, the choice of another quantity A will enable us to complete the square —provided, as usual, that it does not lead to repetitions or negative quantities (Fig. 343 and, author's example, Fig. 344).

s_1	A	s_2+s_3-A
$2s_2+s_3-s_1-A$	s_2	$2s_2-2s_3+A$
s_2-s_3+A	$2s_2-A$	s_3

Fig. 343

60	54	42
34	52	70
62	50	44

Fig. 344

5. The given numbers are in the middle row

This is subject to just the same restrictive condition as before. Here

---

[218] *Un traité médiéval*, p. 88, line 749 of the Arabic text; or below, Appendix 46.

too, an optional quantity will suit our purpose provided it is put in a corner cell (Fig. 345 and, as in the text, Fig. 346).

A	$4s_2-s_1-2A$	s_1-s_2+A
s_1	s_2	s_3
$3s_2-s_1-A$	s_1-2s_2+2A	$2s_2-A$

Fig. 345

49	50	57
60	52	44
47	54	55

Fig. 346

§2. Square of order 5

1. The given numbers are in the first row

s_1	s_2	s_3	s_4	s_5
A				$2r-A$
B		r		$2r-B$
$3r+s_5-s_1-A-B$				$A+B+s_1-s_5-r$
$2r-s_5$	$2r-s_2$	$2r-s_3$	$2r-s_4$	$2r-s_1$

Fig. 347

Take thus $M = \sum s_i$ as the magic sum and suppose it to be divisible by 5 and $r = \frac{1}{5} M$ to be in the central cell (this condition will hold for the subsequent cases as well). First, we can fill the last row with the complements to $2r$ of the given terms. Choosing two quantities enables us to complete the border (Fig. 347). To fill the inner square, we shall either take three arithmetical progressions centred around r (above, pp. 215-216), or we shall choose two of its elements, one being for a corner cell.

92	52	53	41	62
50	59	57	64	70
51	65	60	55	69
49	56	63	61	71
58	68	67	79	28

Fig. 348

80	66	116	7	81
61	64	68	78	79
77	84	70	56	63
73	62	72	76	67
59	74	24	133	60

Fig. 349

The main difficulty will be to avoid repetitions or negative quantities. Both the *Harmonious arrangement* and the *Brief treatise* have an example, with the phrase محمد بن احمد والد حميد (= 300) for the first (Fig. 348) and حسبيا به وكفى الله حسبى (= 350) for the second (Fig. 349).[219]

2. The given numbers are in the second row

We thus again know M and r ($M = 5r$), and, in addition, one row of the inner square, which can then be completed immediately (above, p. 216). It will not, however, be a magic 3×3 square, nor, therefore, will the whole square be a bordered one; indeed (Fig. 350), we shall normally not have $s_2 + s_3 + s_4 = 3r$. As the author of the *Harmonious arrangement* says about the top and bottom rows of the 3×3 square, *it is here not necessary that these two rows be equalized, and you do not need to care about it.*[220] The choice of five quantities enables us to complete the border. See the author's example, with the same numbers as before (Fig. 351).

B	$5r-B-C-D-E$	D	E	C
s_1	s_2	s_3	s_4	s_5
$3r-s_1+C-A-B$	$r+s_4-s_2$	r	$r+s_2-s_4$	$s_1-r+A+B-C$
A	$2r-s_4$	$2r-s_3$	$2r-s_2$	$s_2+s_3+s_4-r-A$
$2r-C$	$B+C+D+E-3r$	$2r-D$	$2r-E$	$2r-B$

Fig. 350

42	77	75	55	51
92	52	53	41	62
57	49	60	71	63
40	79	67	68	46
69	43	45	65	78

Fig. 351

3. The given numbers are in the diagonal

We again know M and r ($M = 5r$), but with $r = s_3$: *the condition is that the central number be a fifth of (the sum of) these five numbers.*[221]

In order to fill the inner square, we shall choose one quantity A and complete the lines and columns to $3s_3$ beginning with the lower row (Fig. 352). As for the cells surrounding it, we shall first place in the end

---

[219] *Un traité médiéval*, pp. 90 & 160; *L'Abrégé*, pp. 129-130 & 152.

[220] *Un traité médiéval*, p. 91, line 786 of the Arabic text; or below, Appendix 47.

[221] *Un traité médiéval*, p. 92, line 799 of the Arabic text; or below, Appendix 48.

cells of the incomplete diagonal the complements (to $2s_3$) of the numbers occupying the opposite corner cells of the inner square. Four chosen quantities will enable us to complete the lateral rows. Fig. 353 is the example of the text, with the phrase محمد بن عبد الله سيدك, which suits the condition.

s_1	B	C	$s_2+s_5+A-B-C$	s_3+s_4-A
D	s_2	A	$3s_3-s_2-A$	$2s_3-D$
E	$2s_3+s_4-s_2-A$	s_3	s_2-s_4+A	$2s_3-E$
$2s_3+s_4+s_5$ $-A-D-E$	s_3-s_4+A	$2s_3-A$	s_4	$A+D+E-s_4-s_5$
s_2-s_3+A	$2s_3-B$	$2s_3-C$	$2s_3-s_2-s_5$ $+B+C-A$	s_5

Fig. 352

92	97	62	72	57
79	52	85	91	73
78	81	76	71	74
70	95	67	66	82
61	55	90	80	94

Fig. 353

4. The given numbers are in opposite lateral rows

Since M need not, or not necessarily, be the sum of the given numbers, we are first to determine $M = 5r$. For that purpose, put in the upper row the complements to $2r$ of the numbers below (Fig. 319). Since this gives us the sum $4r + s_1 + s_3 + s_5 - s_2 - s_4 = M$, then $r = s_1 + s_3 + s_5 - s_2 - s_4$.

s_1	$2r-s_2$	s_3	$2r-s_4$	s_5
B	s_1+s_5-2r+A	$3r-s_1-s_5$	$2r-A$	$2r-B$
C	$3r-s_1-A$	r	s_1-r+A	$2r-C$
$5r-s_1-A-B-C$	$2r-s_5$	s_1+s_5-r	$2r-s_1$	$s_1-3r+A+B+C$
A	s_2	$2r-s_3$	s_4	$4r-s_1-s_5-A$

Fig. 354

A first choice enables us to complete those elements of the last row which are not complements of vertically opposite ones. Then the corner cells of the 3×3 square are filled with the complements to $2r$ of the corner elements in the 5×5 square. The inner square being completed (to the sum $3r$), we are left with choosing two numbers to fill one of the lateral columns, then, with their complements, its opposite. The *Harmonious arrangement* constructs thus the square of Fig. 355 (same numbers as in the first two examples).[222]

92	176	53	187	62
116	67	188	87	112
111	109	114	119	117
110	166	40	136	118
141	52	175	41	161

Fig. 355

5. The given numbers are in the first two rows

Suppose once again that $M = 5r$, with r in the central cell.

s_1	A	s_3	$5r-s_1-s_3-s_5-A$	s_5
B	s_2	$3r-s_2-s_4$	s_4	$2r-B$
C	$r+s_4-s_2$	r	$r+s_2-s_4$	$2r-C$
$3r+s_5-s_1-B-C$	$2r-s_4$	s_2+s_4-r	$2r-s_2$	$s_1-s_5-r+B+C$
$2r-s_5$	$2r-A$	$2r-s_3$	$s_1+s_3+s_5-3r+A$	$2r-s_1$

Fig. 356

92	18	53	15	62
32	52	51	41	64
40	37	48	59	56
42	55	45	44	54
34	78	43	81	4

Fig. 357

With an appropriate choice of r (to avoid later occurrences of repeated and negative values), we shall construct a bordered square, thus with $3r$ as the magic sum of the inner square. We can then directly fill this inner square, of which three elements and the magic sum are known. The choice

---

[222] *Un traité médiéval*, pp. 94-95 & (Arabic) 157-158.

of one element enables us to complete the horizontal rows of the border, and the choice of two further ones to fill the rest of the border (Fig. 356). Fig. 357 is the example of the *Harmonious arrangement*, with the same numbers as before.[223]

Here ends the *Harmonious arrangement*'s treatment of the 5 × 5 square, as well as consideration of odd-order squares. For, according to the author, other cases could be reduced to the above ones: *if you know how to place numbers with irregular differences in the squares of 3 and of 5, you will be able to place them in all squares of odd orders.*[224] Now we have indeed seen the general principles but the fact remains that with growing orders the risk of seeing repeated or negative values will increase as well.

An example of a 7 × 7 square, namely that of Fig. 358, is found in a later text.[225] The words محمد وعلي وفاطمة والحسن والحسين عدة ليوم معادى (with the last two on the left taken together) make the sum 959; this being divisible by 7, the quotient, 137, is put in the central cell. The border of the inner 3 × 3 square is easy to fill: the neighbours of 137 are put in the diagonal, then 1, 2, 3 and their complements are placed in the known way for the usual 3 × 3 square. The outer border is also easily filled: its bottom line is to receive the complements of the given numbers, while for one of the columns we shall choose four numbers (the fifth is determined) and put their complements opposite. As to the border of order 5, we shall fill it with pairs of complements not already employed and making the appropriate sum.

92	116	141	155	165	79	211
164	90	82	172	171	170	110
162	128	2	271	138	146	112
161	241	273	137	1	33	113
160	122	136	3	272	152	114
157	104	192	102	103	184	117
63	158	133	119	109	195	182

Fig. 358

---

[223] *Un traité médiéval*, pp. 96-97 & 156-157.

[224] *Un traité médiéval*, p. 97, lines 857-858 of the Arabic text; or below, Appendix 49.

[225] MS. Mashhad Āstān-i Quds 12235, p. 452; or 14159, fol. 92ʳ.

B. SQUARES OF EVENLY-EVEN ORDERS

§3. Square of order 4

1. General observations

After obtaining, by moving the numbers in the 4×4 natural square, the 4 × 4 usual magic square (Fig. 359 or Fig. 76), the 10th-century author Abū'l-Wafā Būzjānī remarks: *There appears in this square by nature, besides the magic arrangement we were searching for, a nice equalization, namely that the sum of the content of any two middle cells of the border around the central square and of the two opposite cells also gives 34: thus 12 and 8 added to 9 and 5, and 14 and 15 added to 3 and 2. It also appears that when we add the content of two cells adjacent to any corner cell to the content of the two cells adjacent to the opposite corner cell, this also gives 34; indeed, if we add 12 and 15 to 2 and 5, and if we add 14 and 9 to 8 and 3, it gives 34 (in both cases). It further appears that if we add the content of the four cells gathered together in any corner, it also gives 34; thus 1, 15, 12 and 6 make 34.*[226]

1	15	14	4
12	6	7	9
8	10	11	5
13	3	2	16

Fig. 359

Now there are in a magic square of order 4 two types of property to be distinguished: those which, like the last two, arise from the way it is constructed and those which, like the first, appear whatever the construction, that is, are found in any magic square of order 4. It is important to identify these properties in order not to be misled into thinking we are dealing with a particular type of square.

Let M be the magic sum of the square, which is thus found in any line, in any column, and in the two main diagonals.

I. Any four cells placed symmetrically relative to the axes make the magic sum.

a. We may first verify that for the median cells of the outer horizontal rows. Adding these two rows and the median columns, we find $4M$ (Fig.

---

[226] *Le traité d'Abū'l-Wafā'*, p. 161, lines 470-479, or *Magic squares in the tenth century*, B.8.v; or below, Appendix 50.

$360a$); subtracting the sums in the diagonals will leave $2M$ for twice the considered cells, thus M for the four median cells.

b. We may next verify it for the central cells by adding the sums in the median columns and subtracting from that the sum found in *a* (Fig. $360b$).

c. We may finally verify it for the corner cells by adding the sums in the diagonals and subtracting the sum found in *b* (Fig. $360c$).

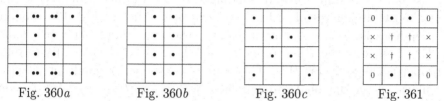

Fig. 360*a*   Fig. 360*b*   Fig. 360*c*   Fig. 361

It thus appears that in any magic square of order 4 summing the four elements as designated in Fig. 361 will give M. It is to be noted that the 4×4 natural square has this property too.

II. The sum of the two extremes in a row equals the sum of the two medians in the opposite row.

If, for instance, the sum in the cells • in the upper line equals α (Fig. $362a$), that of the cells • in the lower line will be $M - \alpha$; so the sum of the cells 0 will be α. To the four aspects of this property obtained by rotating Fig. $362a$ may be added the four analogous aspects for the cells of the middle rows (Fig. $362b$).

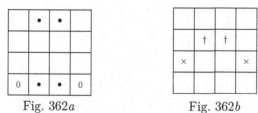

Fig. 362*a*   Fig. 362*b*

III. The sum in a quadrant equals the sum in the opposite quadrant.

This is verified in Fig. $363a$: the sum of the • makes $2M$, that of the † $2M$ too, and removing the common elements will leave the same quantity in the two opposite quadrants.

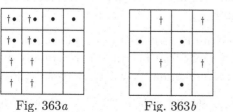

Fig. 363*a*   Fig. 363*b*   Fig. 363*c*

To the two aspects of this property may be added the two aspects of the analogous property for the enlarged quadrants (Fig. 363b); this can be verified as before (Fig. 363c).

These properties are inherent in any magic square of order 4. Others characterize, as said above in discussing Abū'l-Wafā's remark, particular constructions. Such are symmetry (where numbers in diagonally opposite cells make the sum $\frac{M}{2}$) and pandiagonality. Another property, found in almost all squares of order 4 we have met (that of Fig. 135 was an exception), is to have 'equal quadrants' (the sum in any quadrant equals M). This was the third property noted by Abū'l-Wafā'. His second property followed from this: if the sum of two opposite quadrants is $2M$, removal of the diagonal through them will leave M in the pairs of cells enclosing opposite corner cells.

Identifying the properties inherent in any magic square of order 4 will prove to be of considerable help in understanding the constructions presented by the author of the *Harmonious arrangement*.

2. The given numbers are in the upper row

First method

s_1	s_2	s_3	s_4
s_3-s_1+A	s_1+s_4-A	A	s_1+s_2-A
$m-s_3$	$m-s_4$	$m-s_1$	$m-s_2$
$m-A$	$m-s_1-s_2+A$	$m+s_1-s_3-A$	$m-s_1-s_4+A$

121	66	166	97
136	127	91	96
59	128	104	159
134	129	89	98

Fig. 364a Fig. 365

As we are told, *the condition is that the sum of these numbers must have a half in order that the placement of the complements in accordance with it be possible —this being (merely) a condition for making the placement easier.*[227] Let thus $M = \sum s_i = 2m$. With the complements to m of the elements in the first line we fill the corresponding bishop's cells in the third line (Fig. 364a). Choosing then a single quantity A allows us to complete the rows, if need be using the above properties Ib and II. The resulting square will be pandiagonal. Fig. 365 is the author's example ($M = 450 = $ حسبنا الله ونعم الوكيل).

Remark. If M is odd we may fill the square by considering $M = m_1 + m_2$ and completing it in such a way that m_1 and m_2 be equally distributed in each row (Fig. 364b); pandiagonality will, however, be lost.

---

[227] *Un traité médiéval*, p. 98, lines 866-867 of the Arabic text; or below, Appendix 51.

s_1	s_2	s_3	s_4
s_3-s_1+A	s_1+s_4-A	A	s_1+s_2-A
m_2-s_3	m_2-s_4	m_1-s_1	m_1-s_2
m_1-A	$m_1-s_1-s_2+A$	$m_2+s_1-s_3-A$	$m_2-s_1-s_4+A$

Fig. 364*b*

Second method

Starting at one of the upper corner cells, we place a descending sequence, and, from the other, an ascending one, both with the same constant difference α, in the manner seen in Ch. II (pp. 63-64, with the knight's move, then the queen's and again the knight's). It thus appears that the bishop's cells of the median terms in the first upper row are filled, and this will enable us to complete the square. Indeed (Fig. 366), the term $s_1 - 2\alpha$ being associated with s_2, it means that each term of the sequence beginning with s_1 must produce, together with the content of its bishop's cell, the sum $s_1 + s_2 - 2\alpha$. For the same reason, a term of the sequence starting with s_4 and its associate must produce the sum already attained by a pair, namely $s_3 + s_4 + 2\alpha$. This is how the square of Fig. 367 is constructed (same given numbers as before).[228]

s_1	s_2	s_3	s_4
$s_3-\alpha$	$s_4+\alpha$	$s_1-\alpha$	$s_2+\alpha$
$s_4+2\alpha$	$s_3+2\alpha$	$s_2-2\alpha$	$s_1-2\alpha$
$s_2-\alpha$	$s_1-3\alpha$	$s_4+3\alpha$	$s_3+\alpha$

Fig. 366

121	66	166	97
165	98	120	67
99	168	64	119
65	118	100	167

Fig. 367

The 13th-century author Zanjānī fills an example in this way. But in another one he is obliged to resort to trial and error since the previous method leads to repetition of numbers.[229]

Another example places the name G̲h̲afūr = 1286, with the norms 1082 and 204 (Fig. 368).[230]

Remark. There is no need for a descent with regular moves. The sum of the terms added to the four s_is must simply equal zero in the lines,

---

[228] *Un traité médiéval*, pp. 99-100 & 154-155.

[229] *Herstellungsverfahren II & II'*, pp. 261-263, lines 176-219 & 225-240. See below, note 234.

[230] Būnī, *S̲h̲ams al-ma‘ārif*, II, p. 32.

columns and main diagonals (Fig. 370, or Fig. 371). The placing of *kabīr* may be obtained in such a way (Fig. 369).[231]

200	6	80	1000
ر	و	ف	غ
79	1001	199	7
1002	82	4	198
5	197	1003	81

Fig. 368

200	10	2	20
ر	ك	ب	ل
19	3	9	201
8	198	22	4
5	21	199	7

Fig. 369

s_1	s_2	s_3	s_4
$s_4-\alpha$	$s_3+\alpha$	$s_2-\alpha$	$s_1+\alpha$
$s_2-2\alpha$	$s_1-2\alpha$	$s_4+2\alpha$	$s_3+2\alpha$
$s_3+3\alpha$	$s_4+\alpha$	$s_1-\alpha$	$s_2-3\alpha$

Fig. 370

Third method

Method similar to the above except that we start in the middle cells of the upper row. The norms for filling the bishop's cells will be $s_1 + s_2 + 2\alpha$ for the sequence from s_2 and $s_3 + s_4 - 2\alpha$ for the sequence from s_3 (*the second term of the sequence starting with each of them falls in the bishop's cell of the adjacent extremity. You then adopt this as norm for attributing the complement of this (initial term) and the complement of each term of its sequence*).[232] The resulting square is that of Fig. 371; the author's example is in Fig. 372.

s_1	s_2	s_3	s_4
$s_3-\alpha$	$s_4+\alpha$	$s_1-\alpha$	$s_2+\alpha$
$s_4-2\alpha$	$s_3-2\alpha$	$s_2+2\alpha$	$s_1+2\alpha$
$s_2+3\alpha$	$s_1+\alpha$	$s_4-\alpha$	$s_3-3\alpha$

Fig. 371

121	66	166	97
165	98	120	67
95	164	68	123
69	122	96	163

Fig. 372

Fourth method

Our main source and a contemporary one both present a method of placing progressively the complements, but not in the bishop's cells, and

---

[231] Būnī, *Shams al-maʿārif*, II, p. 33.
[232] *Un traité médiéval*, p. 101, lines 899-900; or below, Appendix 52.

this is said by the first author to be *more general* than the previous constructions. It is taught in this way by the second author (Fig. 373).[233] *You put in the two median (cells) of the lower row two numbers with their sum (equal to) the sum of the two numbers in the two extremities of the upper row, and in its two extremities two numbers with their sum (equal to) the content of the two median (cells) of the first row. Next, you put in one of the two median columns the remainder of the sum due. Then one completes each of the two diagonals. There remain two cells in the right-hand column and two cells in the left-hand column. Putting then, in the two cells of one, the remainder of the sum due and completing one of the two horizontal median rows, there will remain one cell common to two rows which is to receive what will complete one and the other. With this being placed the magic square will be finished.*

s_1	s_2	s_3	s_4
D	C	$s_2-s_4+A+C-B$	$s_1+s_3+2s_4$ $+B-A-2C-D$
$s_2+s_3+s_4-B-D$	$s_1+s_3+s_4-A-C$	s_4+B-C	$A+2C+D-s_3-2s_4$
B	A	s_1+s_4-A	s_2+s_3-B

Fig. 373

1	8	40	4
14	24	6	9
13	18	5	17
25	3	2	23

Fig. 374

102	66	329	289
350	286	100	50
87	150	250	299
247	284	107	148

Fig. 375

The example chosen by the two authors (Fig. 374) puts Aḥmad (a, ḥ, m, d = 1, 8, 40, 4) in the first line —which using the previous methods would have led to repetition of numbers.[234] Another notable example is to write in the first line *bism Allah al-raḥman al-raḥīm* = 102, 66, 329, 289 = 786 (Fig. 375).[235]

3. The given numbers are in the second row

---

[233] *L'Abrégé*, p. 130 & (Arabic) 152, 15-22, or below, Appendix 53; *Un traité médiéval*, pp. 130-131 & (Arabic) lines 1217-1228.

[234] It is the choice of this name that had obliged Zanjānī to resort to trial and error (above, p. 227).

[235] MS. Mashhad Āstān-i Quds 12235, p. 450.

As remarked by the author of the *Harmonious arrangement* at the outset, the following methods are analogous to the previous ones.

First method

As we are told, *the condition is that the sum of the numbers have a half*.[236] Thus $M = 2m$. As before, we begin by filling the bishop's cells in the fourth row. Choosing a single quantity (and using, e.g., Property II) will enable us to fill the remainder of the square (Fig. 376, and Fig. 377 for the author's example, لا إله إلا هو, adding up to 110).

$m-A$	$m-s_1-s_2+A$	$m+s_1-s_3-A$	$m-s_1-s_4+A$
s_1	s_2	s_3	s_4
s_3-s_1+A	s_1+s_4-A	A	s_1+s_2-A
$m-s_3$	$m-s_4$	$m-s_1$	$m-s_2$

26	17	25	42
31	36	32	11
30	13	29	38
23	44	24	19

Fig. 376   Fig. 377

Remark. Comparing the square of Fig. 376 with that of Fig. 364a, we note that it is the same square, just shifted by one line. This square is thus also pandiagonal.

Second method

$s_2-\alpha$	$s_1-3\alpha$	$s_4+3\alpha$	$s_3+\alpha$
s_1	s_2	s_3	s_4
$s_3-\alpha$	$s_4+\alpha$	$s_1-\alpha$	$s_2+\alpha$
$s_4+2\alpha$	$s_3+2\alpha$	$s_2-2\alpha$	$s_1-2\alpha$

Fig. 378

We start in the end cells of the second row and proceed with the knight's move, then the queen's and again the knight's. The two norms then appear, which enables us, in theory, to fill the remaining cells (Fig. 378). The same may be done starting with the median cells (Fig. 379).

$s_2+3\alpha$	$s_1+\alpha$	$s_4-\alpha$	$s_3-3\alpha$
s_1	s_2	s_3	s_4
$s_3-\alpha$	$s_4+\alpha$	$s_1-\alpha$	$s_2+\alpha$
$s_4-2\alpha$	$s_3-2\alpha$	$s_2+2\alpha$	$s_1+2\alpha$

Fig. 379

---

[236] *Un traité médiéval*, p. 102, (Arabic) line 917; or below, Appendix 54.

The author has thus adapted the previous method (first row given, Fig. 366 or Fig. 371) to this case, and taken for his example the same numbers as in his first method here, but without filling all the cells (had he done so, he would have noted the repetition of numbers). As a matter of fact he has, relative to the previous method, merely shifted the rows by one. But since the other square was not pandiagonal, the main diagonals will here no longer be magic. Such a filling will work only if we take $\alpha = \frac{1}{4}(s_1 + s_2 - s_3 - s_4) = 6$ in the first case and $\alpha = \frac{1}{4}(s_3 + s_4 - s_1 - s_2) = -6$ in the second. Completing then these squares with the previously given numbers will produce the two (pandiagonal) squares of Fig. 380 and Fig. 381.

30	13	29	38
31	36	32	11
26	17	25	42
23	44	24	19

Fig. 380

18	25	17	50
31	36	32	11
38	5	37	30
23	44	24	19

Fig. 381

Or else we might use the progressive filling already seen; but the author does not mention it (Fig. 382).

$s_2+s_3+s_4-B-C$	$s_1+s_3+s_4-A-D$	$2A+2C+B+D$ $-s_1-s_2-s_3-2s_4$	$s_1+s_2+s_4-A-C$
s_1	s_2	s_3	s_4
B	A	s_1+s_4-A	s_2+s_3-B
C	D	$s_1+2s_2+s_3+2s_4$ $-A-B-2C-D$	$A+B+C-s_2-s_4$

Fig. 382

4. The given numbers are in the end cells of the first row and the median of the second

s_1	C	s_2+s_3-C	s_4
B	s_2	s_3	s_1+s_4-B
s_3+2s_4-A-B	s_1+s_4-A	A	s_2-2s_4+A+B
s_2-s_4+A	s_3+A-C	$s_1+s_4-s_3+C-A$	s_3+s_4-A

Fig. 383

31	22	46	11
15	36	32	27
16	19	23	52
48	33	9	20

Fig. 384

Suppose we take the sum of the four given numbers as the magic sum M. The choice of a quantity A enables us, using property Ib of the 4×4

magic square, to fill the central cells, then the two diagonals. Choosing two further quantities, we can complete the lateral rows (Fig. 383 and Fig. 384, author's example, same numbers as before).

5. The given numbers are in the end cells of the first row and the median of the third

First method

Let us again take $M = s_1 + s_2 + s_3 + s_4$. Just as before we may fill, by choosing A, the inner square then the diagonals and, by choosing B and C, the remainder. See Fig. 385 and (example of the *Harmonious arrangement*) Fig. 386.

s_1	C	s_2+s_3-C	s_4
B	s_1+s_4-A	A	s_2+s_3-B
$s_2+s_4-s_1+A-B$	s_2	s_3	$2s_1-s_2+B-A$
s_1+s_3-A	s_3+A-C	$s_1+s_4-s_3+C-A$	s_2-s_1+A

__31__	22	46	__11__
24	19	23	44
15	__36__	__32__	27
40	33	9	28

Fig. 385   Fig. 386

Second method

We may take two sequences, one descending the other ascending, starting from the ends, then complete their bishop's cells by means of the two norms $s_1 + s_3$ and $s_2 + s_4$, the sum of which is M (Fig. 387). The author of the *Harmonious arrangement* merely mentions this construction in a remark, without giving an example.

s_1	$s_3+2\alpha$	$s_2-2\alpha$	s_4
$s_2-3\alpha$	$s_4+\alpha$	$s_1-\alpha$	$s_3+3\alpha$
$s_4+2\alpha$	s_2	s_3	$s_1-2\alpha$
$s_3+\alpha$	$s_1-3\alpha$	$s_4+3\alpha$	$s_2-\alpha$

Fig. 387

6. The given numbers are in the diagonal

First method

Here the magic sum is given. With the choice of two quantities we may first fill the central cells, then the second diagonal; this leaves us with choosing two quantities to complete the square (Fig. 388, and the author's example in Fig. 389, same numbers as before).

s_1	D	s_4+B-D	s_2+s_3-B
C	s_2	A	$s_1+s_3+s_4-A-C$
$s_2+s_3+s_4-B-C$	s_1+s_4-A	s_3	$A+B+C-s_3-s_4$
B	s_3+A-D	$s_1+s_2+D-A-B$	s_4

Fig. 388

31	17	22	40
26	36	19	29
25	23	32	30
28	34	37	11

Fig. 389

Second method

Here too we have a simple remark, without example. We are to place in the empty end of the upper row $s_2 + \alpha$ and write in from the two end cells two arithmetical progressions with the same constant difference α, one ascending and the other descending, then complete the bishop's cells with the two (given) norms $s_1 + s_3$ and $s_2 + s_4$, the sum of which is the magic sum (Fig. 390). Since there is no example, a reader added one in the manuscript; but as he unfortunately chose $\alpha = 1$, his square displays repetitions.

s_1	$s_3-2\alpha$	$s_4+\alpha$	$s_2+\alpha$
$s_4+2\alpha$	s_2	$s_1+\alpha$	$s_3-3\alpha$
$s_2-\alpha$	$s_4-\alpha$	s_3	$s_1+2\alpha$
$s_3-\alpha$	$s_1+3\alpha$	$s_2-2\alpha$	s_4

Fig. 390

7. The given numbers are in the corner cells

First method

By rule Ic, we know that they add up to the magic sum. An optional quantity A enables us to complete one diagonal, another, B, the first line, a third one, C, the last line then the median columns, and with D the square may be completed (Fig. 391).

s_1	s_2+s_3-B	B	s_4
s_3+s_4-D	A	$s_3-s_4+A+C-B$	$s_1+s_2+s_4-s_3$ $+B+D-2A-C$
D	$s_1+s_4+B-A-C$	s_2+s_4-A	$s_3-s_4+2A+C-B-D$
s_2	C	s_1+s_4-C	s_3

Fig. 391

Remark. The author did not notice that, in his example, two numbers are the same. An unwitting warning that such methods are no guarantee of a successful result.

Second method

Once again, we are told, we shall come down from the upper corner cells with two sequences and use the norms $s_1 + s_2 + \alpha$ and $s_3 + s_4 - \alpha$ (Fig. 392).

s_1	$s_2-\alpha$	$s_3+\alpha$	s_4
$s_3+2\alpha$	$s_4-\alpha$	$s_1+\alpha$	$s_2-2\alpha$
$s_4-2\alpha$	$s_3-\alpha$	$s_2+\alpha$	$s_1+2\alpha$
s_2	$s_1+3\alpha$	$s_4-3\alpha$	s_3

Fig. 392

Remark. There is no treatment of the case with the given numbers in the central cells (Fig. 393 & Fig. 394). If this is not a copyist's omission, it might have been left out because it just requires the same methods as before. Still, it would have been worth a mention.

A	D	$s_1+s_2+s_3+s_4$ $-A-B-D$	B
C	s_1	s_3	s_2+s_4-C
$s_2+s_3+B-A-C$	s_2	s_4	$s_1-s_2+A+C-B$
s_1+s_4-B	s_3+s_4-D	$A+B+D-s_3-s_4$	s_2+s_3-A

Fig. 393

$s_4-\alpha$	$s_3-2\alpha$	$s_2+2\alpha$	$s_1+\alpha$
$s_2+3\alpha$	s_1	s_4	$s_3-3\alpha$
$s_1-\alpha$	s_2	s_3	$s_4+\alpha$
$s_3-\alpha$	$s_4+2\alpha$	$s_1-2\alpha$	$s_2+\alpha$

Fig. 394

8. Writing in the sum as a whole

a. *Cases of impossibility*

As three texts point out, the given numbers might make filling of the square impossible.[237] This problem is discussed only for the square of order 4, for, as the last author observes, it is the most frequently used;

---

[237] *Un traité médiéval*, pp. 129-132, lines 1197-1229 (Arabic); *L'Abrégé*, pp. 136-137, & (Arabic) 154, 7-17; *Herstellungsverfahren II & II'*, pp. 261-264, & lines 173-265.

furthermore, it is supposed that the given numbers will occupy the first upper row and therefore the magic sum, their sum, is given.

Now in certain cases, the construction is impossible according to 'form' (صورة): that is, we cannot attribute a single cell to each term; if so, we shall fill the square according to 'substance' (معنى) and be content with obtaining the magic sum.

(1) Neither construction is possible, thus neither in 'form' nor in 'substance' if we wish to place four numbers with $M = s_1 + s_2 + s_3 + s_4 < M_4 = 34$. Such is the case for Dā'ūd (d, a, u, d = 4, 1, 6, 4) and Ayyūb (a, y, u, b = 1, 10, 6, 2).

(2) Construction according to form is not possible when, although $s_1 + s_2 + s_3 + s_4 = M \geq 34$, for one of them, say s_t, the quantity $M - s_t$ cannot be represented in a second way as the sum of three natural numbers; indeed, s_t is common to two rows. This is for instance the case with ch, h, a, b = 300, 5, 1, 2 (شهاب, star) where *there should have been with 300 in another row three other numbers the sum of which is 8, which is not possible.* As noted by another author, there should even be three distinct representations of 8 since 300 occupies a corner cell.[238] Thus we must fill according to substance.

(3) The same holds if a letter occurs twice in the name. A common example is Muḥam(m)ad (written m, ḥ, m, d).

(4) It may be that we wish to put in a square of order 4 words with a number of letters other than 4. Normally, they could only be placed according to substance, thus to their sum. If, however, we can combine consecutive letters so as to have four entities, we may proceed according to form. Now it is clear that, in order for these consecutive letters to appear together in the same cell, they must not belong to the same decimal order. A name like Ibrāhīm (written a, b, r, a, h, i, m = 1, 2, 200, 1, 5, 10, 40) admits of only one combination (1, 2, 201, 5, 10, 40), and would thus have to be placed according to substance.

b. Filling according to 'substance'

(1) Prior filling of half the cells.

According to the method seen above, we shall take two progressions with constant difference α and start from the end cells of the first row. Dividing then the quantity M to enter into $M = m_1 + m_2$, we complete to m_1 the bishop's cells of one of the sequences and to m_2 the other (Fig. 395). This is the case for the triliteral name 'Alī (', l, i = 70, 30, 10),

---

[238] *Un traité médiéval*, p. 129 & lines 1202-1204; *L'Abrégé*, pp. 136-137 & 154.

with the sum $110 = 60 + 50 = m_1 + m_2$, by putting $A = 1$, $B = 8$, $\alpha = 1$ —that is, by placing the continuous sequence of the first eight numbers (Fig. 396).[239]

A	$m_1-A-2\alpha$	$m_2-B+2\alpha$	B
$m_2-B+3\alpha$	$B-\alpha$	$A+\alpha$	$m_1-A-3\alpha$
$B-2\alpha$	m_2-B	m_1-A	$A+2\alpha$
$m_1-A-\alpha$	$A+3\alpha$	$B-3\alpha$	$m_2-B+\alpha$

Fig. 395

1	57	44	8
45	7	2	56
6	42	59	3
58	4	5	43

Fig. 396

The resulting square will look better if filled with a continuous sequence of numbers. In the museum adjoining the Āstān-i Quds mosque in Mashhad there is a soldier's tunic bearing, for protection, the (pandiagonal) magic square shown in Fig. 397 ('substance' of the name 'Alī = 110 —or perhaps of لا إله إلا هو as in Fig. 377— thus $m = \frac{M}{2} = 55$, with the continuous sequence $20, \ldots, 35$; obtainable also by inverting Fig. 78 and increasing its elements by 19).

27	30	33	20
32	21	26	31
22	35	28	25
29	24	23	34

Fig. 397

(2) Prior filling of three-quarters of the cells.

A	$M-(A+B+C+2\alpha)$	$C+2\alpha$	B
$C+3\alpha$	$B-\alpha$	$A+\alpha$	$M-(A+B+C+3\alpha)$
$B-2\alpha$	C	$M-(A+B+C)$	$A+2\alpha$
$M-(A+B+C+\alpha)$	$A+3\alpha$	$B-3\alpha$	$C+\alpha$

Fig. 398

1	90	11	8
12	7	2	89
6	9	92	3
91	4	5	10

Fig. 399

Here just one set of four cells is left empty, the third set being filled by placing one more progression in the bishop's cells of a previous progression. Next, the four empty cells (one in each type of row) will be filled by putting in them the quantity M minus the sum in each of the rows meeting in it (Fig. 398). See Fig. 399, again for the name 'Alī, with

---

[239] Example of Zanjānī, see *Herstellungsverfahren II & II'*, p. 264 & lines 258-260.

sum 110, the prior filling being that of the first twelve numbers taken consecutively.[240]

(3) Varying the differences in the progressions.

A recurrent difficulty, arising as we fill, is repetition of numbers. To avoid that, the author of the *Harmonious arrangement* reverts to the subject in an appendix to his work. He suggests varying, as far as possible, the constant differences.

In order to better evaluate the situation, let us assume that the two sequences starting in A and B have the constant differences k_1 and k_2 for the first knight's move, l_1 and l_2 for the queen's, then p_1 and p_2 for the last knight's move (Fig. 400).

A	$m-A-k_1-l_1$	$m-B-k_2-l_2$	B
$m-B-k_2-l_2-p_2$	$B+k_2$	$A+k_1$	$m-A-k_1-l_1-p_1$
$B+k_2+l_2$	$m-B$	$m-A$	$A+k_1+l_1$
$m-A-k_1$	$A+k_1+l_1+p_1$	$B+k_2+l_2+p_2$	$m-B-k_2$

Fig. 400

Let then the magic sum be $M = 2m$ (it could be $M = m_1 + m_2$, each m_i being associated with one of the sequences). After completing the bishop's cells to the quantity m, consider the resulting situation. Since, according to property Ib, the magic sum must also be in the central cells, we must take $k_1 = -k_2$; likewise, equality in the lower end cells and the median top ones gives $l_1 = -l_2$; finally, the sum in the last row leads to setting $p_1 = -p_2$ while from the sum in the lateral columns we infer that $p_1 = -k_2$ and $p_2 = -k_1$. The result is the square of Fig. 401, with constant differences α ($= k_1 = -k_2 = p_1 = -p_2$) and β ($= l_1 = -l_2$).

A	$m-A-\alpha-\beta$	$m-B+\alpha+\beta$	B
$m-B+2\alpha+\beta$	$B-\alpha$	$A+\alpha$	$m-A-2\alpha-\beta$
$B-\alpha-\beta$	$m-B$	$m-A$	$A+\alpha+\beta$
$m-A-\alpha$	$A+2\alpha+\beta$	$B-2\alpha-\beta$	$m-B+\alpha$

Fig. 401

This result is clearly stated in the original text:[241] *If you put what*

---

[240] Example of Zanjānī, *ibid.* & lines 253-255.

[241] *Un traité médiéval*, p. 127, lines 1176-1183 of the Arabic text; or below, Appendix 55.

takes the place of the small extreme and what takes the place of the small middle[242] *at their (respective) places in some row, and move from one of them by increasing and from the other by decreasing with the same difference, it is not necessary that this difference be maintained for all the displacements. It is indeed imposed for the displacements which are of the same type, such as the displacements by the knight's move, or those by the queen's move. However, that the displacements by the knight's move between two rows display the same difference as the displacements by the queen's move, which, themselves, are between the union of two rows towards the union of a third and a fourth row, this is by no mean imposed.* In other words, the constant difference must be the same for identical moves within the same sequence, and equal but with the opposite sign for the corresponding moves in the other sequence. The author illustrates this with an example (Fig. 402, $m = 24$).

1	17	9	21
10	20	2	16
15	3	23	7
22	8	14	4

Fig. 402

10	23	4	21
5	20	11	22
25	8	19	6
18	7	24	9

Fig. 403

He continues by saying that the signs of our α and β may differ within the same sequence:[243] *Neither is it imposed that, having begun from one of them with addition, you are to continue always, for all subsequent moves, with addition; nor that, having begun from one of them with subtraction, you are to continue, for all subsequent moves, with subtraction. It is indeed obligatory for displacements of a same type, but no longer if the displacements are of two types. For instance, beginning with addition for the knight's move it is possible to advance with subtraction for the queen's move, but for the knight's move you are to return to addition. The same holds for a beginning with subtraction: it is possible to advance with addition for the queen's move, but you are to return to subtraction for the knight's move.* This is illustrated with the example of Fig. 403 ($m = 29$).

Remark. This square, like the previous one and the general square of Fig. 401, is pandiagonal (provided M is divided into two equal parts).

---

[242] Here our A and B (according to the denominations already used, see p. 67).

[243] *Un traité médiéval*, pp. 127-128, lines 1186-1192; or below, Appendix 56.

§4. Square of order 8

1. The given numbers are in the first row

a. Particular case: method of division

We suppose that s_1, s_2, s_3, s_4, s_5, s_6, s_7, s_8 are the numbers to be written in the first row and that $s_1 + s_2 + s_3 + s_4 = s_5 + s_6 + s_7 + s_8 = 2m$. Since the two upper quadrants are each to contain the same sum, they may be filled separately. This is explained as follows:[244] *In this case you are brought back to the three methods explained for placing different numbers in the first row of the square of four.*[245] *You fill the first half in one square of 4 by a method, then the second half in a second square (of 4), by the same or another method, until half the square (of 8) is filled. One fills then the other half as we shall show.*

There is no example in the *Harmonious arrangement*. There is one in the *Brief treatise*, which by the way considers only this particular case (upper half of Fig. 404, with $2m = 285$; filled by the fourth method — known to the reader of the *Brief treatise* since already taught there, see note 245 here below).

<u>80</u>	<u>66</u>	<u>116</u>	<u>23</u>	<u>61</u>	<u>85</u>	<u>58</u>	<u>81</u>
39	94	88	64	7	90	91	97
84	92	11	98	146	42	62	35
82	33	70	100	71	68	74	72
1	213	40	20	6	50	224	16
41	21	2	214	223	15	5	49
18	38	226	14	3	211	52	8
225	13	17	37	53	9	4	212

Fig. 404

b. General case

Here it is only required that the sum of the given quantities be divisible by 4. Let then $M = \sum s_i = 4m$, where m shall thus represent the sum of a pair of complementary terms. We shall fill the two upper quadrants in such a way that in each the columns and main diagonals display the

---

[244] *Un traité médiéval*, pp. 113-114, lines 1036-1040 of the Arabic text; or below, Appendix 57.

[245] Above, pp. 226-228; 'three methods': at this point the fourth remains to be explained (in the appendix to the work).

sum $2m$ (the sum of the horizontal rows of each may differ in this general case). Let us first fill the third rows with the complements to m of the quantities in the bishop's cells (we must have $m > s_i$). Considering then the first quadrant, we put two chosen quantities A and B in its second row and their complements in the bishop's cells (Fig. 405).

s_1	s_2	s_3	s_4	s_5	s_6	s_7	s_8
A	B	s_1-s_3+A	s_2-s_4+B	C	D	s_5-s_7+C	s_6-s_8+D
$m-s_3$	$m-s_4$	$m-s_1$	$m-s_2$	$m-s_7$	$m-s_8$	$m-s_5$	$m-s_6$
$m+s_3-s_1-A$	$m+s_4-s_2-B$	$m-A$	$m-B$	$m+s_7-s_5-C$	$m+s_8-s_6-D$	$m-C$	$m-D$

Fig. 405

This enables us to determine the content of the still empty cells (here, unlike in the case of Fig. 364a, we are to choose two quantities since the sum in the horizontal rows is not fixed). We proceed in the same way for the second quadrant, except that one of the two quantities will be determined, as seen by equating the sum in the second row to M: we must have $A + B + C + D = s_3 + s_4 + s_7 + s_8$. The *Harmonious arrangement* has the example of Fig. 406 (هوالحى لا اله الا هو له الحمد كله, $m = 86$).

60	31	36	32	11	35	83	56
29	40	53	39	80	58	8	37
50	54	26	55	3	30	75	51
33	47	57	46	78	49	6	28

Fig. 406

As the author remarks, this construction remains valid if the given quantities are in another line of the square; indeed, since the broken diagonals of each quadrant display the required sum $2m$, the horizontal rows of Fig. 405 may be cyclically permuted.

c. Filling the second half

A	$m-A-\alpha-\beta$	$m-B+\alpha+\beta$	B	C	$m-C-\alpha'-\beta'$	$m-D+\alpha'+\beta'$	D
$m-B+2\alpha+\beta$	$B-\alpha$	$A+\alpha$	$m-A-2\alpha-\beta$	$m-D+2\alpha'+\beta'$	$D-\alpha'$	$C+\alpha'$	$m-C-2\alpha'-\beta'$
$B-\alpha-\beta$	$m-B$	$m-A$	$A+\alpha+\beta$	$D-\alpha'-\beta'$	$m-D$	$m-C$	$C+\alpha'+\beta'$
$m-A-\alpha$	$A+2\alpha+\beta$	$B-2\alpha-\beta$	$m-B+\alpha$	$m-C-\alpha'$	$C+2\alpha'+\beta'$	$D-2\alpha'-\beta'$	$m-D+\alpha'$

Fig. 407

Choosing four elements for the upper end cells, we fill half of each quadrant using successively the moves knight, queen, knight; we then write in each empty cell the quantity completing the content of its bishop's cell (Fig. 407). In the *Harmonious arrangement* the square is completed in this way (Fig. 408, with $A = 1$, $B = 22$, $C = 14$, $D = 27$ and $\alpha = \beta = \alpha' = \beta' = 3$, $m = 86$)

1	79	70	22	14	66	65	27
73	19	4	76	68	24	17	63
16	64	85	7	21	59	72	20
82	10	13	67	69	23	18	62

Fig. 408

From this we obtain the whole square of Fig. 409.

60	31	36	32	11	35	83	56
29	40	53	39	80	58	8	37
50	54	26	55	3	30	75	51
33	47	57	46	78	49	6	28
1	79	70	22	14	66	65	27
73	19	4	76	68	24	17	63
16	64	85	7	21	59	72	20
82	10	13	67	69	23	18	62

Fig. 409

Remarks.

1. As already said, this filling requires that $M = 4m$, that is, that M be divisible by 4. As to A, B, C, D, α, β, α', β', the choice is restricted since we are to later avoid repeated or negative numbers.

2. In the example of the *Brief treatise* (Fig. 404), the author, who makes extensive use of the knight's move (Ch. II, §16), has filled the second half as in Fig. 410. For writing in the complements, he has divided the half-sum into $m_1 = 227$ and $m_2 = 58$. The constant differences in his progressions ($\alpha = 1$, $\beta = 3$) have been chosen so as to avoid repetitions.

A	$m_1-B+2\alpha$	$m_2-C-\alpha+\beta$	C	D	$m_2-D+\alpha-\beta$	$m_1-A-2\alpha$	B
$m_2-C+\beta$	$C+\alpha$	$A+\alpha$	$m_1-B+3\alpha$	$m_1-A-3\alpha$	$B-\alpha$	$D-\alpha$	$m_2-D-\beta$
$C+\alpha-\beta$	m_2-C	m_1-A	$B-2\alpha$	$A+2\alpha$	m_1-B	m_2-D	$D-\alpha+\beta$
$m_1-A-\alpha$	$B-3\alpha$	$C-\beta$	$m_2-C-\alpha$	$m_2-D+\alpha$	$D+\beta$	$A+3\alpha$	$m_1-B+\alpha$

Fig. 410

2. The given numbers are in the first two rows, within the quadrants' diagonals

The initial arrangement is thus that of Fig. 411 (*Harmonious arrangement*'s example, same numbers as before). We shall then fill the two quadrants so as to obtain the same sum in the lines and columns of each, which sum must also be that of the main half-diagonal of the whole square. With that in mind, we must have in the first quadrant (Fig. 412):

$$\sum a_i = \sum b_i = \sum c_i = \sum d_i = a_t + b_t + c_t + d_t \quad (t = 1, 2, 3, 4)$$

and this must also equal $a_1 + b_2 + c_3 + d_4$. Note that, since this is just a quadrant, these sums do not need to equal $a_4 + b_3 + c_2 + d_1$. But we may already observe that in such a quadrant, according to the above condition, the sum in the two lateral columns must equal the sum in the two median lines, so that the sum in the four corner cells will equal the sum in the central cells.

60			32	11			56
	31	36			35	83	

Fig. 411

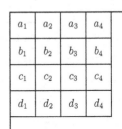

Fig. 412

In order to fill the first half of the square, the author of the *Harmonious arrangement* starts by completing the diagonals of the quadrants so as to have equal sums in the pairs of diagonals symmetrically placed (Fig. 413, with A, B, C, D optional quantities). The two main half-diagonals will thus contain the sum $s_1 + s_2 + s_7 + s_8$, the double of which will be set as the magic sum M (thus it is not the sum of the s_i). The optional quantities E, F, G, H will enable us to complete the two quadrants, each of which is to end up containing, vertically and horizontally, the sum $\frac{M}{2}$. We are then to fill the second half of the whole square as before, as a separate entity.

s_1	F		s_4	s_5	H		s_8
E	s_2	s_3		G	s_6	s_7	
	B	A			D	C	
s_5+s_6-B			s_7+s_8-A	s_1+s_2-D			s_3+s_4-C

Fig. 413

The example given in the *Harmonious arrangement* is, however, faulty. The author failed to observe that the quantities A and B on one side, C and D on the other, are not independent. Indeed, since the sums in the corner and central cells must be equal, we shall have, on the one hand,

$$s_2 + s_3 + A + B = s_1 + s_4 + s_5 + s_6 - B + s_7 + s_8 - A,$$

whence

$$A + B = \tfrac{1}{2}(s_1 - s_2 - s_3 + s_4 + s_5 + s_6 + s_7 + s_8) = \tfrac{1}{2}\sum s_i - (s_2 + s_3),$$

and, on the other hand, likewise,

$$C + D = \tfrac{1}{2}\sum s_i - (s_6 + s_7),$$

so that

$$A + B + s_2 + s_3 = C + D + s_6 + s_7.$$

3. The given numbers are in the first and third rows, within the diagonals

60			32	11			55
	31	36			35	83	

Fig. 414

s_1	F		s_4	s_5	H		s_8
	A	B			C	D	
E	s_2	s_3		G	s_6	s_7	
s_5+s_7-B			s_6+s_8-A	s_1+s_3-D			s_2+s_4-C

Fig. 415

The initial placing in the example of the *Harmonious arrangement* is that of Fig. 414. To fill the square (Fig. 415), we complete as before

the quadrants' diagonals with the help of quantities A, B, C, D meeting
the same requirements as before. This determines the magic sum for the
whole square, namely $M = 2(s_1 + s_3 + s_6 + s_8)$. We are then to choose
E, F, G, H and thus complete the whole arrangement.

The author explains this case but gives no example. He does, however,
tell us why he has modified here the value of s_8: filling the second half
requires, as we have seen (p. 241), that M be divisible by 4.

4. The given numbers are in the first and fourth rows, within the diagonals

60			32	11			56
31			36	35			83

Fig. 416

The example of Fig. 416 is that of the *Harmonious arrangement*. The
author begins by equalizing the upper and lower rows, placing in them
pairs of numbers in such a way that the sum in each half-row is equal to
that opposite in the other quadrant: *You place in the two median cells of
the lower row of the right-hand square of 4 two numbers with a sum equal
to the sum of the two numbers placed in the extremities of the upper row
of the left-hand square of 4, and conversely. You place in the two median
cells of the lower row of the left-hand square of 4 two numbers with a
sum equal to the sum of the two numbers placed in the extremities of the
upper row of the right-hand square of 4, and conversely. This being done,
the first and fourth rows of the square of 8 are equalized.*[246]

s_1	C	s_6+s_7-C	s_4	s_5	D	s_2+s_3-D	s_8
G	E			H		F	
		$2m-s_1-s_3-E$			$2m-s_6-s_8-F$		
s_2	A	s_5+s_8-A	s_3	s_6	B	s_1+s_4-B	s_7

Fig. 417

The magic sum will therefore again be the sum of the given numbers,
thus $M = \sum s_i = 4m$. Then (Fig. 417), the quantities E and F enable

[246] *Un traité médiéval*, pp. 120-121, lines 1120-1125 of the Arabic text; or below,
Appendix 58.

us to complete to $2m$ the main half-diagonals followed by the median half-columns. Then, with G and H, the upper half of the square will be determined, the median half-lines of which will, like the half-columns, contain $2m$ (this does not hold for the other half-lines, which before had been equalized by pairs). An example (slightly modified relative to that of the text, faulty) is seen in Fig. 418.

60	55	63	32	11	30	37	56
69	4	3	96	62	54	38	18
12	80	72	8	64	43	50	15
31	33	34	36	35	45	47	83

Fig. 418

Remark. We would treat in the same way the case with initial placing in the central cells: the lines to be equalized by pairs would be the median ones. This analogy may account for the author's omission of such a case.

5. Writing in the global sum

The *Brief treatise*'s author uses his method of descent by the knight's move (II, § 16) to fill the square of order 8 and, generally, the squares of evenly-even orders $n = 4k$ with a given magic sum M. We place, as already seen several times, $\frac{n}{2}$ sequences with the first natural numbers from 1 to $\frac{n^2}{2}$, of which $\frac{n}{4}$ are ascending sequences and $\frac{n}{4}$ descending ones. Next, we divide the sum to be placed into $\frac{n}{2}$ equal or unequal parts m_i, to each of which one of the $\frac{n}{2}$ sequences of n numbers will be attributed in order to determine the complements in the 4×4 subsquares. Each horizontal or vertical row will then contain the required sum M.

The case of the two diagonals is less simple. Each time one of the sequences of the numbers placed first passes through a cell of a main diagonal, the corresponding complements will appear there too, namely in the bishop's cell. Thus, the sum of these two cells will equal the part of M attributed to the sequence in question. Therefore, the part m_j of M will occur in this diagonal as many times as the corresponding jth sequence crosses one of its cells. Consequently, the sum in each diagonal will depend on its $\frac{n}{2}$ crossings by the various sequences of the numbers placed initially. The author of the *Brief treatise* examines the cases of the first three evenly-even orders.

(1) *Square of order 4.* Each of the two sequences has just one cell in

each diagonal (Fig. 419). The sum in the whole diagonal will then be $m_1 + m_2$, and the two parts of M may be taken either equal or unequal. In his example (Fig. 420), the author uses the word already seen (p. 235): 300, 5, 1, 2, with the breaking up of its sum into 150 and 158. We have already encountered a similar example, with the name 'Alī (Fig. 396) —which cannot appear in a square of order 3.

1	m_1-3	m_2-6	8
m_2-5	7	2	m_1-4
6	m_2-8	m_1-1	3
m_1-2	4	5	m_2-7

Fig. 419

150			158
1	147	152	8
153	7	2	146
6	150	149	3
148	4	5	151

Fig. 420

(2) *Square of order 8.* Let $M = m_1 + m_2 + m_3 + m_4$ with m_1 and m_2 associated with the two ascending sequences and m_3, m_4 with the descending ones. In this case, each diagonal contains the same parts twice, and their sums will be $2(m_1 + m_2)$ and $2(m_3 + m_4)$, respectively (Fig. 421). If the m_i are unequal, we need to have $m_1 + m_2 = m_3 + m_4 = \frac{M}{2}$ (M must thus be even). The example of the *Brief treatise* takes the same quantity as before, with $308 = 54 + 100 + 89 + 65$ (Fig. 422).

m_1			m_2	m_3			m_4
1	m_4-30	m_2-11	9	24	m_3-22	m_1-3	32
m_2-12	10	2	m_4-29	m_1-4	31	23	m_3-21
11	m_2-9	m_1-1	30	3	m_4-32	m_3-24	22
m_1-2	29	12	m_2-10	m_3-23	21	4	m_4-31
28	m_1-7	m_3-18	20	13	m_2-15	m_4-26	5
m_3-17	19	27	m_1-8	m_4-25	6	14	m_2-16
18	m_3-20	m_4-28	7	26	m_1-5	m_2-13	15
m_4-27	8	17	m_3-19	m_2-14	16	25	m_1-6

Fig. 421

54			100	89			65
1	35	89	9	24	67	51	32
88	10	2	36	50	31	23	68
11	91	53	30	3	33	65	22
52	29	12	90	66	21	4	34
28	47	71	20	13	85	39	5
72	19	27	46	40	6	14	84
18	69	37	7	26	49	87	15
38	8	17	70	86	16	25	48

Fig. 422

(3) *Square of order 12.* Let $M = m_1 + m_2 + m_3 + m_4 + m_5 + m_6$, with m_j associated with the jth sequence as numbered from the upper left corner (Fig. 423). The $\frac{n}{2} = 6$ crossings of the $\frac{n}{2} = 6$ sequences placed initially will give, as the magic condition for the diagonals, $m_1 + m_4 + m_5 + 3m_2 = m_2 + m_3 + m_6 + 3m_5 = M$. Either the parts are all equal

(provided M be divisible by $\frac{n}{2} = 6$), or they must obey the conditions $m_3 + m_6 = 2m_2$, $m_1 + m_4 = 2m_5$. Here as before a major difficulty will be to have throughout positive quantities and no repetitions.

m_1			m_2	m_3			m_4	m_5			m_6
1	m_4-46	m_2-15	13	25	m_6-70	m_1-3	48	60	m_5-58	m_3-27	72
m_2-16	14	2	m_4-45	m_1-4	47	26	m_6-69	m_3-28	71	59	m_5-57
15	m_2-13	m_1-1	46	3	m_4-48	m_3-25	70	27	m_6-72	m_5-60	58
m_1-2	45	16	m_2-14	m_3-26	69	4	m_4-47	m_5-59	57	28	m_6-71
44	m_5-54	m_6-66	68	17	m_3-31	m_4-42	56	5	m_1-7	m_2-19	29
m_6-65	67	43	m_5-53	m_4-41	55	18	m_3-32	m_2-20	30	6	m_1-8
66	m_6-68	m_4-44	54	42	m_5-56	m_2-17	31	19	m_3-29	m_1-5	7
m_4-43	53	65	m_6-67	m_2-18	32	41	m_5-55	m_1-6	8	20	m_3-30
52	m_1-11	m_3-35	33	64	m_2-23	m_5-50	9	40	m_4-38	m_6-62	21
m_3-36	34	51	m_1-12	m_5-49	10	63	m_2-24	m_6-61	22	39	m_4-37
35	m_3-33	m_5-52	11	50	m_1-9	m_6-64	23	62	m_2-21	m_4-40	38
m_5-51	12	36	m_3-34	m_6-63	24	49	m_1-10	m_4-39	37	61	m_2-22

Fig. 423

100			175	150			220	160			200
1	174	160	13	25	130	97	48	60	102	123	72
159	14	2	175	96	47	26	131	122	71	59	103
15	162	99	46	3	172	125	70	27	128	100	58
98	45	16	161	124	69	4	173	101	57	28	129
44	106	134	68	17	119	178	56	5	93	156	29
135	67	43	107	179	55	18	118	155	30	6	92
66	132	176	54	42	104	158	31	19	121	95	7
177	53	65	133	157	32	41	105	94	8	20	120
52	89	115	33	64	152	110	9	40	182	138	21
114	34	51	88	111	10	63	151	139	22	39	183
35	117	108	11	50	91	136	23	62	154	180	38
109	12	36	116	137	24	49	90	181	37	61	153

Fig. 424

We present in Fig. 424 an example for $1005 = 100 + 175 + 150 + 220 + 160 + 200$; the extant manuscript of the *Brief treatise* has an incomplete one.

C. SQUARES OF EVENLY-ODD ORDERS

§5. Square of order 6

1. The given numbers are in the first row

For convenience, we shall construct a bordered square (Fig. 425). Suppose then that $M = \sum s_i = 3m$, with $s_i < m$ all different. After filling the last line, we consider the two columns, one of which presents, relative to m, an excess of $\delta = |s_1 - s_6|$ and the other, an equal deficit.

s_1	s_2	s_3	s_4	s_5	s_6
$m-\beta-B$					$\beta+B$
β					$m-\beta$
$m-\alpha-A$					$\alpha+A$
α					$m-\alpha$
$m-s_6$	$m-s_2$	$m-s_3$	$m-s_4$	$m-s_5$	$m-s_1$

Fig. 425

31	36	32	11	25	24
47					6
3					50
48					5
1					52
29	17	21	42	28	22

Fig. 426

We then continue as follows (supposing here $\delta = s_1 - s_6 > 0$):[247] *For their equalization, you consider the row in excess. You divide this excess into two parts $(\delta = A + B)$. You place then in one of its cells some small number (α), and in another of its cells what is less than its complement by one of the two parts of this excess $(m-\alpha-A)$. You place in a third cell*

---

[247] *Un traité médiéval*, pp. 122-123, lines 1148-1153 of the Arabic text; or below, Appendix 59.

of this (same row) some other small number (β), and in a fourth cell what is less than its complement by the second part of this excess ($m - \beta - B$). Next, you place the complements of these numbers in the opposite row. The four rows are thus equalized, and there remains the square of 4 inside, with the filling of which you are familiar. The author's example is that of Fig. 426, where $M = 159$, $m = 53$, $\delta = 7 = A + B = 4 + 3$, $\alpha = 1$, $\beta = 3$ (α and β are taken small in order to avoid later occurrences of negative quantities). The *Brief treatise* proceeds in the same manner for another example, with its inner square filled as in Fig. 374. See Fig. 2 or here Fig. 427a-b.

١٢٤ ٣٦ ١٠٧ ٩٢ ١٠١ ١٣١

124	36	107	92	101	131
6	85	115	100	94	191
194	72	135	50	137	3
5	62	75	134	123	192
196	175	69	110	40	1
66	161	90	105	96	73

جميعا	آله	وعلى	محمد	على	سلام
٦	٨٥	١١٥	١٠٠	٩٤	١٩١
١٩٤	٧٢	١٣٥	٥٠	١٣٧	٣
٥	٦٢	٧٥	١٣٤	١٢٣	١٩٢
١٩٦	١٧٥	٦٩	١١٠	٤٠	١
٦٦	١٦١	٩٠	١٠٥	٩٦	٧٣

Fig. 427a   Fig. 427b

Examples of non-bordered 6 × 6 squares are also found, such as the following one from a Persian manuscript.[248]

201	121	289	329	66	11
60	170	256	137	191	203
196	254	146	139	189	93
32	276	144	141	187	237
172	176	42	143	185	299
356	20	140	128	199	174

هو   الله الرّحمن الرّحيم الملك القدوس

٢٠١	١٢١	٢٨٩	٣٢٩	٦٦	١١
٦٠	١٧٠	٢٥٦	١٣٧	١٩١	٢٠٣
١٩٦	٢٥٤	١٤٦	١٣٩	١٨٩	٩٣
٣٢	٢٧٦	١٤٤	١٤١	١٨٧	٢٣٧
١٧٢	١٧٦	٤٢	١٤٣	١٨٥	٢٩٩
٣٥٦	٢٠	١٤٠	١٢٨	١٩٩	١٧٤

Fig. 428a   Fig. 428b

2. The given numbers are equally distributed in the lateral rows

Two consecutive corner cells must therefore again be occupied by two

---

[248] MS. Mashhad Āstān-i Quds 12235, p. 379. In the manuscripts, the Arabic words determining the magic sum are written either in the upper cells or above them, as in these examples.

of the given numbers. Here too we shall construct a bordered square, and shall thus suppose as before that $M = 3m$; the author of the *Harmonious arrangement* sets also (which is by no means obligatory) $M = \sum s_i$. On the other hand, as he mentions,[249] it is necessary that *the sum of the three numbers placed in the lower row equal the sum of the three numbers placed in the upper row; the reason for that is clear.* Indeed, as shown in Fig. 429, forming the border implies that $s_1 + s_2 + s_3 = s_4 + s_5 + s_6$, a very restrictive situation, analogous to what we had encountered for the square of order 8 with the 'method of division' (p. 239). The lateral columns will be filled as before. The *Harmonious arrangement* has the example of Fig. 430, where $M = 198$, $m = 66$ (Allah?), $\delta = s_1 + s_4 - m = 14 = A + B = 7 + 7$, $\alpha = 1$, $\beta = 2$.

s_1	s_2	s_3	$m-s_5$	$m-s_6$	$m-s_4$
$m-\beta-B$					$\beta+B$
β					$m-\beta$
$m-\alpha-A$					$\alpha+A$
α					$m-\alpha$
s_4	$m-s_2$	$m-s_3$	s_5	s_6	$m-s_1$

Fig. 429

31	36	32	40	42	17
57					9
2					64
58					8
1					65
49	30	34	26	24	35

Fig. 430

The author observes that the method would remain the same if two of the given numbers were to occupy the upper corners. We may add that, if s_1 and s_3 are in those cells, we shall take $\delta = A + B = s_1 - s_3$ (Fig. 431).

---

[249] *Un traité médiéval*, pp. 123-124, lines 1158-1160 of the Arabic text; or below, Appendix 60.

s_1	s_2	$m-s_4$	$m-s_5$	$m-s_6$	s_3
$m-\beta-B$					$\beta+B$
β					$m-\beta$
$m-\alpha-A$					$\alpha+A$
α					$m-\alpha$
$m-s_3$	$m-s_2$	s_4	s_5	s_6	$m-s_1$

Fig. 431

§6. Squares of higher evenly-odd orders

Consider then $n = 4k + 2$, $k \geq 2$. The *Harmonious arrangement* explains very concisely how to fill its border. In the first of the above initial placings, we shall have $M = \sum s_i$, in the second $s_1 + s_2 + \cdots + s_{2k+1} = s_{2k+2} + \cdots + s_{4k+2}$, and we shall fill the horizontal rows as taught earlier. For the lateral columns, we shall again proceed as taught: having equalized one of them by means of four numbers, and placed their complements, we shall be left with a number of empty cells divisible by 4. These we shall fill using the middle and extreme terms of tetrads of numbers either consecutive or in progression, each group of four cells then making the sum due, thus $2m$ if $M = (2k + 1)m$.

Chapter VII

Other magic figures

By 'other magic figures' we mean those which either are squares but do not display the usual magic features or take another form (and thus obey other principles), such as magic triangles, rectangles, circles, crosses, cubes. Starting with the case of square figures, we shall first examine 'literal squares', that is, ones where we are to place different letters, in number equal to the order, in such a way that each line, column and diagonal includes them all. Such an arrangement may lend itself to the construction of magic squares. The second kind examined here will be that of squares with an empty cell, where the magic sum is nevertheless displayed. Finally, we shall consider the squares with 'divided cells', which are in fact two magic squares combined in a single figure.

§ 1. Literal squares

Arabic texts call 'literal squares' (*awfāq ḥarfiya*, as opposed to *awfāq 'adadiya*, numerical squares) those of any order greater than 3 where letters, all different and in number equal to the order, are arranged in such a way that each letter appears only once in each of the lines, columns and main diagonals; obviously, the order will be limited by the number of letters, thus 28 for the Arabic alphabet (pp. 10, 207). Shabrāmallisī will be our main source, his explanations being particularly clear.[250] In what follows, we shall adapt his methods to our alphabet, with its 26 letters taken in the usual order.

1. Squares of odd orders

a. Case of a prime order

Write in the first line the natural sequence of letters starting, say, in the corner. Reproduce these letters from line to line using the knight's move. Shabrāmallisī has the examples of order 5 (Fig. 432) and 7 (Fig. 433). Note that each letter is found a single time not only in the rows and main diagonals but in the broken diagonals as well.

This simple construction is applicable to all prime orders less than 28. Of the remaining odd orders, namely 9, 15, 21, 25, 27, all composite, Shabrāmallisī will examine only 9, 15, 25, suggesting two possibilities for the last.

---

[250] MS. Paris BNF Arabe 2698, fol. 19ʳ-24ʳ (uneven orders) and 49ᵛ-51ʳᵛ (even orders).

© Springer Nature Switzerland AG 2019
J. Sesiano, *Magic Squares*, Sources and Studies in the History of Mathematics and Physical Sciences, https://doi.org/10.1007/978-3-030-17993-9_7

a	b	c	d	e
d	e	a	b	c
b	c	d	e	a
e	a	b	c	d
c	d	e	a	b

Fig. 432

a	b	c	d	e	f	g
f	g	a	b	c	d	e
d	e	f	g	a	b	c
b	c	d	e	f	g	a
g	a	b	c	d	e	f
e	f	g	a	b	c	d
c	d	e	f	g	a	b

Fig. 433

b. Case of the order n=9

We may not proceed as before: each ascending diagonal would end up with the same triads of letters. Considering (Fig. 434) the first triad of lines, put a at the beginning of the first line, then repeat it in the next two lines in the cell following that attained by the knight's move; place after it the remaining letters in succession. For the second triad of lines, put first a in the middle, then for the next two lines again in the cell following that reached by the knight's move, and complete the lines with the other letters taken in succession. For the last triad, place a at the end of the first line and, for the other two lines, next to the corresponding knight's cells; the subsequent letters are written by jumping each time over one cell and advancing in the opposite direction. The main diagonals (rarely the broken ones) will thus contain each letter once.

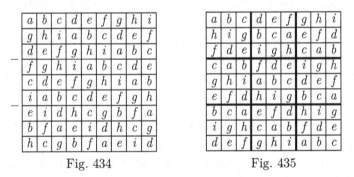

Fig. 434   Fig. 435

Another possibility is found in a Persian manuscript (Fig. 435).[251] Divide the square into 3×3 subsquares and write the sequence of consecutive letters in the first line. Reproduce each triad of letters, shifted back by one place, in the next lines of the 3×3 neighbour. This completes the top subsquares. The columns of each of these subsquares are then repeated in the subsquares immediately below, shifted forward by one place.

---

[251] MS. Āstān-i Quds 12235, p. 453.

c. Case of the order $n=3\cdot z$, with $z\neq3$ (in fact, $n=15$)

a	b	c	d	e	f	g	h	i	j	k	l	m	n	o
n	o	a	b	c	d	e	f	g	h	i	j	k	l	m
l	m	n	o	a	b	c	d	e	f	g	h	i	j	k
j	k	l	m	n	o	a	b	c	d	e	f	g	h	i
h	i	j	k	l	m	n	o	a	b	c	d	e	f	g
e	f	g	h	i	j	k	l	m	n	o	a	b	c	d
c	d	e	f	g	h	i	j	k	l	m	n	o	a	b
o	a	b	c	d	e	f	g	h	i	j	k	l	m	n
m	n	o	a	b	c	d	e	f	g	h	i	j	k	l
k	l	m	n	o	a	b	c	d	e	f	g	h	i	j
i	j	k	l	m	n	o	a	b	c	d	e	f	g	h
g	h	i	j	k	l	m	n	o	a	b	c	d	e	f
f	g	h	i	j	k	l	m	n	o	a	b	c	d	e
d	e	f	g	h	i	j	k	l	m	n	o	a	b	c
b	c	d	e	f	g	h	i	j	k	l	m	n	o	a

Fig. 436

We begin in the corner cell of the first line and write the letters in succession (Fig. 436). We then proceed line by line, putting *a* in the knight's cell followed by the other letters, but with three exceptions for placing *a*: in the first line of the second group of five (thus, as the author puts it, in the line following that in which the third letter, our *c*, has fallen in the diagonal), *a* will be in the cell following the knight's; in the median line (thus, as the author puts it, in the line following that in which the third letter has fallen at the beginning of the line), *a* will again be in the cell following the knight's; in the first line of the last triad of lines, *a* is this time written in the cell preceding the knight's.

a	b	c	d	e	f	g	h	i	j	k	l	m	n	o
n	o	a	b	c	d	e	f	g	h	i	j	k	l	m
l	m	n	o	**k**	b	**h**	d	e	f	g	**c**	i	j	**a**
j	k	l	m	n	o	a	b	c	d	e	f	g	h	i
h	i	j	k	l	m	n	o	a	b	c	d	e	f	g
e	f	g	h	i	j	k	l	m	n	o	a	b	c	d
c	d	e	f	g	h	i	j	k	l	m	n	o	a	b
o	a	b	c	d	e	f	g	h	i	j	k	l	m	n
m	n	o	a	b	c	d	e	f	g	h	i	j	k	l
k	l	m	n	o	a	b	c	d	e	f	g	h	i	j
i	j	k	l	m	n	o	a	b	c	d	e	f	g	h
g	**c**	i	j	**a**	l	m	n	o	**k**	b	**h**	d	e	f
f	g	h	i	j	k	l	m	n	o	a	b	c	d	e
d	e	f	g	h	i	j	k	l	m	n	o	a	b	c
b	**h**	d	e	f	g	**c**	i	j	**a**	l	m	n	o	**k**

Fig. 437

Proceeding thus ensures that all horizontal and vertical rows will contain each letter just once. The first diagonal, however, does not meet that condition, for a and c recur while h and k are missing. Then we shall exchange, in the columns concerned, one of the a with k and one of the c with h. But the problem is now with the lines, and in order to avoid other repetitions we are to modify them, or, rather, a minimal number of them, and without touching the other diagonal, which meets the condition. The author does it in the way seen in Fig. 437 (the letters concerned are in bold and underlined). The smaller order $n = 15$ is in this respect particularly suitable, for it requires few modifications.

d. Case of the square order $n = z^2$, with $z \neq 3$ (thus $n = 25$)

a	b	c	d	e	f	g	h	i	j	k	l	m	n	o	p	q	r	s	t	u	v	w	x	y
u	v	w	x	y	a	b	c	d	e	f	g	h	i	j	k	l	m	n	o	p	q	r	s	t
p	q	r	s	t	u	v	w	x	y	a	b	c	d	e	f	g	h	i	j	k	l	m	n	o
k	l	m	n	o	p	q	r	s	t	u	v	w	x	y	a	b	c	d	e	f	g	h	i	j
f	g	h	i	j	k	l	m	n	o	p	q	r	s	t	u	v	w	x	y	a	b	c	d	e
t	u	v	w	x	y	a	b	c	d	e	f	g	h	i	j	k	l	m	n	o	p	q	r	s
o	p	q	r	s	t	u	v	w	x	y	a	b	c	d	e	f	g	h	i	j	k	l	m	n
j	k	l	m	n	o	p	q	r	s	t	u	v	w	x	y	a	b	c	d	e	f	g	h	i
e	f	g	h	i	j	k	l	m	n	o	p	q	r	s	t	u	v	w	x	y	a	b	c	d
y	a	b	c	d	e	f	g	h	i	j	k	l	m	n	o	p	q	r	s	t	u	v	w	x
n	o	p	q	r	s	t	u	v	w	x	y	a	b	c	d	e	f	g	h	i	j	k	l	m
i	j	k	l	m	n	o	p	q	r	s	t	u	v	w	x	y	a	b	c	d	e	f	g	h
d	e	f	g	h	i	j	k	l	m	n	o	p	q	r	s	t	u	v	w	x	y	a	b	c
x	y	a	b	c	d	e	f	g	h	i	j	k	l	m	n	o	p	q	r	s	t	u	v	w
s	t	u	v	w	x	y	a	b	c	d	e	f	g	h	i	j	k	l	m	n	o	p	q	r
h	i	j	k	l	m	n	o	p	q	r	s	t	u	v	w	x	y	a	b	c	d	e	f	g
c	d	e	f	g	h	i	j	k	l	m	n	o	p	q	r	s	t	u	v	w	x	y	a	b
w	x	y	a	b	c	d	e	f	g	h	i	j	k	l	m	n	o	p	q	r	s	t	u	v
r	s	t	u	v	w	x	y	a	b	c	d	e	f	g	h	i	j	k	l	m	n	o	p	q
m	n	o	p	q	r	s	t	u	v	w	x	y	a	b	c	d	e	f	g	h	i	j	k	l
g	m	s	y	f	l	r	x	e	k	q	w	d	j	p	v	c	i	o	u	b	h	n	t	a
b	h	n	t	a	g	m	s	y	f	l	r	x	e	k	q	w	d	j	p	v	c	i	o	u
v	c	i	o	u	b	h	n	t	a	g	m	s	y	f	l	r	x	e	k	q	w	d	j	p
q	w	d	j	p	v	c	i	o	u	b	h	n	t	a	g	m	s	y	f	l	r	x	e	k
l	r	x	e	k	q	w	d	j	p	v	c	i	o	u	b	h	n	t	a	g	m	s	y	f

Fig. 438

Divide first the square into z groups of z consecutive lines. Then, starting from the corner, put in the first line the letters taken in succession (Fig. 438). For the first group of z lines, we again put the letters in succession but by shifting each time the first one by z cells. The letters are also put in succession for the second group, shifting each time the first one by z cells; but here the point of departure will be the cell $(z+1)+1$ $(= 7\text{th})$ counted from the side. So again for the third and fourth group, but starting from the cells, counted from the side, $2(z+1)+1$ $(= 13\text{th})$ and $3(z+1)+1$ $(= 19\text{th})$. The placing in the last group is slightly different: the first letter will be in the last cell of the first line, and then placed as before in the rows below; but each subsequent letter will be in the $(z-1)\text{th}$ cell from it —thus with a jump over $z - 2$ cells— and in the other direction.

Remark. As noted by the author himself, this construction is in keeping with that of the square of 9 seen before (where each letter is shifted by $z = 3$, and with the points of departure being in the first, fifth, ninth cell from the side; see Fig. 434). The same method is thus applicable to the two odd square orders considered here.

e. Case of the composite order $n = t \cdot z$, with t, z odd $\neq 3$

a	b	c	d	e	f	g	h	i	j	k	l	m	n	o	p	q	r	s	t	u	v	w	x	y
d	e	a	b	c	i	j	f	g	h	n	o	k	l	m	s	t	p	q	r	x	y	u	v	w
b	c	d	e	a	g	h	i	j	f	l	m	n	o	k	q	r	s	t	p	v	w	x	y	u
e	a	b	c	d	j	f	g	h	i	o	k	l	m	n	t	p	q	r	s	y	u	v	w	x
c	d	e	a	b	h	i	j	f	g	m	n	o	k	l	r	s	t	p	q	w	x	y	u	v
k	l	m	n	o	p	q	r	s	t	u	v	w	x	y	a	b	c	d	e	f	g	h	i	j
n	o	k	l	m	s	t	p	q	r	x	y	u	v	w	d	e	a	b	c	i	j	f	g	h
l	m	n	o	k	q	r	s	t	p	v	w	x	y	u	b	c	d	e	a	g	h	i	j	f
o	k	l	m	n	t	p	q	r	s	y	u	v	w	x	e	a	b	c	d	j	f	g	h	i
m	n	o	k	l	r	s	t	p	q	w	x	y	u	v	c	d	e	a	b	h	i	j	f	g
u	v	w	x	y	a	b	c	d	e	f	g	h	i	j	k	l	m	n	o	p	q	r	s	t
x	y	u	v	w	d	e	a	b	c	i	j	f	g	h	n	o	k	l	m	s	t	p	q	r
v	w	x	y	u	b	c	d	e	a	g	h	i	j	f	l	m	n	o	k	q	r	s	t	p
y	u	v	w	x	e	a	b	c	d	j	f	g	h	i	o	k	l	m	n	t	p	q	r	s
w	x	y	u	v	c	d	e	a	b	h	i	j	f	g	m	n	o	k	l	r	s	t	p	q
f	g	h	i	j	k	l	m	n	o	p	q	r	s	t	u	v	w	x	y	a	b	c	d	e
i	j	f	g	h	n	o	k	l	m	s	t	p	q	r	x	y	u	v	w	d	e	a	b	c
g	h	i	j	f	l	m	n	o	k	q	r	s	t	p	v	w	x	y	u	b	c	d	e	a
j	f	g	h	i	o	k	l	m	n	t	p	q	r	s	y	u	v	w	x	e	a	b	c	d
h	i	j	f	g	m	n	o	k	l	r	s	t	p	q	w	x	y	u	v	c	d	e	a	b
p	q	r	s	t	u	v	w	x	y	a	b	c	d	e	f	g	h	i	j	k	l	m	n	o
s	t	p	q	r	x	y	u	v	w	d	e	a	b	c	i	j	f	g	h	n	o	k	l	m
q	r	s	t	p	v	w	x	y	u	b	c	d	e	a	g	h	i	j	f	l	m	n	o	k
t	p	q	r	s	y	u	v	w	x	e	a	b	c	d	j	f	g	h	i	o	k	l	m	n
r	s	t	p	q	w	x	y	u	v	c	d	e	a	b	h	i	j	f	g	m	n	o	k	l

Fig. 439

With the limited number of alphabetical letters, the only possible order here is $n = 25$. Shabrāmallisī begins by placing the successive letters in the first line (Fig. 439). He then divides his square into subsquares of order 5, and fills each upper one with the letters of its first line, repeated with the knight's move, as in Fig. 432. This arrangement in the top subsquares is then repeated in each of those below distant by a knight's move made in the other direction. Note (as Shabrāmallisī does not) that each pair of broken diagonals also contains each letter just once.

General remark. Shabrāmallisī does not say (or does not know) that magic squares can be obtained using such odd-order configurations. Suppose we replace the literal square by a numerical square by attributing to each of the n letters each of the numbers from 1 to n, then take the literal square in inverse direction and attribute this time to each letter a multiple of the order equal to the number of this letter less 1. Adding the numbers in the corresponding cells of the two squares will produce a magic square.[252]

1	2	3	4	5
4	5	1	2	3
2	3	4	5	1
5	1	2	3	4
3	4	5	1	2

Fig. 440a

20	15	10	5	0
10	5	0	20	15
0	20	15	10	5
15	10	5	0	20
5	0	20	15	10

Fig. 440b

21	17	13	9	5
14	10	1	22	18
2	23	19	15	6
20	11	7	3	24
8	4	25	16	12

Fig. 440c

1	2	3	4	5	6	7	8	9
7	8	9	1	2	3	4	5	6
4	5	6	7	8	9	1	2	3
6	7	8	9	1	2	3	4	5
3	4	5	6	7	8	9	1	2
9	1	2	3	4	5	6	7	8
5	9	4	8	3	7	2	6	1
2	6	1	5	9	4	8	3	7
8	3	7	2	6	1	5	9	4

Fig. 441a

72	63	54	45	36	27	18	9	0
45	36	27	18	9	0	72	63	54
18	9	0	72	63	54	45	36	27
36	27	18	9	0	72	63	54	45
9	0	72	63	54	45	36	27	18
63	54	45	36	27	18	9	0	72
0	45	9	54	18	63	27	72	36
54	18	63	27	72	36	0	45	9
27	72	36	0	45	9	54	18	63

Fig. 441b

Thus, from the configuration of Fig. 432 we derive the figures 440a–c, which represent the two auxiliary squares and the resulting (pandiagonal) square of order 5. Using Fig. 434 we obtain the figures 441a–c of the auxiliary squares and their sum (ordinary magic square of order 9). For

---

[252] Method attributed in modern times to Philippe de la Hire (1705).

order 15, the configuration of Fig. 437 leads to the two auxiliary squares represented together in Fig. 442a, their sum giving Fig. 442b. Finally, from the configurations of the figures 438 and 439 for order 25 we infer the auxiliary squares of Fig. 443a and 444a, which, using the key of Fig. 443c, produce the squares of Fig. 443b and 444b.

73	65	57	49	41	33	25	17	9
52	44	36	19	11	3	76	68	60
22	14	6	79	71	63	46	38	30
42	34	26	18	1	74	66	58	50
12	4	77	69	61	53	45	28	20
72	55	47	39	31	23	15	7	80
5	54	13	62	21	70	29	78	37
56	24	64	32	81	40	8	48	16
35	75	43	2	51	10	59	27	67

Fig. 441c

1	2	3	4	5	6	7	8	9	10	11	12	13	14	15
210	195	180	165	150	135	120	105	90	75	60	45	30	15	0
14	15	1	2	3	4	5	6	7	8	9	10	11	12	13
180	165	150	135	120	105	90	75	60	45	30	15	0	210	195
12	13	14	15	11	2	8	4	5	6	7	3	9	10	1
0	135	120	30	90	75	60	45	105	15	150	210	195	180	165
10	11	12	13	14	15	1	2	3	4	5	6	7	8	9
120	105	90	75	60	45	30	15	0	210	195	180	165	150	135
8	9	10	11	12	13	14	15	1	2	3	4	5	6	7
90	75	60	45	30	15	0	210	195	180	165	150	135	120	105
5	6	7	8	9	10	11	12	13	14	15	1	2	3	4
45	30	15	0	210	195	180	165	150	135	120	105	90	75	60
3	4	5	6	7	8	9	10	11	12	13	14	15	1	2
15	0	210	195	180	165	150	135	120	105	90	75	60	45	30
15	1	2	3	4	5	6	7	8	9	10	11	12	13	14
195	180	165	150	135	120	105	90	75	60	45	30	15	0	210
13	14	15	1	2	3	4	5	6	7	8	9	10	11	12
165	150	135	120	105	90	75	60	45	30	15	0	210	195	180
11	12	13	14	15	1	2	3	4	5	6	7	8	9	10
135	120	105	90	75	60	45	30	15	0	210	195	180	165	150
9	10	11	12	13	14	15	1	2	3	4	5	6	7	8
105	90	75	60	45	30	15	0	210	195	180	165	150	135	120
7	3	9	10	1	12	13	14	15	11	2	8	4	5	6
75	60	45	105	15	150	210	195	180	165	0	135	120	30	90
6	7	8	9	10	11	12	13	14	15	1	2	3	4	5
60	45	30	15	0	210	195	180	165	150	135	120	105	90	75
4	5	6	7	8	9	10	11	12	13	14	15	1	2	3
30	15	0	210	195	180	165	150	135	120	105	90	75	60	45
2	8	4	5	6	7	3	9	10	1	12	13	14	15	11
150	210	195	180	165	0	135	120	30	90	75	60	45	105	15

Fig. 442a

211	197	183	169	155	141	127	113	99	85	71	57	43	29	15
194	180	151	137	123	109	95	81	67	53	39	25	11	222	208
12	148	134	45	101	77	68	49	110	21	157	213	204	190	166
130	116	102	88	74	60	31	17	3	214	200	186	172	158	144
98	84	70	56	42	28	14	225	196	182	168	154	140	126	112
50	36	22	8	219	205	191	177	163	149	135	106	92	78	64
18	4	215	201	187	173	159	145	131	117	103	89	75	46	32
210	181	167	153	139	125	111	97	83	69	55	41	27	13	224
178	164	150	121	107	93	79	65	51	37	23	9	220	206	192
146	132	118	104	90	61	47	33	19	5	216	202	188	174	160
114	100	86	72	58	44	30	1	212	198	184	170	156	142	128
82	63	54	115	16	162	223	209	195	176	2	143	124	35	96
66	52	38	24	10	221	207	193	179	165	136	122	108	94	80
34	20	6	217	203	189	175	161	147	133	119	105	76	62	48
152	218	199	185	171	7	138	129	40	91	87	73	59	120	26

Fig. 442*b*

Ya	Xb	Wc	Vd	Ue	Tf	Sg	Rh	Qi	Pj	Ok	Nl	Mm	Ln	Ko	Jp	Iq	Hr	Gs	Ft	Eu	Dv	Cw	Bx	Ay
Tu	Sv	Rw	Qx	Py	Oa	Nb	Mc	Ld	Ke	Jf	Ig	Hh	Gi	Fj	Ek	Dl	Cm	Bn	Ao	Yp	Xq	Wr	Vs	Ut
Op	Nq	Mr	Ls	Kt	Ju	Iv	Hw	Gx	Fy	Ea	Db	Cc	Bd	Ae	Yf	Xg	Wh	Vi	Uj	Tk	Sl	Rm	Qn	Po
Jk	Il	Hm	Gn	Fo	Ep	Dq	Cr	Bs	At	Yu	Xv	Ww	Vx	Uy	Ta	Sb	Rc	Qd	Pe	Of	Ng	Mh	Li	Kj
Ef	Dg	Ch	Bi	Aj	Yk	Xl	Wm	Vn	Uo	Tp	Sq	Rr	Qs	Pt	Ou	Nv	Mw	Lx	Ky	Ja	Ib	Hc	Gd	Fe
St	Ru	Qv	Pw	Ox	Ny	Ma	Lb	Kc	Jd	Ie	Hf	Gg	Fh	Ei	Dj	Ck	Bl	Am	Yn	Xo	Wp	Vq	Ur	Ts
No	Mp	Lq	Kr	Js	It	Hu	Gv	Fw	Ex	Dy	Ca	Bb	Ac	Yd	Xe	Wf	Vg	Uh	Ti	Sj	Rk	Ql	Pm	On
Ij	Hk	Gl	Fm	En	Do	Cp	Bq	Ar	Ys	Xt	Wu	Vv	Uw	Tx	Sy	Ra	Qb	Pc	Od	Ne	Mf	Lg	Kh	Ji
De	Cf	Bg	Ah	Yi	Xj	Wk	Vl	Um	Tn	So	Rp	Qq	Pr	Os	Nt	Mu	Lv	Kw	Jx	Iy	Ha	Gb	Fc	Ed
Xy	Wa	Vb	Uc	Td	Se	Rf	Qg	Ph	Oi	Nj	Mk	Ll	Km	Jn	Io	Hp	Gq	Fr	Es	Dt	Cu	Bv	Aw	Yx
Mn	Lo	Kp	Jq	Ir	Hs	Gt	Fu	Ev	Dw	Cx	By	Aa	Yb	Xc	Wd	Ve	Uf	Tg	Sh	Ri	Qj	Pk	Ol	Nm
Hi	Gj	Fk	El	Dm	Cn	Bo	Ap	Yq	Xr	Ws	Vt	Uu	Tv	Sw	Rx	Qy	Pa	Ob	Nc	Md	Le	Kf	Jg	Ih
Cd	Be	Af	Yg	Xh	Wi	Vj	Uk	Tl	Sm	Rn	Qo	Pp	Oq	Nr	Ms	Lt	Ku	Jv	Iw	Hx	Gy	Fa	Eb	Dc
Wx	Vy	Ua	Tb	Sc	Rd	Qe	Pf	Og	Nh	Mi	Lj	Kk	Jl	Im	Hn	Go	Fp	Eq	Dr	Cs	Bt	Au	Yv	Xw
Rs	Qt	Pu	Ov	Nw	Mx	Ly	Ka	Jb	Ic	Hd	Ge	Ff	Eg	Dh	Ci	Bj	Ak	Yl	Xm	Wn	Vo	Up	Tq	Sr
Gh	Fi	Ej	Dk	Cl	Bm	An	Yo	Xp	Wq	Vr	Us	Tt	Su	Rv	Qw	Px	Oy	Na	Mb	Lc	Kd	Je	If	Hg
Bc	Ad	Ye	Xf	Wg	Vh	Ui	Tj	Sk	Rl	Qm	Pn	Oo	Np	Mq	Lr	Ks	Jt	Iu	Hv	Gw	Fx	Ey	Da	Cb
Vw	Ux	Ty	Sa	Rb	Qc	Pd	Oe	Nf	Mg	Lh	Ki	Jj	Ik	Hl	Gm	Fn	Eo	Dp	Cq	Br	As	Yt	Xu	Wv
Qr	Ps	Ot	Nu	Mv	Lw	Kx	Jy	Ia	Hb	Gc	Fd	Ee	Df	Cg	Bh	Ai	Yj	Xk	Wl	Vm	Un	To	Sp	Rq
Lm	Kn	Jo	Ip	Hq	Gr	Fs	Et	Du	Cv	Bw	Ax	Yy	Xa	Wb	Vc	Ud	Te	Sf	Rg	Qh	Pi	Oj	Nk	Ml
Ag	Tm	Ns	Hy	Bf	Ul	Or	Ix	Ce	Vk	Pq	Jw	Dd	Wj	Qp	Kv	Ec	Xi	Ro	Lu	Fb	Yh	Sn	Mt	Ga
Ub	Oh	In	Ct	Va	Pg	Jm	Ds	Wy	Qf	Kl	Er	Xx	Re	Lk	Fq	Yw	Sd	Mj	Gp	Av	Tc	Ni	Ho	Bu
Pv	Jc	Di	Wo	Qu	Kb	Eh	Xn	Rt	La	Fg	Ym	Ss	My	Gf	Al	Tr	Nx	He	Bk	Uq	Ow	Id	Cj	Vp
Kq	Ew	Xd	Rj	Lp	Fv	Yc	Si	Mo	Gu	Ab	Th	Nn	Ht	Ba	Ug	Om	Is	Cy	Vf	Pl	Jr	Dx	We	Qk
Fl	Yr	Sx	Me	Gk	Aq	Tw	Nd	Hj	Bp	Uv	Oc	Ii	Co	Vu	Pb	Jh	Dn	Wt	Qa	Kg	Em	Xs	Ry	Lf

Fig. 443*a*

601	577	553	529	505	481	457	433	409	385	361	337	313	289	265	241	217	193	169	145	121	97	73	49	25
496	472	448	424	400	351	327	303	279	255	231	207	183	159	135	111	87	63	39	15	616	592	568	544	520
366	342	318	294	270	246	222	198	174	150	101	77	53	29	5	606	582	558	534	510	486	462	438	414	390
236	212	188	164	140	116	92	68	44	20	621	597	573	549	525	476	452	428	404	380	356	332	308	284	260
106	82	58	34	10	611	587	563	539	515	491	467	443	419	395	371	347	323	299	275	226	202	178	154	130
470	446	422	398	374	350	301	277	253	229	205	181	157	133	109	85	61	37	13	614	590	566	542	518	494
340	316	292	268	244	220	196	172	148	124	100	51	27	3	604	580	556	532	508	484	460	436	412	388	364
210	186	162	138	114	90	66	42	18	619	595	571	547	523	499	475	426	402	378	354	330	306	282	258	234
80	56	32	8	609	585	561	537	513	489	465	441	417	393	369	345	321	297	273	249	225	176	152	128	104
600	551	527	503	479	455	431	407	383	359	335	311	287	263	239	215	191	167	143	119	95	71	47	23	624
314	290	266	242	218	194	170	146	122	98	74	50	1	602	578	554	530	506	482	458	434	410	386	362	338
184	160	136	112	88	64	40	16	617	593	569	545	521	497	473	449	425	376	352	328	304	280	256	232	208
54	30	6	607	583	559	535	511	487	463	439	415	391	367	343	319	295	271	247	223	199	175	126	102	78
574	550	501	477	453	429	405	381	357	333	309	285	261	237	213	189	165	141	117	93	69	45	21	622	598
444	420	396	372	348	324	300	251	227	203	179	155	131	107	83	59	35	11	612	588	564	540	516	492	468
158	134	110	86	62	38	14	615	591	567	543	519	495	471	447	423	399	375	326	302	278	254	230	206	182
28	4	605	581	557	533	509	485	461	437	413	389	365	341	317	293	269	245	221	197	173	149	125	76	52
548	524	500	451	427	403	379	355	331	307	283	259	235	211	187	163	139	115	91	67	43	19	620	596	572
418	394	370	346	322	298	274	250	201	177	153	129	105	81	57	33	9	610	586	562	538	514	490	466	442
288	264	240	216	192	168	144	120	96	72	48	24	625	576	552	528	504	480	456	432	408	384	360	336	312
7	488	344	200	31	512	368	224	55	536	392	248	79	560	416	272	103	584	440	296	127	608	464	320	151
502	358	214	70	526	382	238	94	575	406	262	118	599	430	286	142	623	454	310	166	22	478	334	190	46
397	228	84	565	421	252	108	589	445	276	132	613	469	325	156	12	493	349	180	36	517	373	204	60	541
267	123	579	435	291	147	603	459	315	171	2	483	339	195	26	507	363	219	75	531	387	243	99	555	411
137	618	474	305	161	17	498	329	185	41	522	353	209	65	546	377	233	89	570	401	257	113	594	450	281

Fig. 443b

a	b	c	d	e	f	g	h	i	j	k	l	m	n	o	p	q	r	s	t	u	v	w	x	y
1	2	3	4	5	6	7	8	9	10	11	12	13	14	15	16	17	18	19	20	21	22	23	24	25

A	B	C	D	E	F	G	H	I	J	K	L	M	N	O	P	Q	R	S	T	U	V	W	X	Y
0	25	50	75	100	125	150	175	200	225	250	275	300	325	350	375	400	425	450	475	500	525	550	575	600

Fig. 443c

Ya	Xb	Wc	Vd	Ue	Tf	Sg	Rh	Qi	Pj	Ok	Nl	Mm	Ln	Ko	Jp	Iq	Hr	Gs	Ft	Eu	Dv	Cw	Bx	Ay
Wd	Ve	Ua	Yb	Xc	Ri	Qj	Pf	Tg	Sh	Mn	Lo	Kk	Ol	Nm	Hs	Gt	Fp	Jq	Ir	Cx	By	Au	Ev	Dw
Ub	Yc	Xd	We	Va	Pg	Th	Si	Rj	Qf	Kl	Om	Nn	Mo	Lk	Fq	Jr	Is	Ht	Gp	Av	Ew	Dx	Cy	Bu
Xe	Wa	Vb	Uc	Yd	Sj	Rf	Qg	Ph	Ti	No	Mk	Ll	Km	On	It	Hp	Gq	Fr	Js	Dy	Cu	Bv	Aw	Ex
Vc	Ud	Ye	Xa	Wb	Qh	Pi	Tj	Sf	Rg	Lm	Kn	Oo	Nk	Ml	Gr	Fs	Jt	Ip	Hq	Bw	Ax	Ey	Du	Cv
Jk	Il	Hm	Gn	Fo	Ep	Dq	Cr	Bs	At	Yu	Xv	Ww	Vx	Uy	Ta	Sb	Rc	Qd	Pe	Of	Ng	Mh	Li	Kj
Hn	Go	Fk	Jl	Im	Cs	Bt	Ap	Eq	Dr	Wx	Vy	Uu	Yv	Xw	Rd	Qe	Pa	Tb	Sc	Mi	Lj	Kf	Og	Nh
Fl	Jm	In	Ho	Gk	Aq	Er	Ds	Ct	Bp	Uv	Yw	Xx	Wy	Vu	Pb	Tc	Sd	Re	Qa	Kg	Oh	Ni	Mj	Lf
Io	Hk	Gl	Fm	Jn	Dt	Cp	Bq	Ar	Es	Xy	Wu	Vv	Uw	Yx	Se	Ra	Qb	Pc	Td	Nj	Mf	Lg	Kh	Oi
Gm	Fn	Jo	Ik	Hl	Br	As	Et	Dp	Cq	Vw	Ux	Yy	Xu	Wv	Qc	Pd	Te	Sa	Rb	Lh	Ki	Oj	Nf	Mg
Tu	Sv	Rw	Qx	Py	Oa	Nb	Mc	Ld	Ke	Jf	Ig	Hh	Gi	Fj	Ek	Dl	Cm	Bn	Ao	Yp	Xq	Wr	Vs	Ut
Rx	Qy	Pu	Tv	Sw	Md	Le	Ka	Ob	Nc	Hi	Gj	Ff	Jg	Ih	Cn	Bo	Ak	El	Dm	Ws	Vt	Up	Yq	Xr
Pv	Tw	Sx	Ry	Qu	Kb	Oc	Nd	Me	La	Fg	Jh	Ii	Hj	Gf	Al	Em	Dn	Co	Bk	Uq	Yr	Xs	Wt	Vp
Sy	Ru	Qv	Pw	Tx	Ne	Ma	Lb	Kc	Od	Ij	Hf	Gg	Fh	Ji	Do	Ck	Bl	Am	En	Xt	Wp	Vq	Ur	Ys
Qw	Px	Ty	Su	Rv	Lc	Kd	Oe	Na	Mb	Gh	Fi	Jj	If	Hg	Bm	An	Eo	Dk	Cl	Vr	Us	Yt	Xp	Wq
Ef	Dg	Ch	Bi	Aj	Yk	Xl	Wm	Vn	Uo	Tp	Sq	Rr	Qs	Pt	Ou	Nv	Mw	Lx	Ky	Ja	Ib	Hc	Gd	Fe
Ci	Bj	Af	Eg	Dh	Wn	Vo	Uk	Yl	Xm	Rs	Qt	Pp	Tq	Sr	Mx	Ly	Ku	Ov	Nw	Hd	Ge	Fa	Jb	Ic
Ag	Eh	Di	Cj	Bf	Ul	Ym	Xn	Wo	Vk	Pq	Tr	Ss	Rt	Qp	Kv	Ow	Nx	My	Lu	Fb	Jc	Id	He	Ga
Dj	Cf	Bg	Ah	Ei	Xo	Wk	Vl	Um	Yn	St	Rp	Qq	Pr	Ts	Ny	Mu	Lv	Kw	Ox	Ie	Ha	Gb	Fc	Jd
Bh	Ai	Ej	Df	Cg	Vm	Un	Yo	Xk	Wl	Qr	Ps	Tt	Sp	Rq	Lw	Kx	Oy	Nu	Mv	Gc	Fd	Je	Ia	Hb
Op	Nq	Mr	Ls	Kt	Ju	Iv	Hw	Gx	Fy	Ea	Db	Cc	Bd	Ae	Yf	Xg	Wh	Vi	Uj	Tk	Sl	Rm	Qn	Po
Ms	Lt	Kp	Oq	Nr	Hx	Gy	Fu	Jv	Iw	Cd	Be	Aa	Eb	Dc	Wi	Vj	Uf	Yg	Xh	Rn	Qo	Pk	Tl	Sm
Kq	Or	Ns	Mt	Lp	Fv	Jw	Ix	Hy	Gu	Ab	Ec	Dd	Ce	Ba	Ug	Yh	Xi	Wj	Vf	Pl	Tm	Sn	Ro	Qk
Nt	Mp	Lq	Kr	Os	Iy	Hu	Gv	Fw	Jx	De	Ca	Bb	Ac	Ed	Xj	Wf	Vg	Uh	Yi	So	Rk	Ql	Pm	Tn
Lr	Ks	Ot	Np	Mq	Gw	Fx	Jy	Iu	Hv	Bc	Ad	Ee	Da	Cb	Vh	Ui	Yj	Xf	Wg	Qm	Pn	To	Sk	Rl

Fig. 444a

These literal squares were apparently intended for use as talismans, with the letters of the same *word* (and thus the corresponding sum) appearing in each line, column and main diagonal as in the 4 × 4 square of Fig. 445 (Ja'far = 353, arrangement as in Fig. 446, reading right to left).[253] Similar arrangements are found with the words of sentences.

ر	ف	ع	ج
ع	ج	ر	ف
ج	ع	ف	ر
ف	ر	ج	ع

Fig. 445

---

[253] MS. Mashhad Āstān-i Quds 12235, p. 453.

601	577	553	529	505	481	457	433	409	385	361	337	313	289	265	241	217	193	169	145	121	97	73	49	25
554	530	501	602	578	434	410	381	482	458	314	290	261	362	338	194	170	141	242	218	74	50	21	122	98
502	603	579	555	526	382	483	459	435	406	262	363	339	315	286	142	243	219	195	166	22	123	99	75	46
580	551	527	503	604	460	431	407	383	484	340	311	287	263	364	220	191	167	143	244	100	71	47	23	124
528	504	605	576	552	408	384	485	456	432	288	264	365	336	312	168	144	245	216	192	48	24	125	96	72
236	212	188	164	140	116	92	68	44	20	621	597	573	549	525	476	452	428	404	380	356	332	308	284	260
189	165	136	237	213	69	45	16	117	93	574	550	521	622	598	429	405	376	477	453	309	285	256	357	333
137	238	214	190	161	17	118	94	70	41	522	623	599	575	546	377	478	454	430	401	257	358	334	310	281
215	186	162	138	239	95	66	42	18	119	600	571	547	523	624	455	426	402	378	479	335	306	282	258	359
163	139	240	211	187	43	19	120	91	67	548	524	625	596	572	403	379	480	451	427	283	259	360	331	307
496	472	448	424	400	351	327	303	279	255	231	207	183	159	135	111	87	63	39	15	616	592	568	544	520
449	425	396	497	473	304	280	251	352	328	184	160	131	232	208	64	40	11	112	88	569	545	516	617	593
397	498	474	450	421	252	353	329	305	276	132	233	209	185	156	12	113	89	65	36	517	618	594	570	541
475	446	422	398	499	330	301	277	253	354	210	181	157	133	234	90	61	37	13	114	595	566	542	518	619
423	399	500	471	447	278	254	355	326	302	158	134	235	206	182	38	14	115	86	62	543	519	620	591	567
106	82	58	34	10	611	587	563	539	515	491	467	443	419	395	371	347	323	299	275	226	202	178	154	130
59	35	6	107	83	564	540	511	612	588	444	420	391	492	468	324	300	271	372	348	179	155	126	227	203
7	108	84	60	31	512	613	589	565	536	392	493	469	445	416	272	373	349	325	296	127	228	204	180	151
85	56	32	8	109	590	561	537	513	614	470	441	417	393	494	350	321	297	273	374	205	176	152	128	229
33	9	110	81	57	538	514	615	586	562	418	394	495	466	442	298	274	375	346	322	153	129	230	201	177
366	342	318	294	270	246	222	198	174	150	101	77	53	29	5	606	582	558	534	510	486	462	438	414	390
319	295	266	367	343	199	175	146	247	223	54	30	1	102	78	559	535	506	607	583	439	415	386	487	463
267	368	344	320	291	147	248	224	200	171	2	103	79	55	26	507	608	584	560	531	387	488	464	440	411
345	316	292	268	369	225	196	172	148	249	80	51	27	3	104	585	556	532	508	609	465	436	412	388	489
293	269	370	341	317	173	149	250	221	197	28	4	105	76	52	533	509	610	581	557	413	389	490	461	437

Fig. 444*b*

2. Squares of even orders

First method

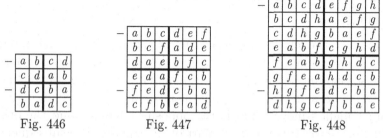

Fig. 446   Fig. 447   Fig. 448

We begin by placing the letters in their order in the top line and, in inverted order, in the penultimate line (Fig. 446-450). These two lines thus arranged will enable us to fill the two halves of the square, except for their last two lines. Consider first the upper half of the squares in figures 447-450 ($n > 4$). We shall fill successively all their lines (except the last) in the same way: the letters in the end cells of a line are written in the median cells of the next, but each in the other quadrant; the remaining letters are placed by queen's moves in direction of the (nearest) side of the square. Just the same is done for the second half, but going upwards from line to line beginning with the penultimate line. We are thus left with filling the last lines in each of the halves.

a	b	c	d	e	f	g	h	i	j	k	l
b	c	d	e	f	l	a	g	h	i	j	k
c	d	e	f	l	k	b	a	g	h	i	j
d	e	f	l	k	j	c	b	a	g	h	i
e	f	l	k	j	i	d	c	b	a	g	h
g	a	b	c	d	h	e	i	j	k	l	f
h	g	a	b	c	d	i	j	k	l	f	e
i	h	g	a	b	c	j	k	l	f	e	d
j	i	h	g	a	b	k	l	f	e	d	c
k	j	i	h	g	a	l	f	e	d	c	b
l	k	j	i	h	g	f	e	d	c	b	a
f	l	k	j	i	e	h	d	c	b	a	g

Fig. 449

a	b	c	d	e	f	g	h	i	j	k	l	m	n
b	c	d	e	f	g	n	a	h	i	j	k	l	m
c	d	e	f	g	n	m	b	a	h	i	j	k	l
d	e	f	g	n	m	l	c	b	a	h	i	j	k
e	f	g	n	m	l	k	d	c	b	a	h	i	j
f	g	n	m	l	k	j	e	d	c	b	a	h	i
h	a	b	c	d	e	i	f	j	k	l	m	n	g
i	h	a	b	c	d	e	j	k	l	m	n	g	f
j	i	h	a	b	c	d	k	l	m	n	g	f	e
k	j	i	h	a	b	c	l	m	n	g	f	e	d
l	k	j	i	h	a	b	m	n	g	f	e	d	c
m	l	k	j	i	h	a	n	g	f	e	d	c	b
n	m	l	k	j	i	h	g	f	e	d	c	b	a
g	n	m	l	k	j	f	i	e	d	c	b	a	h

Fig. 450

The last line of the upper half is filled using the preceding line: the latter's end cells become the median ones of the other, as before, then the remaining cells are filled by writing the letters above in inverse order and by changing quadrant. For the last line of the lower half, we place the medians of the penultimate line in the end cells, but on the other side; the remaining letters are placed, starting from the side, using the queen's move, until only the median cells remain empty; they will then receive the missing letters, using this time the knight's move.

Shabrāmallisī's examples are a square of order 4 (particular case) and one of order 6; we have added those of orders 8, 12 and 14.[254]

Second method

We place the letters in direct order in the first line of the upper half, and in inverse order in the first line of the lower half (Fig. 451-452). We fill the second line of the upper half by moving the letters of the first line: its extreme ones, inverted, become the medians of the second, its medians make a knight's move towards the opposite quadrant and the others, a queen's move towards the outer part of the square. The subsequent lines are filled as in the first method by transposing the letters at the ends of

---

[254] Another configuration for order 4 is to take the lines in the order 1-3-4-2; this second possibility is found in the MS. Mashhad Āstān-i Quds 12235, p. 453. The configurations of order 8 and order 14 are found there, p. 454.

the preceding line to the median cells, each in the other quadrant, the other letters by a queen's move towards the outer part of the square; which completes the upper half of the square. The second half is filled in the same way except that we start there with filling the very last line by means of the first; the remaining lines are filled one by one using the last, just as we proceeded in the other half but now the movement is upward.

Shabrāmallisī has the figure of order 4, identical to the preceding one, and that of order 6. Here too, we have decided to add a further example, that of order 12.

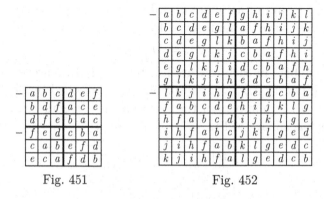

Fig. 451   Fig. 452

§2. Squares with one empty cell

1. Squares of odd orders with central cell empty

A square 'with empty centre' (khālī al-wasaṭa) is possible only from order 5 on and not for all odd orders. Shabrāmallisī explains how to fill such a square as follows.[255] *You fill it by the method of the (move of the) knight, beginning first with the central cell, but without putting anything in it; from there you proceed by the knight's move, upwards or downwards, towards the left or towards the right, and you write 1 in the corresponding knight's cell. From there, using the same move as from the central cell, you place the numbers in succession until you have placed numbers in quantity equal to the order of the square less one. For placing the other sequences of n numbers, we shall remain in the column reached and either go down four cells, as seen earlier (pp. 37-38), or go up one cell, as taught by the author at this point.* The figures 453 and 454 show a square of order 7 filled in these two ways (both are pandiagonal).

---

[255] MS. Paris BNF Arabe 2698, fol. 15ʳ, 14-19 & 16ʳ, 2-5; or below, Appendix 61.

45	14	39	8	33	2	27
12	30	6	24	42	18	36
21	46	15	40	9	34	3
37	13	31		25	43	19
4	22	47	16	41	10	28
20	38	7	32	1	26	44
29	5	23	48	17	35	11

Fig. 453

29	46	7	24	41	2	19
21	38	6	16	33	43	11
20	30	47	8	25	35	3
12	22	39		17	34	44
4	14	31	48	9	26	36
45	13	23	40	1	18	28
37	5	15	32	42	10	27

Fig. 454

The magic sum of a square of order n with empty centre filled with the first $n^2 - 1$ natural numbers will be

$$M_v = M_n - n$$

since the numbers in each row are less by a unit. If, however, we wish to attain the quantity M_n, we shall merely, as mentioned by Shabrāmallisī, increase by n the numbers of the last sequence (*supra*, p. 213). Such is the case in the square of Fig. 455 (pandiagonal).

52	14	39	8	33	2	27
12	30	6	24	49	18	36
21	53	15	40	9	34	3
37	13	31		25	50	19
4	22	54	16	41	10	28
20	38	7	32	1	26	51
29	5	23	55	17	35	11

Fig. 455

This is a remarkable and admirable method, which enables the construction of all squares of this kind with empty centre, except the square of order 3 and those with side divisible by 3, adds Shabrāmallisī. Indeed, filling squares with central cell left empty merely amounts to beginning with 0 in the middle; now, as we have seen (Ch. II, pp. 41-42, 44), the starting cell is fixed for an odd square with order divisible by 3, and therefore cannot be the central cell.

2. Case of the square of order 3

It is not possible to arrange a square of order 3 with an empty central cell and still obtain the usual magic conditions. Indeed (Fig. 456), we would then have $c+f=c+e+h$, thus $f=e+h$, along with $a+h=a+d+f$, thus $h=d+f$ and therefore $d+e=0$. However, a magic arrangement is possible if the empty cell is a lateral one. Shabrāmallisī expresses his gratitude to God for having led him to this construction —which seems to indicate that he either invented it or believed that he had. This rule, explained by means of the usual magic square of order 3, is as follows.[256] *You take the difference between its magic sum (wafq, $M_3 = 15$) and its equalizing number ('idl, $n^2 + 1 = 10$) and divide it into two unequal parts; put them in two corners of one side of the square, and put in the median cell of this side the quantity of the equalizing number; then put the whole aforesaid difference in the central cell.* That is, 5 will be in the central cell, 1 and 4 or 2 and 3 in two consecutive corners and 10 between them (Fig. 457 et 458). It is easy to complete to the known sum $M = 15$ the two lateral rows, leaving empty the bottom cell between the diagonals (its column already contains the magic sum).

Fig. 456

Fig. 457

Fig. 458

We may also have the square occupied by consecutive numbers, which was not the case above. We shall then obtain the smallest possible sum 12 (Fig. 459).[257] (Here too, it is as if we begin with zero in the usual magic square of 3). Obtaining other magic sums is possible: multiples of 15 or 12 by multiplying each number of the corresponding square (Fig. 460, derived from Fig. 457); or quantities of the form $M + 3z$ if the numbers of the second diagonal are increased by z and those of the last sequence, by $2z$. Thus Fig. 461a, with $z = 2$, is derived from Fig. 460, and gives the sum 66, corresponding to the name 'Allah' (a, 1, 1, h = 1, 30, 30, 5) —which may be emphasized by its explicit appearance in the empty cell (Fig. 461b).

66

Fig. 459

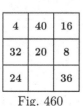
Fig. 460

Fig. 461a

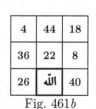
Fig. 461b

[256] MS. Paris BNF Arabe 2698, fol. 16ᵛ, 11-14; or below, Appendix 62. Also *Quelques méthodes*, p. 74.

[257] *Quelques méthodes*, p. 74. Also seen, added by a reader, in the MS. Mashhad Āstān-i Quds 12235, p. 481.

The preferred configuration was, however, with the empty cell in the centre, which could then therefore be occupied by some symbol or word. That is possible, says Sha̲brāmallisī, but at the cost of two magic conditions: it will not be each diagonal but the sum of both which will give the magic sum. To obtain such a square, we shall divide the considered sum M into four different quantities

$$M = m_1 + m_2 + m_3 + m_4$$

which are put in the corners. We are then only to complete the median lateral cells (Fig. 462).

m_1	$m_2 + m_4$	m_3
$m_3 + m_4$		$m_1 + m_2$
m_2	$m_1 + m_3$	m_4

Fig. 462

It is of course necessary for all quantities to be different. A later (18th-century) follower of Sha̲brāmallisī, Ki̲s̲h̲nāwī, chose the examples of our figures 463a-b and 464, with $66 = 13 + 19 + 25 + 9$ and $60 = 9 + 28 + 20 + 3$. Another form of the same arrangement appears in Fig. 465, from which we infer the simpler filling, always applicable for $M \geq 12$, of Fig. 466.[258]

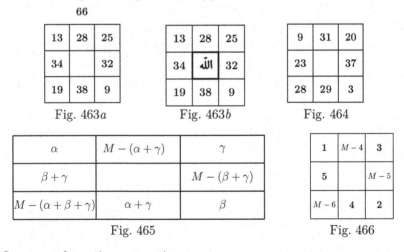

66

13	28	25
34		32
19	38	9

Fig. 463a

13	28	25
34	ﷲ	32
19	38	9

Fig. 463b

9	31	20
23		37
28	29	3

Fig. 464

α	$M - (\alpha + \gamma)$	γ
$\beta + \gamma$		$M - (\beta + \gamma)$
$M - (\alpha + \beta + \gamma)$	$\alpha + \gamma$	β

Fig. 465

1	$M - 4$	3
5		$M - 5$
$M - 6$	4	2

Fig. 466

3. Squares of evenly-even orders

In order to obtain a square of evenly-even order with an empty cell, we shall use one of the methods already taught, where each subsquare of order 4 was a magic entity (p. 79 $seqq.$).

---

[258] *Quelques méthodes*, p. 76.

○	●		
		○	●
●	○		
		●	○

Fig. 467

		○	●
○	●		
		●	○
●	○		

Fig. 468

2	7	12	13
11	14	1	8
5	4	15	10
16	9	6	3

Fig. 469

1	6	11	12
10	13		7
4	3	14	9
15	8	5	2

Fig. 470

Consider thus a square of order 4, to be filled with the sixteen first numbers. The placing of the first two consecutive sequences will follow the arrangement of the figures 467 or 468. We know that any cell may be the starting point; the second number will then occupy the cell bearing the same mark in the adjacent (upper or lower) line, and this will determine the direction of the line-by-line movement, with knight's and queen's moves, while the second sequence will proceed in the other vertical direction; the last two sequences, with the complements to $n^2+1 = 17$, are then put in the corresponding bishop's cells (Fig. 469). Now we shall proceed in just the same way to obtain a square of order 4 with an empty cell (Fig. 470): we shall choose a cell to be left unoccupied and then fill the remaining seven according to the above pattern and, with their complements to $n^2 - 1 = 15$, find what to put in the eight remaining cells (counting the empty cell as zero). The resulting square, with sum 30, will be pandiagonal.

60	59	6	1	52	51	14	9
7		61	58	15	8	53	50
57	62	3	4	49	54	11	12
2	5	56	63	10	13	48	55
44	43	22	17	36	35	30	25
23	16	45	42	31	24	37	34
41	46	19	20	33	38	27	28
18	21	40	47	26	29	32	39

Fig. 471

	7	58	61	8	15	50	53
59	60	1	6	51	52	9	14
5	2	63	56	13	10	55	48
62	57	4	3	54	49	12	11
16	23	42	45	24	31	34	37
43	44	17	22	35	36	25	30
21	18	47	40	29	26	39	32
46	41	20	19	38	33	28	27

Fig. 472

We shall do just the same for a square of order $4k$: we choose the cell to be left empty in any subsquare, which we fill as above, then the other subsquares with the following numbers; lastly, all the complements to $n^2 - 1$ will be put in the bishop's cells of the corresponding subsquares. We shall thus have placed the $n^2 - 1$ first numbers, and the resulting magic sum will be $M_n - n$. In the figures 471 and 472, of order 8 and therefore with magic sum 252, we have left an inner cell and a corner cell empty, and adopted for the first two sequences the configuration of Fig.

468 with, first, an upward movement then that of Fig. 467 with, first, a downward movement. These squares are, like those from which they differ by 1 in each cell, pandiagonal (see p. 83).

Shabrāmallisī teaches the construction of evenly-even order squares with an empty cell, but briefly;[259] after all, these constructions are fairly easy and, with their empty cell not in the centre, must have been considered less attractive than the odd-order squares with the central cell left empty.

Remark. He does not mention evenly-odd orders. A 6 × 6 square with an empty corner cell could be obtained by subtracting 1 from each number in Fig. 331 (p. 214).

§ 3. Squares with divided cells

Shabrāmallisī also constructs squares in which the cells are halved by their diagonals, each cell having then to contain a pair of numbers.[260] We fill the upper halves with the numbers from 1 to n^2 according to some magic arrangement, then do the same in the lower halves with the subsequent numbers, from $n^2 + 1$ to $2n^2$ (Fig. 473 & 474, the latter following the arrangement seen in Fig. 81, p. 46). The resulting squares are not only magic considering similar half-cells but also as a whole since summing two magic squares of the same order cell by cell also makes a magic square. These three magic sums are, respectively,

$$\tfrac{n}{2}\left(n^2 + 1\right), \quad \tfrac{n}{2}\left(3n^2 + 1\right), \quad n\left(2n^2 + 1\right).$$

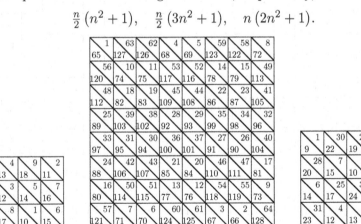

Fig. 473   Fig. 474   Fig. 475

We may also, if the order is evenly-even, start by filling half of the divided cells with the first half of the numbers leaving empty the corre-

[259] MS. Paris BNF Arabe 2698, fol. 48^{v}-49^{v}.

[260] MS. Paris BNF Arabe 2698, fol. 63^{r}-63^{v}.

sponding bishop's cells, then fill the latter with the complements of the half-cells already filled (Fig. 475; 1 is thus associated with 32, 2 with 31, and so on). In that case, the magic sum in each row of half-cells is

$$\tfrac{n}{2}\left(2n^2 + 1\right)$$

and twice that for the rows of whole cells.

These three examples are those of Shabrāmallisī; in the last, the magic sum in each row of half-cells is 66 ('Allah'). Since the two half-cell squares are pandiagonal, each broken diagonal of corresponding half-cells will also display the sum 66, and twice that if we consider all eight cells. This square led to a more elaborate pattern (Fig. 476a), successful enough even to be found in a 14th-century Sanskrit text by Nārāyaṇa (Fig. 476b).[261]

Fig. 476a   Fig. 476b

Note. Both the French and Russian editions have at this point a short history of the knight's move on the chessboard, here omitted since now superseded by our edition of Euler's manuscript notes on the subject.

§4. Magic triangles

Consider an equilateral (or isosceles) triangle and divide its sides into n equal parts, then draw parallels to the sides through each point of section. This gives, starting from the top, quantities of cells according to the succession of the odd numbers, thus altogether n^2 cells.

Accordingly, in the simplest case, a magic triangle of order 3 comprises nine cells in which the nine first numbers must be arranged in such a way that the sum on each side and in each bisector be the same. In Fig. 477 this means fulfilling the six conditions $a + b + c = a + f + i = c + g + i = a + d + g = c + e + f = i + h + b$. This is less than what is required for a magic square of order 3 since the conditions $h + a + e$ and $b + d + f$ are omitted (Fig. 478). Such are the usual conditions. Now Fig. 479 and 480 show that we could have a magic triangle fulfilling all the conditions of

---

[261] Another, similar pattern of his is reproduced in Abe's study, p. 41.

the 3 × 3 magic square. However, the rôle of the triangle is precisely to enlarge the possibility of a magic placement, in particular for three-letter names or triads of words not representable in a 3 × 3 square.

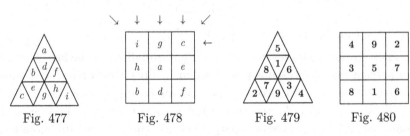

Fig. 477   Fig. 478   Fig. 479   Fig. 480

The following three numerical examples occur in Persian texts already mentioned.[262] They are more computing exercises than constructions: three of the elements must be determined in accordance with the six magic conditions of the triangle.

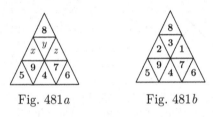

Fig. 481a   Fig. 481b

Example 1 (Fig. 481a). Here six elements and the magic sum $5+4+6 = 15$ are known. We immediately infer that $x = 2$, $y = 3$, $z = 1$ (Fig. 481b). They are thus the elements b, d, f in the figures 477 and 478.

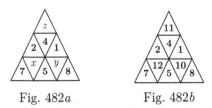

Fig. 482a   Fig. 482b

Example 2 (Fig. 482a; z common to three rows). Here too six elements and the magic sum are known, the latter being $7 + 5 + 8 = 20$ (هادی = 5, 1, 4, 10, not suitable for a 3 × 3 or a 4 × 4 square since the sum is not divisible by 3 and less than 34). So we shall have $z = 11$, $x = 12$, $y = 10$ (Fig. 482b). This time we have determined the elements h, a, e in the figures 477 and 478 (thus the other non-magic triad).

---

[262] MSS. Istanbul Ayasofya 2794, fol. 24ᵛ-25ᵛ, and Mashhad Āstān-i Quds 12235, pp. 446-447, & 14159, fol. 90ʳ.

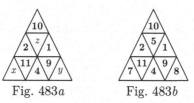

Fig. 483a   Fig. 483b

Example 3 (Fig. 483a). Six elements are known, but not the magic sum S. Since

$$S = x + 4 + y = x + 12 = y + 11 = z + 14,$$

we shall have $y = 8$, $x = 7$, $z = 5$, $S = 19$ (Fig. 483b).

We may also construct triangular talismans, which are of particular appeal. Thus the name 'Alī (with sum $70 + 30 + 10 = 110$) being set and put, for example, in the central column (Fig. 484), we may complete the triangle as in Fig. 485a-b.[263] For this, choose one element, say c, whence $b = 100 - c$, $i = 40 - c$, $f = 60 + c$, $e = 50 - 2c$, $h = 2c - 30$ (thus $15 < c < 25$). Another example, this time for a phrase, is the placing of هو الغنى الحميد (= 11, 1091, 93, adding up to 1195, not divisible by 3) in the base of the triangle (Fig. 486).[264] Here too one appropriately chosen element enables us to complete the triangle.

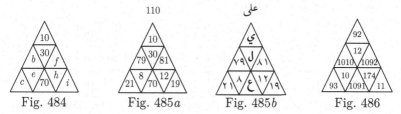

Fig. 484   Fig. 485a   Fig. 485b   Fig. 486

Remark. Some manuscripts feature two triangles combined to make a star.

§5. Magic crosses

We have already seen an example of such a figure (p. 97). Filled with the first natural numbers, the smallest possible cross of this kind will be as in Fig. 487, and display the same sum 63 in the six longer rows, and 21 in any pair of adjacent cells (vertically or horizontally opposite; in the centre, diagonally). Now the branches of such a cross can be repeatedly extended by four cells, to be filled then by neutral placings, the numbers in the original cross being accordingly increased, just as for the construction of the successive frames in a bordered square (in this case, adding $2t$ cells

---

[263] Example in the MS. Mashhad Āstān-i Quds 12235, p. 449.

[264] MS. Mashhad Āstān-i Quds 12235, p. 449, or 14159, fol. 90ᵛ.

means an increment of t). Then, if the cross has altogether n cells filled
with the numbers from 1 to n, each tranversal pair of cells will contain
the quantity $n + 1$, and $\frac{m}{2}(n + 1)$ for the longer rows with m cells.

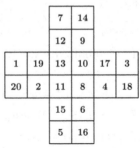

Fig. 487

This configuration would hardly be worth mentioning were its most
elementary example not frequently encountered in magical and astrologi-
cal treatises. It consists of five elements a, b, c, d, e, with c in the centre
of the cross. We must then have $a + b = d + e$ (Fig. 488). With the first
natural numbers, we may put in the middle 1 (Fig. 489), 3 or 5. We can
also place quantities representing a name, like 'Alī with 70, 30, 10 (Fig.
490). Thus here again it became possible to represent a three-letter name
unsuitable for the 3×3 square.

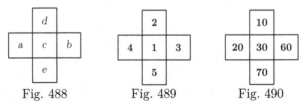

Fig. 488   Fig. 489   Fig. 490

§6. Magic circles

1. First type

Shabrāmallisī describes the following arrangement, which changes a
given magic square (Fig. 491) into a magic circle (*wafq mustadīr*).[265]
*You draw, around the same centre, circles in a quantity equal to the
order of the square and equidistant from one another. Then you divide
the circumference of the large circle into as many parts as the order of
the square, and you join the centre to the places of section with segments
of straight line. After that, you place*
*— the numbers of the first row of the square, either line or column, in
the cells around the centre which are within the small circle;*

---

[265] MS. Paris BNF Arabe 2698, fol. 64^r, 19 - 64^v, 18; or below, Appendix 63. As in
the Arabic text the rows are numbered from right to left.

— the numbers of the second row, next to the first, in the cells located between the small circle and the next, which is the second circle;

— the numbers of the third row between the second circle and the third circle;

you proceed likewise until you have placed the numbers of the last row in the cells between the large circle and its neighbour on the side of the centre.

22	47	16	41	10	35	4
5	23	48	17	42	11	29
30	6	24	49	18	36	12
13	31	7	25	43	19	37
38	14	32	1	26	44	20
21	39	8	33	2	27	45
46	15	40	9	34	3	28

Fig. 491

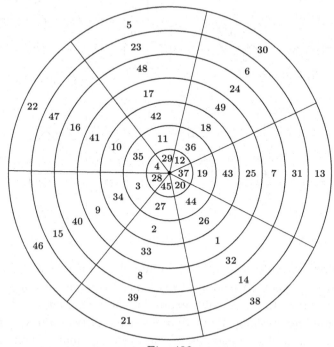

Fig. 492

Thus, in Fig. 492 (Shabrāmallisī's example), the numbers in the columns of the magic square have been placed successively, starting from the first right-hand column of the square, in successive rings from the inner circle. Therefore, to the columns of the square correspond the rings while to the lines correspond the sectors. The magic sum is thus found circularly and radially.

Remark. In fact, since the condition of the two main diagonals is now irrelevant, we may begin in the central circle with numbers taken from any column in the square, and take the columns in any order.

2. Second type

Whereas in the previous case the rings corresponded to the columns and thus all had the same number of cells, they will now contain the successive borders and thus have unequal numbers of cells. If the squares are of odd order, the circle in the centre will contain a single element, or none if the magic square at the outset was a square with an empty central cell. For an even square, we shall have four numbers in the centre. Shabrāmallisī's three examples are those of the figures 493 to 495.

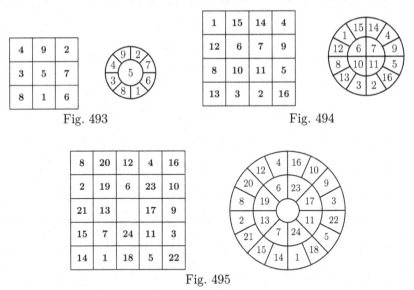

Fig. 493   Fig. 494

Fig. 495

The first example is clear: three numbers diametrically aligned make the magic sum (they correspond to the diagonals and medial rows of the original square). In the second example, the magic sum is found, first, in the diameters crossing the median elements of two opposite quadrants (corresponding to the diagonals of the original square), next along the separations of the quadrants (corresponding to the median rows of the square) and finally in the four outer cells situated on both sides of an axis (corresponding to the lateral rows of the square); to this may be added that the sum in each quadrant of the circle is magic, as was the case in the quadrants of the original square (which had 'equal quadrants', see p. 226). Shabrāmallisī's third figure illustrates, better than the other two, why such arrangements are of rather limited interest.

§7. Magic rectangles

We are to place, in a rectangle with n cells in length and m cells in width, $m \cdot n$ different natural numbers in such a way that each of the m lines contains the same sum and each of the n columns likewise. If the numbers taken are the first natural ones, thus from 1 to $m \cdot n$, the sum on the whole rectangle will be

$$\frac{mn(mn + 1)}{2}$$

and we shall then find the 'magic' sums

$$M_{\text{vert.}} = \frac{m\,(mn + 1)}{2} \quad \text{in each column}$$

$$M_{\text{hor.}} = \frac{n\,(mn + 1)}{2} \quad \text{in each line.}$$

Since both $m\,(mn + 1)$ and $n\,(mn + 1)$ must be even, a rectangle can be magic only if m and n have the same parity. Indeed, if m and n are both even, the two products will be divisible by 2, which will also be the case if m and n are both odd since mn will be odd and thus $mn + 1$ even. On the other hand, if m and n have different parity, mn will be even, hence $mn + 1$ odd, thus one of its products by m and n also.

Filling such rectangles is briefly explained in Thābit's (and Antākī's) text.[266] Since this has to do with the construction of evenly-odd squares by means of a cross and a central square (above, Ch. II, §§ 26 & 28), all the rectangles considered have their sides even. Thus, the first case we shall consider below would have been out of place there, and is treated here merely for the sake of completeness.

1. Both sides are odd

Constructing such a rectangle by trial and error is not difficult. We first arrange in a rectangle of the size desired, more or less equitably, the $m \cdot n$ numbers to be placed; moving them will first enable us to reach the

---

[266] *Magic squares in the tenth century*, A.II.49, and MS. London BL Delhi Arabic 110, fol. 99^{v}, 22 - 100^{r}, 4.

required sums in the columns, then we shall equalize the lines without changing the sums in the columns.

Example (Fig. 496a-c): $m = 3$, $n = 5$, $M_{\text{vert.}} = 24$, $M_{\text{hor.}} = 40$.

21	24	21	27	27
1	7	10	4	13
14	5	2	8	11
6	12	9	15	3

Fig. 496*a*

4	7	13	1	10	35
14	5	2	8	11	40
6	12	9	15	3	45

Fig. 496*b*

4	12	13	1	10
14	5	2	8	11
6	7	9	15	3

Fig. 496*c*

2. Both sides are evenly even

Since each side is divisible by 4, we shall fill the rectangle by pairs of opposite rows, horizontal or vertical, with tetrads or groups of $\frac{n}{4}$ or $\frac{m}{4}$ numbers serving as extremes and means (Fig. 497-498).

1	2	30	29	28	27	7	8
9	10	22	21	20	19	15	16
24	23	11	12	13	14	18	17
32	31	3	4	5	6	26	25

Fig. 497

1	5	9	13	20	24	28	32
31	27	23	19	14	10	6	2
30	26	22	18	15	11	7	3
4	8	12	16	17	21	25	29

Fig. 498

Remark. Reversing in Fig. 497 the median lines we obtain (Fig. 499) another form of magic rectangle, namely the easy placing by continuous filling (see Fig. 100*b*, p. 51).

1	2	30	29	28	27	7	8
16	15	19	20	21	22	10	9
17	18	14	13	12	11	23	24
32	31	3	4	5	6	26	25

Fig. 499

3. One side evenly even and the other evenly odd

1	7	6	4
8	2	3	5

Fig. 500

1	5	9	16	20	24
23	19	15	10	6	2
22	18	14	11	7	3
4	8	12	13	17	21

Fig. 501

Then one of the sides will contain $4r$ cells and the other 2 or $4s + 2$ cells. This case is reducible to the previous one by using the rows of length $4r$ for placing tetrads, or groups of r numbers taken as extremes and means (Fig. 500-501).

4. Both sides are evenly odd

Suppose this time that one of the sides contains 2 or $4r + 2$ cells and the other $4s + 2$ cells. Consider first the simplest case, that of a rectangle with dimensions $m = 2$ and $n = 6$. What we have seen in the method of the cross (p. 95 *seqq.*) is applicable here, except that we may now disregard the particular placing in the median cells since there is no condition for the diagonals. We shall thus fill accordingly the strip of Fig. 502 (cf. the vertical strip in the cross of Fig. 173, where 93 takes the place of 1).

1	11	9	8	7	3
12	2	4	5	6	10

Fig. 502

This solves, as for the cross, the case of a strip $4s + 2$ long with $s > 1$, since we are left with putting in only tetrads (Fig. 503).

1	19	5	15	13	12	11	7	18	4
20	2	16	6	8	9	10	14	3	17

Fig. 503

The case of a rectangle with length $4s + 2$ with $s \geq 1$ and width $4r + 2$, $r \neq 0$, is inferred thereby, for we need only to fill the central strip separately, with pairs taken from the median numbers, the remainder then being filled with tetrads (Fig. 504).

1	5	9	28	32	36
35	31	27	10	6	2
13	23	21	20	19	15
24	14	16	17	18	22
34	30	26	11	7	3
4	8	12	25	29	33

Fig. 504

§8. Magic cubes

A magic cube (*wafq mujassam*) of (say) order n is the arrangement of n^3 numbers in such a way that the magic sum is found in all the lines and columns. With the first natural numbers the magic sum will be $M_n^c = \frac{n}{2}(n^3 + 1)$. Several constructions are reported, perhaps all by Khāzinī (twelfth century).[267] In one of them, for $n = 4$, thus with $M_4^c = 130$, the vertical layers are as in Fig. 505-508 (Fig. 505 is thus the front). The horizontal layers, starting from the upper one, are thus as in Fig. 509-512 and those in depth as in Fig. 513-516.

1	48	31	50		54	27	44	5		59	22	37	12		16	33	18	63
32	49	2	47		43	6	53	28		38	11	60	21		17	64	15	34
46	3	52	29		25	56	7	42		24	57	10	39		35	14	61	20
51	30	45	4		8	41	26	55		9	40	23	58		62	19	36	13

Fig. 505   Fig. 506   Fig. 507   Fig. 508

16	33	18	63		17	64	15	34		35	14	61	20		62	19	36	13
59	22	37	12		38	11	60	21		24	57	10	39		9	40	23	58
54	27	44	5		43	6	53	28		25	56	7	42		8	41	26	55
1	48	31	50		32	49	2	47		46	3	52	29		51	30	45	4

Fig. 509   Fig. 510   Fig. 511   Fig. 512

16	59	54	1		33	22	27	48		18	37	44	31		63	12	5	50
17	38	43	32		64	11	6	49		15	60	53	2		34	21	28	47
35	24	25	46		14	57	56	3		61	10	7	52		20	39	42	29
62	9	8	51		19	40	41	30		36	23	26	45		13	58	55	4

Fig. 513   Fig. 514   Fig. 515   Fig. 516

Since each of the 48 quadrants of these twelve squares makes the sum 130, each of the eight $2 \times 2 \times 2$ cubes forming the whole cube adds up to 260.

Remark. Here, unlike in the magic cubes considered today, the transverse diagonals do not make the magic sum.

---

[267] MS. London BL Delhi Arabic 110, fol. 109^{v} - 116^{v}.

Appendices

(1) كان اول معرفتى بامر العدد الوفق جدول الثلثة الذى ذكر نيقوماخس فى الارثماطيقى. ثمّ وقع الى ابى القاسم الحجازى جدول الاربعة الذى يبتدئ من الواحد بتفاضل واحد واحد الى ستّة عشر وكان تعجّب منه. ثمّ وجدتُ جدول الستّة على ظهر كتاب اقليدس نسخَه اسحاق. ثمّ وقع الىّ كتاب فيه ثلثة او اربعة جداول ممّا دون العشرة. ثمّ وجدتُ فى خزانة من كتب شيوخنا كتابين قد آتت الارضة على اكثرهما حتى لا يفهم منهما الا اليسير وكان المختصر منهما بخط الماهانى والورقة الاولى من الاكثر بخط الحسين بن موسى النوبختى. فنظرتُ فيهما فوجدتُ استخراجهما متعبًا جدًّا ولاح لى انّه يتهيّأ ان يستخرج بعض ما تلف من كلّ واحد منهما بما سلم من صاحبه.

(2) Figura Iovis quadrata est, et sunt 4^{or} multiplicata per 4, et sunt in quolibet latere 34. Cum volueris operari per eam, facies laminam argenteam in die et in hora Iovis, et sit Iupiter in bono statu; et in dicta lamina sculpes figuram et suffumigabis eam cum ligno aloes et cum ambra, et portabis tecum. Et diligent te qui viderunt te et impetrabis ab eis quod quesieris. Et si posueris eam in repositorio mercatoris, mercatura eius augmentabitur. Et si posueris eam in columbario vel ubi sunt apes, congregabuntur ibi columbe et apes. Et si quis portaverit eam qui sit infortunatus, fortunabitur, de bono in melius proficiet. Et si posueris eam in sede prelati, prelatio eius durabitur et suos inimicos non timebit, sed proficiet apud eos.

16	3	2	13
5	10	11	8
9	6	7	12
4	15	14	1

14	10	1	22	18
20	11	7	3	24
21	17	13	9	5
2	23	19	15	6
8	4	25	16	12

Figura Martis infortunati significat bellum et graverias, et est figura quadrata, et sunt 5 multiplicata per 5, et sunt in quolibet latere 65.

J. Sesiano, *Magic Squares*, Sources and Studies in the History of Mathematics and Physical Sciences, https://doi.org/10.1007/978-3-030-17993-9

Cum volueris operari per eam, accipe laminam cupri in die et hora
Martis cum Mars fuerit diminutus numero vel lumine, vel infortunatus
et retrogradus vel in aliquo malo statu, et sculpes ipsam in lamina; et
suffumigabis ipsam cum stercoribus murium vel murilegorum. Et si posueris eam in edificio novo, non complebitur. Et si posueris eam in sede
prelati, deteriorabitur cottidie et infortunabitur. Et si posueris eam
in operatorio mercatoris, totum destruetur. Et si feceris predictam figuram cum nomine duorum mercatorum et subterraveris eam in domo
unius illorum, cadet inter eos odium et inimicitia. Et si forte timorem
de rege vel de aliquo magnato habueris, vel de inimicis, vel ad iudicium
vel ante iudicem intrare volueris, sculpe hanc figuram ut supra dictum
est, et sit Mars fortunatus, directus, auctus numero et lumine, et suffu-
miga eam cum dragma 1 de lapide corneola; et ponatur predicta lamina
in panno serico rubeo et portabis eam tecum; tunc vinces in iudicio et
in bello adversarios tuos, et fugient a facie tua et timebunt te et revere-
buntur. Et si posueris eam super crus mulieris, continuo fluxum sangui-
nis patietur. Et si scripseris eam in pergameno die et hora Martis et
suffumigabis eam cum aristologia et eam posueris in loco apum, fugient
omnes apes.

(3) فمن الناس مَن يبتدىء بوضع هذه الاعداد على توالى النظم الطبيعى من
الواحد الى حيث بلغت عدّة مربّعات هذه الصورة التى يريد ان يرسم اعداد
الوفق فيها ثمّ ينقل اعدادها ابدًا بالزيادة فى بعض السطور والنقصان فى السطور
المضادّة لها ثمّ ينتسق جميع السطور على أمر واحد. وهذا باب فيه صعوبة على
المبتدىء.   ومن الناس مَن يعمله على وجه آخر اسهل من هذا.

(4) فامّا ما ذكره ابو علي بن الهيثم (...) فان كان عدد ضلع المربّع فردًا فانّه اذا
أُثبت فيه العدد على النظم الطبيعى من الواحد الى آخر عدد يحتوى عليه المربّع
الاعظم الذى هو العدد من المربّعات الصغار الذى حازها فانّه يوجد ما فى
قطريه متساويًا ويوجد ما فى كلّ قطرين عن جنبتى القطر الاعظم عدد بيوتهما
مثل عدد بيوته مثل ما فى القطر الاعظم وهذا خاصّة من خواصّ العدد الطبيعية
لهذه الاعداد اذا أُثبتت فى هذه المربّعات على النظم الطبيعى.
وهذا مثال ما ذكر فى هذا الجدول قد قُسم ضلعه بخمسة اقسام فانقسم
بخمسة وعشرين مربّعًا وكتب فيها العدد من الواحد الى خمسة وعشرين على
النظم الطبيعى فاذا عدّ ما فى احد القطرين الاعظمين الذى فيه أ ز يج يط

كه كان مجموعها خمسة وستّين واذا عدّ ما فى القطر الآخر وهو ه ط يج يز
كا كان ايضًا خمسة وستّين فاذا اعتمد على كلّ قطرين عن جنبتى هذين القطرين
عدّتهما مثل عدّته وجد فيهما مثل ما فيه وذلك اتّه اذا جُمع ما فى القطر الذى
فيه ب ح يد كا وهو اربعة بيوت واضيف اليه ما فى بيت الزاوية التى فيها
كا كان مجموع البيوت خمسة وهى كعدّة ابيات القطر فكان العدد ايضًا خمسة
وستّين وكذلك اذا جُمع ما فى القطر الذى فيه ج ط يه وهو ثلثة ابيات
واضيف اليه من الجانب الآخر ما فى القطر الذى هو بيتان الذى فيهما يو
كب كان ذلك ايضًا خمسة وستّين وكذلك ان أخذ ما فى القطر الذى فيه د ى
واضيف اليه من الجانب الآخر ما فى القطر الذى هو ثلثة ابيات الذى فيها يا
يز كج كان مجموع ذلك خمسة وستّين وكذلك اذا اعتبرتَ الابيات التى عن جنبتى
القطر الآخر وجدتَ كذلك.

ه	د	ج	ب	ا
ي	ط	ح	ز	و
يه	يد	يج	يب	يا
ك	يط	يح	يز	يو
كه	كد	كج	كب	كا

فهو لمّا وجد هذه الخاصّة لازمة فى هذه الجداول التى اضلاعها عدد فرد أمر
ان يُرسم مربّعان ويثبت فى احدهما العدد على النظم الطبيعى ويثبت ما فى
السطرين الاوسطين من سطور الطول والعرض فى القطرين من المربّع الآخر
ويعتمد نقل ما فى الاقطار الباقية الى نظائرها بشرائط طويلة يطول ذكرُها
ويصعب على المبتدىء عملُها.

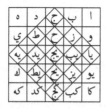

(5) فاستنبطتُ من هذه الطريقة طريقة كلّيّة لكلّ مربّع ضلعه عدد فرد كائنًا ما
كان وهى هذه وذاك ان كلّ مربّع عدد ضلعه فرد فانّه اذا قُسم ضلعه بنصفين
فان القسمة تقع فى وسط البيت الاوسط فنخطّ خطوطًا من كلّ نقطة فى نصف

الضلع الى الاخرى من الضلع القائم عليه لحدث مربّع على السوك ونجد اضلاع
هذا المربّع الذى على السوك تقاطع اضلاع المربّعات التى فى المربّع الاوّل فنخطّ
خطوطاً من كلّ نقطة الى نظيرها من الضلع المقابل له فيحدث فى المربّع الذى
على السوك ايضًا خمسة وعشرون مربّعًا فاذا كنّا قد كتبنا فى المربّع الاوّل العدد
على النظم الطبيعى نجد قد وقع منها فى بيوت المربّع الذى على السوك ثلثة عشر
عددًا وقد بقى من بيوته اثنا عشر بيتًا لها اقطار ونجد قد فصل من المربّع الاوّل
اربعة مثلّثات كلّ واحد منها يحيط به ضلع من اضلاع المربّع الذى على السوك
ونصفا ضلعين من المربّع الاكبر فاذا توهّمنا اتّا قد رفعنا كلّ مثلّث من هذه
المثلّثات ووضعنا ضلعه الذى هو احد اضلاع المربّع الاصغر على ضلعه المقابل له
فاتّا نجد الثلثة الاحرف المثبتة فيه تقع على ثلثة بيوت من بيوت المربّع الاصغر
التى لها الاقطار فاذا اثبتنا ما فى كلّ بيت فى البيت الذى ينطبق عليه تمّ فى
المربّع الاصغر العدد الوفق. وهذا قانون مطرد فى جميع المربّعات التى ضلعها عدد
فرد.

(6) Mr. Vincent dont j'ay souvent parlé dans ma Relation, me voyant un jour, dans le Vaisseau pendant nôtre retour, ranger par amusement des quarrés magiques à la maniere de Bachet, me dit que les Indiens de Suratte les rangeoient avec bien plus de facilité, & m'enseigna leur methode pour les quarrés impairs seulement, ayant, disoit-il, oublié celle des pairs.

Le premier quarré qui est celuy de 9 cases revenoit au quarré d'Agrippa, il estoit seulement renversé; mais les autres quarrés impairs estoient essentiellement differents de ceux d'Agrippa. Il rangeoit les nombres dans les cases tout d'un coup, & sans hésiter, & j'espere qu'on ne desapprouvera pas que je donne les regles, & la demonstration de cette methode, qui est surprenante par son extreme facilité à executer une chose, qui a paru difficile à tous nos Mathematiciens.

1°. Aprés avoir divisé le quarré total en ses petits quarrez, on y place les nombres selon leur ordre naturel, je veux dire en commençant par l'unité, & en continüant par 2, 3, 4, & par tous les autres nombres de suite, & l'on place l'unité, ou le premier nombre de la progression arithmetique donnée, à la case du milieu du gisant d'en haut.

2°. Quand on a mis un nombre dans la plus haute case d'un montant, on met le nombre suivant dans la plus bas(s)e case du montant qui suit vers la droite; c'est à dire que du gisant d'en haut on descend tout d'un coup à celuy d'en bas.

3°. Quand on a placé un nombre dans la derniere case d'un gisant, on place le suivant dans la premiere case du gisant immediatement superieur, c'est à dire que du dernier montant à droit on revient tout d'un coup à gauche au premier montant.

4°. En toute autre rencontre aprés avoir placé un nombre, on place les suivants dans les cases qui suivent diametralement ou en écharpe de bas en haut & de la gauche à la droite, jusqu'à ce qu'on arrive à l'une des cases du gisant d'en haut, ou du dernier montant à droit.

5°. Quand on trouve le chemin bouché par quelque case déja remplie de quelque nombre, alors on prend la case immediatement au dessous de celle qu'on vient de remplir, & l'on continuë comme auparavant diametralement de bas en haut & de la gauche à la droite.

Ce peu de regles aisées à retenir suffisent à ranger tous les quarrés impairs generalement.

(7) Μέθοδος ἑτέρα, διὰ δὲ τῶν τριῶν καὶ πέντε, οὕτως. Ἀναγράφομεν τετράγωνον, καὶ περιγράφομεν αὐτῷ τοὺς τόπους τοῦ τετραγώνου ἀριθμοῦ· εἶτα τιθέαμεν τὴν μονάδα ἀεὶ ἐπὶ τοῦ μέσου τῶν ἀνωτάτω τόπων· καὶ μετροῦμεν τόπους τρεῖς, ἕνα τὸν ἔχοντα τὴν μονάδα, καὶ δύο κατωτέρω τούτου ἐφεξῆς· καὶ τιθέαμεν ἐπὶ τῷ δεξιῷ τοῦ τρίτου κατ' εὐθεῖαν τὰ β̄· εἶτα πάλιν ἐκεῖθεν μετροῦμεν τρεῖς τόπους ὁμοίως· καὶ τιθέαμεν ἐπὶ τῷ δεξιῷ τὰ γ̄. Εἰ δὲ μὴ ἔχομεν ἐπὶ τῶν δεξιῶν τόπον, ἀναστρέφομεν ἐπὶ τὰ ἀριστερὰ κατ' εὐθεῖαν, ὥσπερ ἐπὶ τῆς προτέρας μεθόδου (...). Καὶ τοῦτο ποιοῦμεν ἕως ἂν ἔλθωμεν ἐπὶ τὴν πλευρὰν τοῦ προχειμένου τετραγώνου· ἐπ' ἐκείνην γὰρ ἀφιγμένοι, μετροῦμεν πέντε τόπους, ἕνα τὸν ἔχοντα τὴν πλευράν, καὶ τέσσαρας κατωτέρω τούτου· εἶτα τιθέαμεν ἐπὶ τῷ πέμπτῳ τόπῳ, μὴ παρεκκλίνοντες, τὸν ἐφεξῆς ἀριθμὸν τῆς πλευρᾶς· εἶτα πάλιν μετροῦμεν διὰ τῶν τριῶν μέχρι τῆς πλευρᾶς, ἀνακυκλοῦντες τοὺς τόπους, ὥσπερ ἐπὶ τῆς προτέρας μεθόδου· καὶ τοῦτο μέχρι τέλους ποιοῦμεν. Ἔστι δὲ αὕτη ἡ μέθοδος κατὰ πάντα ὁμοία τῇ προτέρᾳ, πλὴν ὅτι ἐκεῖ μὲν ἡ μονὰς ἐν ἄλλῳ καὶ ἄλλῳ τόπῳ ἐτίθετο, ἐνταῦθα δὲ ἀεὶ ἐπὶ τοῦ μέσου τῶν ἀνωτάτω τόπων· καὶ ὅτι ἐκεῖ μὲν ἐμετροῦμεν διὰ τῶν δύο καὶ τριῶν, ἐνταῦθα δὲ διὰ τῶν τριῶν καὶ πέντε.

(8) فامّا العدد الزوج قال ابن الهيثم فيه قولاً كلّيًا وجعله لجميع انواع الزوج وذاك ان لهذه المربّعات التى اضلاعها مقسومة بعدد زوج خاصّة اخرى غير تلك وهو انّه اذا أُثبت فيها العدد على النظم الطبيعى وُجد ما فى نصفى كلّ سطرين بعدُهما عن الوسط بعد واحد من سطور الطول والعرض اذا جُمعا على التبادل

وجد ما فيهما مثل ما فى كلّ واحد من القطرين.

(9) ونبتدىء بنقلها بالاعداد التى فى مربّع الاربعة. وننقل كـ الى بيت كا وكا الى بيت مد ومد الى بيت مه ومه الى بيت كـ ثمّ ننقل ايضًا لـ الى بيت كز وكز الى بيت لح ولح الى بيت لـ ول الى بيت له. ثمّ ننقل ايضًا يب الى بيت نج ونج الى بيت يج ويج الى بيت نب ونب الى بيت يب ونقلنا ايضًا كو الى بيت لا ولا الى بيت لد ولد الى بيت لط ولط الى بيت كو ثمّ نقلنا يا الى بيت ند وند الى بيت نا ونا الى بيت يد ويد الى بيت يا ونقلنا يح الى بيت مب ومب الى بيت كج وكج الى بيت مز ومز الى بيت يح فتصير اعداد مربّع الستّة الذى فى وسط مربّع الثمنية كلّها وفقًا ومبلغ ما فى كلّ واحد من الصفوف مائة وخمسة وتسعون (...).

ح	سج	سب	د	ه	نط	نح	ا
مط	ى	نا	نج	يد	يا	نب	يو
كد	مز	يط	كا	كب	مه	كو	يز
لج	لا	له	لط	لد	ل	لو	لب
كه	لح	لز	كح	كط	لز	لد	م
يز	كه	مد	كج	كا	مو	كب	مح
نو	يه	يا	نب	نا	يد	ن	مط
سد	ب	س	سا	و	نط	ز	نز

ح	ز	و	ه	د	ج	ب	ا
يو	يه	يد	يج	يب	يا	ى	ط
كد	كج	كب	كا	ك	يط	يح	يز
لب	لا	ل	كط	كح	كز	كو	كه
م	لط	لح	لز	لو	له	لد	لج
مح	مز	مو	مه	مد	مج	مب	ما
نو	نه	ند	نج	نب	نا	ن	مط
سد	سج	سب	سا	س	نط	نح	نز

ثمّ نقلنا ه الى بيت دـ ودـ الى بيت ه وتركنا س وسا فى موضعهما ثمّ نقلنا بـ الى بيت سج وسج الى بيت زـ وزـ الى بيت نح ونح الى بيت بـ ثمّ نقلنا على هذا القياس جـ الى بيت سب وسب الى بيت وـ ووـ الى بيت نط ونط الى بيت جـ فيصير جميع صفوف الطول الاوساط وفقًا عددها موافق لما ينبغى ان يكون فى كلّ صفّ منها وهو مائتان وستّون ويصير ايضًا الصفّ الاوّل والصفّ الثامن من صفوف العرض وفقًا عدد ما فيهما مائتان وستّون. ثمّ نقلنا كه الى بيت لج ولج الى بيت كه وتركنا لب م فيما بيتهما ثمّ نقلنا البيوت فى الصفّ الاوّل والثامن من صفوف الطول على قياس ما نقلنا البيوت فى الصفّ الاوّل والثامن من صفوف العرض وذلك انّا نقلنا مط الى بيت يو ويو الى بيت ط وط الى بيت نو ونو الى بيت مط ثمّ نقلنا ايضًا ما الى بيت كد وكد الى بيت يز ويز الى بيت مح ومح الى بيت ما. فتعتدل اعداد جميع الصفوف

من هذا المربّع.

(10) تضع الطرف الاصغر والواسطة الصغرى فى طرفى الصفّ الاعلى وما يليانهما صاعدًا من الطرف ونازلاً من الواسطة فى بيتى فرسيهما من الصفّ الثانى ثمّ من الصفّ الثالث ثمّ من الصفّ الرابع واذا تعذّر الانتقال من صفّ الى صفّ اسفل منه بنقل الفرس انتقلتَ منه اليه بنقل الفرزان ثمّ تعود الى نقل الفرس وينعكس الوراب من التيامن الى التياسر ومن التياسر الى التيامن واذا انتهيتَ بهما الى الصفّ الاخير انتقلتَ بكلّ واحد منهما الى الصفّ الاوّل الى القطر الثانى من مربّع ٤ الذى وضعتَ مبدأه فى قطره الاوّل وتمضى بهما على عكس وراب المبدأ الاوّل الى الصفّ الاخير فان انتهت الاعداد القليلة فذاك والّا عدتَ بهما الى القطر الاوّل من مربّع ٤ الثانى الذى يلى المربّع الاوّل وتمضى بهما على الوراب الاوّل الى الصفّ الاخير فان انتهت الاعداد القليلة فذاك والّا عدتَ بهما الى القطر الثانى من مربّع ٤ الثانى (...). ولا تزال تفعل كذلك حتّى تأتى على نصف بيوت المربّع وكلّما زاد فى طول المربّع اربعة بيوت زاد الانتقال من اسفل الى فوق ثمّ تضع مقابلاتها فى بيوت افيالها من مربّعات ٤ فيتمّ الوفق. (...) واذا وضعتَ الاعداد القليلة حيث تضع الاعداد الكثيرة المقابلة لها فى بيوت افيالها واردتَ صحّة وضع الاعداد القليلة عن فساده فانظر الى الصفوف العرضية والطولية فان كانت الصفوف العرضية متساوية وكان كلّ صفّ من الصفوف الطولية مساويًا للصفّ الذى تنتقل بينهما بنقل الفيل فوضع الاعداد القليلة صحيح والّا فلا.

(11) اذا ابتدأتَ بوضع الطرف الاصغر والواسطة الصغرى فى الصفّ الاوّل فالقريب من الطبع ان تخلط الصفّ الاوّل بالصفّ الثانى وتنتقل فيهما بسير الفرس وقد علمتَ ذلك والبعيد من الطبع ان تخلط الاوّل بالصفّ الاخير وتقدّرهما كالمّتصلين وتنتقل فيهما بسير الفرس ثمّ تخلط ما يلى احدهما بما يلى الآخر وتنتقل فيهما بسير الفرس ويكون الانتقال من كلّ صفّين الى ما يليانهما بسير الفرزان (...) واذا ابتدأتَ من الصفّ الثانى فالقريب من الطبع ان يخلط بالصفّ الثالث والبعيد ان يخلط بالصفّ الاوّل وينتقل فيهما بسير الفرس ثمّ يخلط الرابع بالثالث فهلمّ جرّاً ويكون الانتقال من كلّ صفّين الى صفّين آخرين بعد تقدير

اتّصال الصفّ المنتقل منه الى الصفّ المنتقل اليه بسير الفرزان.

(12) يمكن ان يوضع الوفق من هذا القسم بدورين فقط وهى طريقة حسنة.
وكيفيّة ذلك ان تضع الواحد فى احدى زوايا المربّع وتمشى مشى الفرس بالاعداد
على تواليها فى البيوت على تواليها من الصفّ الذى اوّله بيت الزاوية والصفّ
الملاصق له الى ان ينقطع المشى فتضع تالى ما انتهيتَ اليه من الاعداد فى البيت
التالى لبيت المنتهى اليه من صفّه (...) ثمّ ارجع فى الصفّين بمشى الفرس ايضًا
الى ان ينقطع المشى فضع تالى ما انتهيتَ اليه من الاعداد فى البيت الموازى
للبيت المنتهى اليه من اوّل الصفّين المتلاصقين المجاورين للصفّين الاوّلين وامشى
فيهما كالاوّلين مستقيمًا وراجعًا الى ان ينقطع النقل فيهما ثمّ انتقل للصفّين
الملاصقين لهما ايضًا كما انتقلتَ اليهما وهكذا بحيث يكون مبدأ الرجوع فى كلّ
صفّين من البيت التالى للمنتهى اليه مرّة بعده ومرّة قبله وهكذا الى ان تأتى الى
آخر الشكل فيكون قد كمل الدور الاوّل ووضع من بيوت الوفق نصفها ومن كلّ
صفّ منه وقطر نصف بيوته ويكون العدد الذى انتهيتَ اليه فى الوضع هو عدد
نصف بيوت الشكل.

ثمّ ابتدىء بوضع الدور الثانى بوضع ما انتهيتَ اليه من الاعداد فى بيت
فرزانه من الصفّين الموضوعين اخرًا من الدور الاوّل ثمّ امشى فيهما كالمشى الاوّل
مستقيمًا وراجعًا الى ان يتمّ وضعهما فضع تالى ما انتهيتَ اليه من الاعداد فى
البيت المقابل له من اوّل الصفّين المجاورين للصفّين اللذين تمّ وضعهما ثمّ امشى
فى هذين الصفّين كذلك الى ان يكملا فانتقل منهما للتاليين لهما وهكذا الى ان
يكمل الوفق.

(13) جميع الطرق التى ذكرناها فى وضع الاعداد المتوالية على هذا الوجه آتٍ
هنا فى وضع الافراد ووضع الازواج المقابلة لها فى بيوت افيالها وبالعكس من
غير فرق فلا معنى لتطويل الكتاب بسببه. (...) واذا تأمّلتَ علمتَ انّ الافراد
متميّزة عن الازواج فى كلّ مربّع من مربّعات ٤ وضع بيوتها على مثال مسدّس
ضلعان منه فقط متقابلان طويلان.

(14) وقد يمكن ان نثبت العدد الوفق فى المربّعات التى اضلاعها رُبع بوجه
احسن من الذى بيّنّاه وهو ان يكون المربّع مقسومًا بعدد مربّعات تكون الاعداد
التى فى صفوفها كلّها وفقًا شبيهًا بعضها ببعض. فاذا اردنا ذلك قسمنا المربّع على

ستّة عشر فما خرج من القسمة فهو عدد المربّعات التى يُقسم بها ذلك المربّع ويكون كلّ مربّع منها فيه ستّة عشر بيتًا فاذا اثبتنا فى نصف مربّع منها كيف ما اتّفق الاعداد المتوالية من الواحد على النظم الطبيعى على الترتيب الذى رتّبنا إثباته فى مربّع الاربعة بأحد الوجوه التى قد بيّنّاها فى النوع الثانى الى ثمانية وتمّمناها بالعدد العدل لذلك المربّع انتقلنا الى مربّع آخر كيف ما اتّفق واثبتنا العدد فى نصفه من تسعة الى ستّة عشر وتمّمناها بالعدد العدل وكذلك نفعل الى ان نأتى بجميع المربّعات فيه فيصير حينئذٍ الاعداد التى فى جميع صفوف المربّع وفقًا ويكون ايضًا الاعداد التى فى جميع صفوف كلّ مربّع منها وفقًا.

(15) وامّا اذا كان المربّع زوج الفرد مثل ستّة وعشرة واربعة عشر وغيرها فانّا نقسم المربّع ارباعًا ونعيّن ربع المعيار. ونثبت فى اوّل بيوته نقطةً فان كان المربّع مربّع الستّة نثبت فى البيت الثانى صفرًا وفى البيت الثالث صليبًا وتتمّ للنقطة بيوت القطر ولكلّ واحد من الصفر والصليب عن جنبتى القطر ما يتمّ به عدّة بيوت القطر فيصير مربّع الستّة على هذا المثال (below, right). ثمّ نطبق هذا الربع الذى هو المعيار على الربع الذى يليه على الايسر للنقطة والصليب وعلى الربع الذى يليه على الاسفل للنقطة والصفر وعلى الربع الرابع للنقطة فقط فيصير المربّع على هذه الصورة (below, centre).

وان كان المربّع مربّع العشرة اثبتنا فى اوّل بيوت المعيار النقطة ثمّ تركنا بيتًا خاليًا واثبتنا الصفر فى البيت الثالث والصليب فى البيت الرابع والنقطة فى البيت الخامس وتتمّم لكلّ واحد من النقطة والصفر والصليب عن جنبتى القطر الاوّل من ربع المعيار ما يساوى عدّة بيوت القطر الاوّل ثمّ نطبق الربع الذى هو المعيار على سائر الارباع للصفر على الربع الذى يليه من جهة الاسفل وللصليب

على الربع الذى يليه على الايسر وللنقطة على سائر الارباع فيصير المربع على هذه الصورة (above, left).

وان كان المربّع هو مربّع الاربعة عشر نثبت النقطة فى البيت الاوّل وتركنا البيت الثانى خاليًا واثبتنا الصفر فى البيت الثالث والصليب فى الرابع والنقطة فى الخامس ونترك السادس خاليًا واثبتنا النقطة فى البيت السابع ولتعلم انّه لا يُثبت الصفر والصليب الّا واحدًا فى السطر الاوّل من سطور العرض من الربع الاوّل وانّما يزاد عند زيادة عدد الضلع فى النقطة ثمّ نتمّم ذلك عن جنبتى القطر ثمّ اطبقناه على سائر الارباع.

مفتاح الصفر   مفتاح النقط

ا	صط	ح	صد	ه	و	صز	صج	صب	ى
يا	يب	يخ	يز	فه	فو	فد	فج	يط	كه
عا	كب	كج	عز	كو	عه	عد	كج	كط	ف
م	سب	لج	لد	سو	سه	لز	لح	سط	سا
س	مط	نج	مد	مه	مو	مز	لح	نب	نا
ن	نط	نج	ند	نه	نو	نز	نج	مب	ما
ع	لط	سج	سد	لو	له	سز	سح	لب	لا
ل	عب	عج	كز	عو	كه	عح	كد	عط	كا
فا	قب	يج	فز	يو	يه	يد	يج	فط	ص
صا	ط	صج	ز	صه	صو	د	ج	ب	ق

مفتاح الخوالى   مفتاح الصليب

مفتاح الصفر   مفتاح النقط

و	لب	لد	ج	ه	ا
ل	يا	كز	ى	ح	كه
يط	كج	يو	يه	ك	يج
يج	يد	كب	كا	يز	كد
ز	كط	ط	كح	كو	يب
لو	ب	ج	د	ه	لا

مفتاح الخوالى   مفتاح الصليب

ثمّ اذا اردنا ان نثبت العدد بدأنا بالبيت الاوّل وسمّيناه مفتاح النقط وجرينا الى اليسار نازلاً واثبتنا العدد فى كلّ بيت فيه نقطة مبتدئين بالواحد الى آخر القطر الاوّل ثمّ نرجع من ذلك وسمّيناه اعنى آخر القطر مفتاح الخوالى ونجرى على السطر الاخير الى اليمين صاعدًا وبدأنا بالواحد واثبتنا فى كلّ بيت خال العدد الذى انتهينا اليه الى ان ننتهى الى اوّل بيوت المربّع ثمّ نبدأ بالطرف الاوّل من طرفى القطر الثانى وسمّيناه مفتاح الصفر ونجرى على اوّل سطور العرض الى اليمين واثبتنا العدد فى كلّ بيت فيه صفر مبتدئين بالواحد وهكذا نفعل فى سائر السطور الى ان ننتهى الى الطرف الاخير من طرفى القطر الثانى ثمّ نرجع من ذلك وسمّيناه مفتاح الصليب ونجرى على آخر سطور العرض الى اليسار صاعدًا واثبتنا العدد فى كلّ بيت فيه صليب مبتدئين بالواحد الى ان ننتهى الى اوّل القطر الثانى فيصير مربّع العشرة هكذا (above, right). وامّا مربّع الستة فقد صار

هكذا (above, left). واعتبر ما ذكرناه فى سائر المربّعات.

(16) اعلم ان الفرد لا يعمل فيه ملفّق الّا ما كان ضلعه مركّبًا تركيبًا ضربيًا كمربّع التسعة ومربّع الخمسة عشر ومربّع الاحد والعشرين ونحو ذلك. وطريقه ان تقسم المربّع الكبير بمربّعات صغار ضلع كلّ واحد منها هو العدد العادل لاقسام ضلع ذلك المربّع وعدّة تلك المربّعات كمربّع العدد الذى به يعدّ ضلع تلك المربّعات الصغار لضلع المربّع الكبير فيُقسم مربّع تسعة بتسعة مربّعات ثلاثة ومربّع الخمسة عشر بتسعة مربّعات خمسة وان شئتَ بخمسة وعشرين مربّع ثلاثة ويُقسم مربّع الاحد والعشرين بتسعة مربّعات سبعة او تسعة واربعين مربّع ثلاثة وعلى هذا فقس ثمّ تضع فى المربّع الاوّل من المربّعات الصغار الاعداد الطبيعية من الواحد الى عدد بيوت المربّع الصغير ويكون الوضع بأحد الطرق التى يوضع بها جنس ذلك المربّع ثمّ تضع فى الثانى من المربّعات الصغار الاعداد الطبيعية مبتدئًا بما يلى آخر الاعداد الموضوعة فى المربّع الاوّل ثمّ تضع فى الثالث من المربّعات الصغار الاعداد الطبيعية مبتدئًا بما يلى آخر الاعداد الموضوعة فى المربّع الثانى ثمّ تضع الاعداد الطبيعية فى المربّع الرابع مبتدئًا بما يلى آخر الاعداد الموضوعة فى المربّع الثالث ولا تزال تملأ المربّعات مربّعًا بعد مربّع على هذا النسق حتّى تملأ المربّع الكبير والاوّل من المربّعات الصغار هو الواقع موقع الواحد لو فرضتَ المربّعات الصغار بيوتًا والثانى هو الواقع موقع الاثنين والثالث هو الواقع موقع الثلاثة والرابع هو الواقع موقع الاربعة وعلى هذا القياس. فاذا سلكتَ هذا الطريق فى عمل الاوفاق الصغار وترتيبها صار المربّع الكبير ايضًا وفقًا. وهذا مثال ذلك فى وفق تسعة فى تسعة يكمل التعمير لتقيس عليه ما شئتَ.

(17) اذا عرفتَ واتقنتَ وضع مربّع ٤ ووضع مربّع ٦ ووضع الصليب المذكور امكنك تقسيم كلّ مربّع يمكن تقسيمه بحيث لا يخرج عن مربّع ٤ ومربّع ٦ وفصل الصليب على المثال المذكور.

(18) وليس ينبغى ان نظنّ ان العددين اللذين وضعناهما فى الزاويتين المتواليتين منها انّهما وقعا فيهما بالاتّفاق فانّ كثيرًا من الاعداد التى فى السطرين المقترنين اذا أخذنا اثنين منها ووضعناهما فى هاتين الزاويتين لم نتمّ بهما تعديل العدد الوفق

فى هذا المربّع. فمن ذلك ان هاتين الزاويتين اذا اُثبت فيهما بدلَ الاربعة والستّة واحد واربعة او واحد واثنان او اربعة وخمسة او ستّة وسبعة لم يوجد فى السطرين المقترنين اعداد تكون جملتها مع ما فى بيتى الزاويتين خمسة وستّين. فان اُثبت فيهما واحد وثلثة او ثلثة وسبعة او اثنان وثمانية او اربعة وستّة او خمسة وسبعة او ستّة وثمانية كان يمكن ان يوجد لهما ثلثة اعداد يكون جملتها مع ما اُثبت فيهما خمسة وستّين. فلنعيّن ان نثبت العدد الذى قبل ضلع المربّع فى الزاوية الاولى والعدد الذى يلى ضلع المربّع بعده فى الزاوية الثانية فانّ السطر يعتدل بهما ويمكن ان يوجد الاعداد الباقية فى باقى الصفوف.

(19) قد رتّبنا العدد الوفق فى مربّع التسعة. وينبغى ان نذكر فى هذا الموضع طريقًا للتقريب على المتعلّم ولمن يختار ان لا يتعب نفسه فى طلب الاعداد التى ترتّب فى المربّع.

(20) تضع ١ وهو الطرف الاصغر فى البيت الاوسط من الصفّ الايمن وما يليه فيما يليه نازلاً الى الزاوية اليمنى السفلى ولا تضع فيها وما يليه فى الزاوية اليسرى السفلى وما يليه فيما يليها من الصفّ الاسفل الى البيت الاوسط منه وما يليه فى الاوسط من الصفّ الاعلى وما يليه فيما فوق الاوسط من الصفّ الايسر وما يليه فيما يليه فصاعدًا الى الزاوية اليسرى العليا فتضع فيها وما يليه فيما يلى الاوسط من الصفّ الاعلى الى مجاور الزاوية اليمنى العليا ثمّ تضع مقابلاتها بازائها كما علمتَ وهكذا تفعل بكلّ حاشية ويقيم ما انتهيتَ اليه من الاعداد القليلة مقام الطرف الاصغر حتّى تنتهى الى مربّع ٣ فتخرّجه على هذه القاعدة.

(21) Ambitus primus. Vides ut dimidia pars terminorum minimorum imparium in latere infimo progrediatur. Alteram partem dimidiam horum vides ponere primum terminum suum intra cellulam mediam sinistri lateris & reliquos in supremo latere. Deinde vides ut pars dimidia terminorum minimorum parium descendat in latere dextro. Vides etiam alteram partem dimidiam horum descendere in sinistro latere & maiorem suum ponere intra cellulam primam. Secundus ambitus. Satis vides ut unica sit regula in quolibet ambitu replendo si habeat latera imparium cellularum.

(22) تُقسم الاعداد القليلة التى تريد وضعها فى الطوق اربعةَ اقسام على الولاء الّا ان اوّل القسم الثالث تجعله اوّل القسم الرابع واوّل القسم الرابع تجعله آخر القسم الثالث ثمّ تضع القسم الاوّل اسفل فى غيرِ الزاوية والقسم الثانى ايمن واوّله فى الزاوية العليا منه والقسم الثالث اعلى وآخره فى الزاوية اليسرى منه والقسم الرابع ايسر فى غير الزاوية ثمّ تضع مقابلاتها بازائها ثمّ تفعل بطوق المربّع الداخل كذلك حتّى تصل الى مربّع ٣ فتضعه كما عرفتَ وفى كلّ طوق تجعل الاعداد القليلة التى تبقى معك كانّها هى اوّل الاعداد القليلة وتقسم الاعداد التى توضع فى الطوق اقسامًا اربعةً كما ذكرنا وعدّة الاعداد القليلة التى توضع فى الطوق هى نصف بيوت الطوق.

(23) تأخذ من اوّل الاعداد القليلة التى تريد وضعها فى الطوق اربعة اعداد متوالية فتضع طرفيها اسفل فى غير الزاويتين ووسطيها فى الزاويتين فوق ثمّ تأخذ عددين يليان تلك الاربعة فتضع اقلّهما فى جهة الاكثر ممّا فى الزاويتين يمينًا كان او يسارًا واكثرهما فى جهة الاقلّ ثمّ تضع مقابلاتها بازائها فتعتدل الصفوف ثمّ تأخذ لكلّ اربعة بيوت من البيوت الباقية اربعة اعداد متوالية فتضع طرفيها فى صفّ ووسطيها فى الصفّ المقابل له حتّى يتمّ الصفوف بعد وضع المقابلات.

(24) Quando ambitus numerum habet cellularum per 8 numerabilem, tunc termini descendunt in sinistro latere, atque dextro, hinc inde donec tot cellulæ repleantur quot unum latus dimidium cellulas habet. Et tunc intermisso descensu illo transitus sit ad latus supremum, & sit progressus per supremum latus atque infimum sicut descensus fieri solet per latus dextrum atque sinistrum. Scilicet semper duo termini, par & impar, ex uno latere ponuntur immediate. Excipiuntur quatuor cellulæ: infima cellula dextri lateris, ea est primi termini cellula; item suprema eiusdem lateris; item secunda cellula supremi lateris; et penultima infimi lateris. Finito autem progressu prædicto per latus supremum & infimum, repetitur descensus ille prius intermissus. Intermittitur autem semper in sinistro latere, & illic iterum repetitur. Hinc fit, ut quatuor cellulas continue videas repleri, & ex alio latere quatuor vacuas (...).

Quando ambitus numerum habet cellularum numerabilem per 4 & non per 8, tunc eadem fiunt per omnia quæ in ambitu fiunt cuius latus unum per octo numeratur —ut fuit primus, & quintus—, hoc solo excepto quod intermissio descensus non fit cum duobus terminis in uno &

eodem latere ponendis, sed prius ponitur unus solitarie in dextro latere, & deinde finit sequens terminus solus partem illam descensus, videlicet in sinistro latere. Idem fit dum descensus repetitur. Ponitur enim unus primo in sinistro, & deinde sequens ponitur in dextro latere; atque ita octo ponuntur solitarie. Satis autem vides quæ sint illæ octo cellulæ quæ solitariæ excipiant terminos minores. Scilicet, infima & suprema dextri lateris; item secunda supremi lateris, & penultima infimi lateris; deinde quatuor illæ de quibus dictum est paulo superius.

(٢٥) تضع الواحد فى القطر الاوّل والاثنين فى السطر الاخير من سطور الطول فى بيت غير القطر والثلثة فى بيت غير القطر من السطر الاخير من سطور العرض والاربعة فى القطر الاخير من السطر الاوّل من سطور العرض والخمسة فى السطر الذى فيه الثلثة فى بيت غير القطر والستّة فى السطر الاوّل من سطور الطول فى بيت غير القطر والسبعة فى السطر الذى فيه الثلثة فى بيت غير القطر والثمانية فى السطر الذى فيه الستّة فى بيت غير القطر والتسعة فى السطر الذى فيه الاثنان فى بيت غير القطر والعشرة فى السطر الاوّل من سطور العرض وينبغى ان لا يقع عدد فى مقابلة عدد (...).

فان كان الوفق وفق الستّة فقد تمّ دوره فتضع فى كلّ بيت خالٍ من الدور ما يتمّ مع ما فى البيت المقابل له مربّع الستّة مزيدًا عليه واحد اعنى سبعة وثلثين. (...) وان كان الوفق بعدد آخر غير الستّة كالعشرة ونحوها فالطريق فيه ان تضع الاعداد العشرة وتتمّم مقابلاتها كما بيّنّا فيبقى فى كلّ سطر بيوت لها رُبع صحيح ابدًا فاملأ رُبع بيوت احد السطور مبتدأً من العدد الذى انتهيتَ اليه وهو احد عشر على الترتيب ثمّ نصف بيوت السطر المقابل له ثمّ رُبع السطر المبدأ به ثمّ املأ رُبع بيوت السطر الثالث ثمّ نصف بيوت الرابع ثمّ رُبع بيوت الثالث بشرط ان لا يقع عدد فى مقابلة عدد ثمّ كمّل مقابلاتها الى تمام الوفق.

(٢٦) واذا تعادلت الصفوف الاربعة بالاعداد العشرة فان كان عملك فى مربّع ٦ فقد تمّ العمل بعد وضع مقابلاتها بازائها وان كان عملك فى غيره يبقى من كلّ صفّ اربعة بيوت او ثمانية وهكذا بزيادة اربعة فتأخذ اربعة اعداد متوالية فتضع طرفيها فى صفّ ووسطيها فى صفّ مقابل له.

(٢٧) نثبت الواحد او ما يقوم مقام الواحد الى جنب الزاوية اليمنى من الصفّ الاسفل وما يليه فى البيت الثالث من الصفّ الاعلى وما يليه فى ثالث بيت آ من

الصفّ الاسفل وما يليه فى الصفّ الايسر وما يليه فى الصفّ الاعلى وما يليه فى
الايمن وما يليه فى الاسفل وما يليه فى الايسر الى ان يبلغ العدد المثبت مثل ضلع
المربّع الذى يلى المربّع المفروض فى الوسط ثمّ اثبتنا ما يليه فى الزاوية اليمنى من
الصفّ الاعلى وما يليه فى الزاوية اليسرى وهو مثل ضلع المربّع المفروض ثمّ اثبتنا
عددين ممّا يليه نحو الايمن وعددًا واحدًا نحو الاسفل وعددًا نحو الايسر الى ان
يمتلىء نصف الصفوف المحيطة باثبات عدد على الايسر ومبلغها مثل مجموع ضلعى
المربّع المفروض والذى يليه فى الوسط ثمّ نتمّمها بالعدد العدل للمربّع المفروض.

(28) Quando ambitus habet numerum cellularum qui est impariter par,
tunc fit descensus hinc inde —ut superius dictum est— donec una so-
lummodo cellula supersit ex iis quæ minoribus terminis debentur in
dextro & sinistro lateribus. Finitur autem descensus ille in sinistro latere tribus terminis immediate positis. Quo sic finito fit transitus ad
infimum latus. Et inde fit progressus per supremum & infimum latera
hinc inde ut in superiori ambitu dictum est, atque ut hic vides. Dum
autem ad finem illius progressus perveneris, id est ad penultimam cellulam infimi lateris, non pones intra illam terminum illum quem ordo
progressionis tangit, sed pones eum intra cellulam in descensu intermis-
sam, id est, intra penultimam cellulam dextri lateris; sequentem vero
terminum pones intra penultimam illam cellulam infimi lateris. Vides
igitur hic ut quinque cellulæ solitarios numeros excipiant, videlicet infima, penultima, suprema dextri lateris; secunda & penultima infimi lateris.

(29) وايضًا فانّ المربّعات التى تحصل فى وسط المربّع تكون على وجهين.
احدهما ان يكون المربّع الذى زواياه مركّبة على موضع النصف من اضلاع المربّع
الكبير وعدد بيوته يكون فردًا وفى بيوته ينبغى ان تكون الاعداد الافراد. واذا
أخذنا صفوفها الموازية للاضلاع التى اطرافها مركّبة على انصاف الاضلاع من
المربّع الكبير وجدنا اضلاعها مساوية للاعداد المتوالية على النظم الطبيعى المبتدئة
من الاثنين. الآ ترى انّ ضلع المربّع الذى يقع فى مربّع الثلثة هو اثنان وضلع
المربّع الذى يقع فى مربّع الخمسة هو ثلثة وضلع المربّع الذى يقع فى مربّع
السبعة هو اربعة وضلع المربّع الذى يقع فى مربّع التسعة هو خمسة وكذلك على
نظم الاعداد المتوالية. فمتى اردنا ان نعرف ضلع هذا المربّع الموزب زدنا على
ضلع المربّع واحدًا وأخذنا نصف ما اجتمع فما كان فهو ضلع المربّع الموزب

وصفوها أنّما تكون مختلفة فمنها صفّ يكون بيوته مساوية للضلع وصفّ ناقص منه ببيت. وامّا الوجه الثانى فان يكون المربّع فى وسطه على رسم ما بيّنا فى المربّعات الفرد والزوج. لكنّ هذا المربّع يقع كلّ واحد منها على نظم توالى الاعداد الافراد المتوالية فى مربّعين. فانّ مربّع الخمسة ومربّع السبعة يقع فيهما مربّع واحد وهو مربّع الثلثة ومربّع التسعة ومربّع الاحد عشر يقع فيهما مربّع واحد وهو مربّع الخمسة وكذلك مربّع ثلثة عشر ومربّع خمسة عشر يقع فى وسطهما مربّع السبعة.

(30) وتجد المربّعات حينئذٍ تقسم اقساماً يحتاج كلّ قسم منها فى ترتيب الاعداد الزوجيّة فى الزوايا الى عمل خلاف عمل الآخر فتكون ٩ و١٣ و١٧ وما زاد عليها من الاعداد ٤ ٤ والسبعة والاحد عشر والخمسة عشر والتسعة عشر وما زاد عليها من الاعداد ٤ ٤.

(31) فامّا رسم الخمسة وما كان من نوعها فانّه يبقى من الصفّ الاوّل المحيط بالمربّع الداخل اربعة ابيات الزوايا وثمنية ابيات تتّصل بها من كلّ جانب بيتان فارغة كلّها ويبقى من الثانى ابيات الزوايا واربعة وعشرون بيتًا تتّصل بها فارغة كلّها ومن الصفّ الثالث ابيات الزوايا واربعون بيتًا تتّصل بها فارغة كلّها. وكذلك ابدًا كلّ صفّ يزيد على الذى قبله بعد ابيات الزوايا ١٦.

(32) وامّا رسم السبعة وما كان من نوعها فانّه يبقى من الصفّ الاوّل المحيط بالمربّع الداخل ٤ ابيات الزوايا فقط فارغة ومن الصفّ الثانى ابيات الزوايا و١٦ بيتًا تتّصل بها من كلّ جانب ٤ ومن الصفّ الثالث ابيات الزوايا و٣٢ بيتًا تتّصل بها فارغة. وكذلك ابدًا كلّ صفّ يزيد على الذى قبله بعد ابيات الزوايا ١٦.

(33) فالصفّ الاوّل الاعلى من مربّع ٩ اذا اثبت الاعداد الفرديّة فيه على هذا الترتيب يزيد على حقّه الواجب له بعشرين وينقص الاسفل عشرين والصفّ الايمن الاوّل منه يزيد على حقّه ١٨ وينقص الايسر عنه ١٨ ويزيد الثانى الاعلى ٣٦ وينقص الاسفل مثلها ويزيد الثانى الايمن ٣٤ وينقص الايسر مثلها. وفى مربّع ١٣ يزيد الاوّل الاعلى ٢٨ وينقص الاسفل مثلها ويزيد الاوّل الايمن ٢٦ وينقص الايسر مثلها ويزيد الثانى الاعلى ٥٢ وينقص الاسفل مثلها ويزيد الثانى

الايمن ٥٠ وينقص الايسر مثلها ويزيد الثالث الاعلى ٧٦ وينقص الاسفل مثلها ويزيد الثالث الايمن ٧٤ وينقص الايسر مثلها. وكذلك سائرها.

وتكون التسعة يزيد الثانى على الذى قبله ١٦. والثلثة عشر يزيد الاول على اول التسعة ٨ ثمّ كلّ صفّ على الذى قبله ٢٤. و١٧ يزيد الاول على الاول من الثلثة عشر ٨ ثمّ كلّ صفّ على الذى قبله ٣٢. ثمّ كذلك ابدًا كلّ صفّ على الذى قبله.

(34) ويكون الصفّ الاول الاعلى من المربعات فى هذا النوع يزيد على حقّه ٤ وينقص الاسفل مثلها ويزيد الاول الايمن على حقّه ٢ وينقص الايسر مثلهما ويكون الصفّ الثانى الاعلى من السبعة يزيد على حقّه ٢٠ والثانى الايمن ١٨. ويزيد الثانى الاعلى فى مربّع ١١ على حقّه ٢٨ والثانى الايمن ٢٦ والثالث الاعلى ٥٢ والثالث الايمن ٥٠ وينقص ما يقابل جميع ذلك مثل الزيادة. والخمسة عشر يزيد الثانى الاعلى على الاول ٣٢ والثالث على الثانى ٣٢. وكذلك سائرها يزيد ٨ ٨.

(35) اذا رسمتَ العدد الاول القليل فى جهة او اىّ عدد شئتَ من القليلة ثمّ رسمتَ الذى يتلوه فى الجهة الاخرى التى تقابلها ورسمتَ بازاء كلّ واحد منهما ما يقابله من الكثيرة قصّرت الجهة التى رسمتَ فيها القليل الاول عن حقّها باثنين وزادت الاخرى باثنين. فان رسمتَ عددًا من القليلة فى جهة ورسمتَ فى الجهة الاخرى التى تقابلها العدد القليل الثالث من ذلك العدد قصّرت الجهة التى فيها القليل الاول عن حقّها ٤. وكذلك كلّما باعدتَ بينهما بعدد زاد النقصان ٢ ابدًا.

(36) فاتّك اذا فعلتَ ذلك وجدتَ اخير الاعداد القليلة والذى قبله اذا صيّرا فى جهة وصيّر بازائهما ما يقابلهما قصّرت الجهة التى فيها القليلان عن حقّها ٤ وزادت الاخرى ٤ واذا صيّر العددان اللذان يتلوانهما يقصّران ١٢ ١٢ واللذان يتلوانهما يقصّران ٢٠ ٢٠ ثمّ ٢٨ ثمّ ٣٦ ثمّ ٤٤ ثمّ ٥٢ ثمّ ٦٠ وكذلك ابدأ الى الاربعة والاثنين.

(37) واذا اُثبت اربعة اعداد فى الزوايا كلّ ٢ منها على نسق فيقابلهما طرفاهما قطرًا واثبتتَ القليلين على قدر ترتيبهما فى الجهة العليا فقصّرت الجهة العليا عن

حقّها ان كانا الاخيرين ٤ وان كانا اللذين قبلهما ١٢ وكذلك ابدًا بزيادة ٨ ٨
حتّى تنتهى الى ٢ و٤ وتكون الجهة اليمنى تزيد على حقّها او تنقص عنه ٢ ٢
ابدأ لا زيادة فيها ولا نقصان. فافهم هذا كلّه فانّه ممّا تحتاج اليه فى تأليف
الازواج فى هذا النوع.

(38) فان بقى من الابيات شىء فارغ فانّما يبقى اربعة ابيات تقابل ٤ او ٨
تقابل ٨ او ١٢ تقابل ١٢ تعدّل كلّ ٤ منها باربعة اعداد منسّقة من الباقية
(...) او باربعة اعداد يكون كلّ عددين منها منسّقين فتثبت اوّل الاوّلين وثانى
الآخرين فى جهة وثانى الاوّلين واوّل الآخرين فى الجهة التى تقابلها وبازاء كلّ
واحد منها العدد الذى يقابله.

(39) فاثبت آخر الاطراف القليلة من الازواج فى الزاوية العليا اليسرى من الصفّ
الاوّل وهو الذى يلى المربّع الداخل الذى حشوتَه بالافراد وبازائه قطرًا فى الزاوية
اليمنى السفلى من الصفّ الاوّل ما يقابله واثبت الطرف القليل الذى يليه قبله فى
الزاوية اليمنى العليا من الصفّ الاوّل وبازائه قطرًا فى الزاوية اليسرى السفلى
الطرف الذى يقابله. (...) فاثبت ٢ فى الصفّ الاعلى وانظر الى كم زيادة الصفّ
الاعلى فخذ نصفها ثمّ عدّ بعد الاثنين مثل ذلك النصف من الازواج القليلة
فحيث انتهيتَ فاثبت العدد الذى تنتهى اليه فى الصفّ الاسفل واثبت بازاء كلّ
واحد منهما الطرف الذى يقابله. واثبت ٤ فى الجهة اليمنى ثمّ عدّ بعد ٤ مثل
نصف الزيادة من الازواج القليلة فحيث انتهيتَ فاثبت العدد الذى تنتهى اليه
فى الصفّ الايسر واثبت بازاء كلّ واحد منهما الطرف الذى يقابله.

(40) فاذا فعلتَ ذلك فقد عدّلتَ الصفّ الاوّل فى جميع المربّعات من هذا
النوع.

(41) وكان الوفق الذى يقع فى المربّع ايضًا مساويًا لضرب عدده العدل فى
نصف ضلع المربّع كما تقدّم ذكره.

(42) فان كانت الاعداد التى نرتّب فى المربّع متفاضلة باعداد متساوية فى
صفوف العرض وكان العدد الاوّل من صفوف الطول لا يزيد على العدد الاخير
من الصفّ الذى قبله بمثل تفاضل صفوف العرض قيل لتلك الاعداد انّها متناسبة

مناسبةً منفصلةً (...) فاذا اردنا ان نعرف عدده الوفق ضربنا تفاضل صفوف الطول وتفاضل صفوف العرض مجموعين فى ضلع المربّع الّا واحدًا فما حصل زدنا عليه ضعف العدد الاوّل فما كان فهو العدد العدل فاذا ضُرب فى نصف ضلع المربّع كان ما حصل هو العدد الوفق.

(43) وقد وقع لبعض المستعملين انّه لا يمكن ان يقع فى المربّعات الّا العدد الوفق الذى هو الاصل او ما يتركّب من تضاعيفه وزيادة عليه من ضلع المربّع وليس الامر كذلك.

(44) فيشترط فى تلك الاعداد امور. احدها ان يكون لمجموعها ثُلث لتخرج منها حصّة البيت الاوسط. الثانى ان يكون العددان الموضوعان فى الطرفين بحيث ينقص كلّ واحد منهما مع مقابل الآخر عن مجموع تلك الاعداد حتّى يبقى بعد وضع مقابليهما ما يوضع فى البيتين الاوسطين يمينًا ويسارًا. الثالث ان لا تكون تلك الاعداد متكرّرة بالفعل او بالقوّة.

(45) تأخذ عددًا اذا ضممناه الى ما فى البيتين اللذين فى الطرفين كان للمجموع ثُلث فتقدّر ذلك العدد المأخوذ للبيت الاوسط من الصفّ الاعلى وثُلث ذلك المجموع للبيت الاوسط فان توافق الصفّان فذاك والّا فنظرنا ما بينهما من التفاوت.

(46) فشرطه ان يكون العدد الاوسط ثُلث الجملة.

(47) وليس يلزم ههنا ان يعتدل الصفّان ولا عليك من ذلك.

(48) فشرطه ان يكون العدد الاوسط خُمس تلك الاعداد.

(49) واذا عرفتَ وضع الاعداد المتفاوتة فى تفاوتها فى مربّعى ٣ وﮪ تمكّنتَ من وضعها فى جميع مربّعات الفرد.

(50) وقد حصل فى هذا المربّع سوى ما كان مطلوبنا من العدد الوفق بالاتّفاق اعتدال حسن وذلك ان كلّ بيتين اوسطين من الصفّ المحيط بالمربّع الاوسط مع البيتين النظيرين لهما اذا جُمع ما فيها كان ايضًا لَدَ مثل اثنى عشر وثمنية اذا جُمعا

مع تسعة وخمسة ومثل اربعة عشر وخمسة عشر اذا جُمعا مع ثلثة واثنين. وحصل ايضًا اتّا متى جمعنا ما فى البيتين اللذين عن جنبتى كلّ احدى من الزوايا الى ما فى البيتين اللذين عن جنبتى الزاوية المقابلة لها كان ايضًا لدّ وذلك اتّا متى زدنا اثنى عشر وخمسة عشر على اثنين وخمسة ومتى زدنا اربعة عشر وتسعة على ثمانية وثلثة كان لدّ. وحصل ايضًا اتّا متى جمعنا ما فى كلّ اربعة بيوت متلاصقة عند احدى زواياه كان لدّ مثل واحد وخمسة عشر واثنى عشر وستّة كان لدّ.

(51) شرطه ان يكون لجملة تلك الاعداد نصف ليمكن ان تضع المقابلات بحسبه وهذا شرط فى سهولة الوضع.

(52) فقد وقع الناشىء الثانى من كلّ واحد منهما فى بيت فيل الطرف الذى يجاوره فتجعل ذلك معيارًا لمقابلته ومقابلة كلّ ناشىء منه.

(53) تضع فى وسطى الصفّ الاسفل عددين مجموعهما مجموع العددين الموضوعين فى طرفى الصفّ الاعلى وفى طرفيه عددين مجموعهما مجموع الموضوع فى الوسطين من الصفّ الاوّل ثمّ تضع فى احد الصفّين الوسطين الطوليين تمام ما يستحقّه ثمّ يتمّم كلّ واحد من القطرين يبقى بيتان فى الصفّ الايمن وبيتان فى الصفّ الايسر فاذا وضعتَ فى بيتى احدهما تمام ما يستحقّه وتمّمتَ احد الصفّين الوسطين العرضيين بقى بيت مشترك بين صفّين يستحقّه ما يتمّ به كلّ واحد منهما فاذا وضعتَه فيه تمّ الوفق.

(54) شرطه ان يكون لجملة الاعداد نصف.

(55) اذا وضعتَ ما يقوم مقام الطرف الاصغر وما يقوم مقام الواسطة الصغرى فى موضعهما من صفّ من الصفوف وانتقلتَ من احدهما بالزيادة ومن الآخر بالنقصان على نسبة واحدة فلا يشترط ان يكون تلك النسبة محفوظة فى جميع الانتقالات بل اتّما يشترط ذلك فى الانتقالات التى هى من جنس واحد كالانتقالات التى هى بسير الفرس او الانتقالات التى هى بسير الفرزان وامّا ان يكون الانتقالات التى هى بسير الفرس وهى خلط صفّ بصفّ على نسبة الانتقالات التى هى بسير الفرزان وهى الانتقالات من خلط صفّ بصفّ الى خلط صفّ ثالث بصفّ رابع فلا.

(56) ولا يشترط ايضًا انّك اذا ابتدأتَ من احدهما بالزيادة تمضى دائمًا فى جميع انتقالاته بالزيادة واذا ابتدأتَ منه بالنقصان تمضى فى جميع انتقالاته بالنقصان بل اتّما يشترط ذلك فى الانتقالات التى هى من جنس واحد وامّا فى الانتقالات التى هى من جنسين فلا. مثلاً اذا ابتدأتَ فى نقل الفرس بالزيادة جاز ان تمضى بنقل الفرزان بالنقصان لكن فى نقل الفرس تعود الى الزيادة وكذلك الامر فى الابتداء بالنقصان جاز ان تمضى بنقل الفرزان بالزيادة لكن فى نقل الفرس تعود الى النقصان.

(57) وحينئذٍ تعود الى الطرق الثلثة المذكورة فى وضع الاعداد المختلفة فى الصفّ الاوّل من مربّع ٤ فتضع النصف الاوّل فى مربّع ٤ بطريق والنصف الآخر فى مربّع آخر امّا بذلك الطريق وامّا بغيره حتّى يتمّ نصف المربّع ثمّ تضع النصف الآخر كما نذكره.

(58) تضع فى البيتين الوسطين من الصفّ الاسفل من مربّع ٤ الايمن عددين مجموعهما مثل مجموع العددين الموضوعين فى طرفى الصفّ الاعلى من مربّع ٤ الايسر وبالعكس وتضع فى وسطى الصفّ الاسفل من مربّع ٤ الايسر عددين مجموعهما مثل مجموع العددين الموضوعين فى طرفى الصفّ الاعلى من مربّع ٤ الايمن وبالعكس فاذا فعلتَ ذلك اعتدل الصفّ الاوّل والرابع من مربّع ٨.

(59) فتعديلهما ان تعمد الى الصفّ الزائد فتقسم ذلك الزائد بقسمين فتضع فى بيت منه عددًا من الاعداد القليلة وما ينقص عن مقابله باحد قسمى ذلك الزائد فى بيت آخر منه وتضع فى بيت ثالث منه عددًا آخر من الاعداد القليلة وفى البيت الرابع ما ينقص عن مقابله بالقسم الثانى من ذلك الزائد ثمّ تضع مقابلات هذه الاعداد فى الصفّ المقابل له فيعتدل الصفوف الاربعة يبقى مربّع ٤ الداخل وانت بوضعه من العارفين.

(60) فشرطه (...) ان تكون الاعداد الثلثة الموضوعة فى الصفّ الاسفل مجموعًا مثل الاعداد الثلثة الموضوعة فى الصفّ الاعلى مجموعًا ولّيّته ظاهرة.

(61) تضعه بطريق الفرس وتبدأ منه اوّلاً ببيت الوسط ولا تضع فيه شيئاً ثمّ

تمشى منه مشى الفرس متصاعدًا او متسافلًا ومتياسرًا او متيامنًا وتضع فى بيت فرسه الواحد ثمّ تمشى منه كمشيك من بيت الوسط اليه وتضع الاعداد على تواليها الى ان تضع اعدادًا بقدر عدّة اقسام الضلع الّا واحدًا (...) وهذه طريقة نفيسة عزيزة يمكن ان ينزل بها كلّ من مربّعات هذا النوع ما عدا المثلّث وما يعدّه ضلعه الثالثة خالى الوسط.

(62) تأخذ الفضل بين وفقه وعدله وتقسمه بقسمين مختلفين وضعهما فى زاويتين من المربّع يكونا طرفى ضلع منه وضع فى البيت الوسط من ذلك الضلع مقدار العدل ثمّ ضع مقدار الفضل كاملاً فى بيت الوسط.

(63) تدير على مركز واحد دوائر وعدّتها كعدّة ضلع المربّع وابعاد ما بينها متساوية ثمّ تقسم محيط الدائرة العظمى باقسام عدّتها كعدّة ضلع المربّع وتصل بين المركز ومواضع القسمة بخطوط مستقيمة ثمّ تضع الاعداد التى فى الصقّ الاوّل من الصفوف الطوليّة او العرضيّة من صفوف المربّع فى البيوت التى حول المركز وهى داخل الدائرة الصغرى والاعداد التى فى الصفّ الثانى الملاصق للاوّل فى البيوت التى هى بين الدائرة الصغرى والتى تليها وهى الدائرة الثانية والاعداد التى فى الصفّ الثالث بين الدائرة الثانية والدائرة الثالثة وعلى هذا حتّى تضع الاعداد التى فى الصفّ الاخير فى البيوت التى هى بين الدائرة العظمى والتى تليها لجهة المركز.

Bibliography

G. Abe: "Magic squares made by Nārāyana Pandita" (in japanese), *Sū-gakushi kenkyū* 104 (1985), pp. 32-44.

Abū'l-Wafā': *see* Sesiano.

C. Agrippa, *De occulta philosophia*. Cologne 1533. Reprinted with appendices by K. A. Nowotny, Graz 1967.

C. G. Bachet de Méziriac, *Problemes plaisans et delectables qui se font par les nombres*. Lyons 1624.

al-Būnī: *Shams al-maʿārif al-kubra wa-laṭā'if al-ʿawārif*, 4 vol. Cairo 1291H.

G. Cardano: *Practica arithmeticæ, & mensurandi singularis*. Milan 1539.

Baron B. Carra de Vaux, "Une solution arabe du problème des carrés magiques", *Revue d'histoire des sciences* 1 (1947), pp. 206-212.

P. Fermat: *Varia opera mathematica*. Toulouse 1679.

——: *Œuvres complètes*, ed. P. Tannery & Ch. Henry (4 vols). Paris 1891-1912.

Fihrist: *see* Ibn al-Nadīm.

M. Folkerts: "Zur Frühgeschichte der magischen Quadrate in Westeuropa", *Sudhoffs Archiv* 65 (1981), pp. 313-338.

B. Frénicle de Bessy, "Table generale des quarrez de quatre", in *Divers ouvrages de mathematique et de physique*, Paris 1693, pp. 484-507.

T. Hayashi, "The *Caturacintāmaṇi* of Giridharabhaṭṭa: a sixteenth-century Sanskrit mathematical treatise", *SCIAMVS* 1 (2000), pp. 133-208.

H. Hermelink: "Die ältesten magischen Quadrate höherer Ordnung und ihre Bildungsweise", *Sudhoffs Archiv* 42 (1958), pp. 199-217.

P. de la Hire: "Nouvelles constructions et considerations sur les quarrés magiques avec les demonstrations", *Memoires de l'Academie Royale des sciences* 1705, pp. 127-171.

Ibn al-Nadīm: *Kitāb al-Fihrist*, edd. with notes by G. Flügel, J. Roediger and A. Müller (2 vol.). Leipzig 1871-1872.

© Springer Nature Switzerland AG 2019
J. Sesiano, *Magic Squares*, Sources and Studies in the History of Mathematics and Physical Sciences, https://doi.org/10.1007/978-3-030-17993-9

W. Kutsch: *Ṯābit b. Qurra's arabische Übersetzung der Ἀριθμητικὴ Εἰσαγωγή des Nikomachos von Gerasa*. Beirut 1959.

S. de la Loubère, *Du royaume de Siam* (2 vols). Amsterdam 1691.

M. Moschopoulos: *see* Tannery, Sesiano.

Nicomachos: *see* Kutsch.

J. Sesiano: "Herstellungsverfahren magischer Quadrate aus islamischer Zeit (II)", *Sudhoffs Archiv* 65 (1981), pp. 251-265.

———: "Herstellungsverfahren magischer Quadrate aus islamischer Zeit (II′)", *Sudhoffs Archiv* 71 (1987), pp. 78-89.

———: "An Arabic treatise on the construction of bordered magic squares", *Historia scientiarum* 42 (1991), pp. 13-31.

———: "Quelques méthodes arabes de construction des carrés magiques", *Bulletin de la Société vaudoise des sciences naturelles* 83 (1994), pp. 51-76.

———: "Herstellungsverfahren magischer Quadrate aus islamischer Zeit (III)", *Sudhoffs Archiv* 79 (1995), pp. 193-226.

———: *Un traité médiéval sur les carrés magiques*. Lausanne 1996.

———: "L'Abrégé enseignant la disposition harmonieuse des nombres", in *De Bagdad a Barcelona* [= Anuari de filologia XIX B-2], Barcelona 1996 (2 vols), I, pp. 103-157.

———: "Le traité d'Abū'l-Wafā' sur les carrés magiques", *Zeitschrift für Geschichte der arabisch-islamischen Wissenschaften* 12 (1998), pp. 121-244.

———: "Les carrés magiques de Manuel Moschopoulos", *Archive for history of exact sciences* 53 (1998), pp. 377-397.

———: "Une compilation arabe du XIIe siècle sur quelques propriétés des nombres naturels", *SCIAMVS* 4 (2003), pp. 137-189.

———: "Magic squares for daily life", in *Studies in the history of the exact sciences in honour of David Pingree* [= *Islamic philosophy, theology and science*, vol. LIV], Leiden 2004, pp. 715-734.

———: *Les carrés magiques dans les pays islamiques*. Lausanne 2004. Russian edition: *Magicheskie kvadraty na srednevekovom vostoke*, Saint Petersburg 2014.

———: *Euler et le problème du cavalier*. Lausanne 2015.

———: *Magic squares in the tenth century, two Arabic treatises by Anṭākī and Būzjānī*. Cham 2017.

M. Stifel: *Arithmetica integra*. Nuremberg 1544.

P. Tannery: "Le traité de Manuel Moschopoulos sur les carrés magiques, texte grec et traduction", *Annuaire de l'Association pour l'encouragement des études grecques en France* 20 (1886), pp. 88-118. Reprinted in Tannery's *Mémoires scientifiques*, IV, pp. 27-60.

——: *Mémoires scientifiques* (17 vols). Paris-Toulouse 1912-1950.

N. Vinel: "Un carré magique pythagoricien?", *Archive for history of exact sciences* 59 (2005), pp. 545-562.

E. Wiedemann: "Beiträge zur Geschichte der Naturwissenschaften, V", *Sitzungsberichte der physikalisch-medizinischen Sozietät zu Erlangen* 37 (1905), pp. 392-455.

Index

© Springer Nature Switzerland AG 2019
J. Sesiano, *Magic Squares*, Sources and Studies in the History of Mathematics
and Physical Sciences, https://doi.org/10.1007/978-3-030-17993-9

23, 36-39, 43-44, 63-85, 227, 230, 237, 238, 241, 245, 253-255, 258, 264, 265, 269 ◇ (queen) 23, 36, 63-65, 67, 69, 71, 74, 227, 230, 237, 238, 241, 264, 265, 269 ◊ (knight's move on the chessboard) 271.

D

E

F

magic cubes: 253, 280.

magic products: 214.

magic rectangles: 105, 253, 277-279.

magic squares: (def.) 1 ◇ (talismanic use) *see* talismans ◇ (origin of the name) vi, 12 ◇ (transformations of) 5-8, 31-32 ◇ (number of configurations) 5-6, 44-45, 143-146, 209 ◇ (sum of two magic squares) 270 ◊ ordinary : 1 (def., possibility, magic sum); 9, 10, 13-15, 19, 21-106 ◇ (odd orders) 21-44, 258 ◇ (evenly-even orders) 44-88, 99, 104, 116, 119, 245-248 ◇ (evenly-odd orders) 88-106, 116, 119 ◇ (separation by parity) 14, 16, 34-36, 43, 57-58, 77-79, 112-113 ◇ (pandiagonal) 3 (def., possibility); 6n, 10, 39, 42, 44, 45, 63, 64, 71, 73, 74, 78, 79, 83, 86, 116, 213, 214, 226, 230, 231, 238, 266, 269-271 ◇ (symmetrical) 6, 8, 25, 31, 32, 38, 45, 49, 50, 54, 56, 57, 93, 142, 214n, 226; (not symmetrical) 38, 51, 56, 60, 93 ◊ bordered : 2 (def., possibility, magic sum); 3, 8, 9, 13-15, 21, 102-104, 141-175 ◇ (odd orders) 142-158 ◇ (evenly-even orders) 158-167 ◇ (evenly-odd orders) 167-174 ◇ (separation by parity) 13, 146-147, 150, 158, 170, 177-210 ◊ composite: 4, 108 (def., possibility); 9, 13, 14, 84n, 107-139 ◊ with non-consecutive numbers: 10, 11, 14, 15, 110, 211-251 ◇ (odd orders) 215-223 ◇ (evenly-even orders) 224-248 ◇ (evenly-odd orders) 248-251 ◊ small orders: (square of order 2) 1-2, 4-5, 85, 108, 116-120, 142 ◇ (square of order 3) 5-6, 9, 12, 16, 25, 44, 177, 210, 215-219, 266-268, 272, 274 ◇ (square of order 4) 6, 9, 16, 44-45, 63-64, 85, 104, 224-238, 245-246, 272 ◇ (square of order 5) 16-17, 22-23, 143-145, 208-210, 219-223 ◇ (square of order 6) 4, 6, 9, 17, 88-91, 248-251 ◇ (square of order 8) 6, 18, 85, 158, 165, 239-246 ◊ history: (antiquity) v-vi, 8-10, 13n, 45, 84, 85, 95, 100, 104, 107, 111-112, 120, 121, 127, 142, 147, 161, 170, 177, 179, 185; *see also* Thābit ibn Qurra ◇ (Islamic times): (ninth century) v-vi, 9; *see also* Thābit ibn Qurra ○ (tenth century) v, 10, 13, 18-19, 21-23, 25, 60, 84, 95, 107, 111, 142, 147, 170; *see also*: Anṭākī, Būzjānī ○ (eleventh and twelfth centuries) 10, 14, 15, 25, 35, 46, 51, 71, 84, 88, 215; *see also*: *Harmonious arrangement, Brief treatise* ○ (thirteenth to eighteenth centuries) 11, 15-16; *see also* Shabrāmallisī, Kishnāwī ◇ (mediaeval and Renaissance Europe) vi, 11, 12, 15-18, 25, 45, 47, 91, 149-150, 161-163, 170-171.

See also: construction methods, literal squares, magic circles, magic crosses, magic cubes, magic rectangles, magic sum, magic triangles, natural square, order, squares with divided cells, squares with one empty cell, sum due, talismans.

N

Nārāyaṇa: 34, 271, 303.

natural square: 21, 22, 24-29, 35, 40-44, 47-49, 51-54, 56, 58, 88, 93, 94, 225.

al-Nawbakhtī: 9.

'neutral placings': 98, 100n, 102, 104, 105, 159, 163-164, 166, 179, 189, 192-193, 200, 202, 205, 251, 273.

Nicomachos: 9, 13, 304.

K. A. Nowotny: 303.

numerical system, alphabetical: 10-11, 215.

O

order: (def.) 1 ◇ (categories of) 4, 6 ◇ (divisibility) 4, 42-44, 107-108, 111, 122.

ordinary magic squares: *see* magic squares.

P

pandiagonal magic squares: *see* magic squares.

perfect number: 12, 207.

planets: vi, 11, 12 ('planetary squares'), 15, 19.

Pythagoreans: 10, 12.

Q

al-Qaswīnī: 8.

R

Rasā'il ikhwān al-ṣafā': see *Epistles of the Brothers of Purity*.

S

al-Shabrāmallisī: 16, 30, 35, 50n, 60, 70, 71, 73, 84, 95, 101n, 108n, 116, 117, 120, 145, 147n, 172, 253, 255-258, 264-268, 270, 271, 274-277.

'small' number, 'small' middle, 'small' extreme: (def.) 67n.

squares with divided cells: 253, 270-271.

squares with one empty cell: 253, 265-270.

M. Stifel: 15, 149-150, 154, 161-163, 166, 170-171, 173, 214, 304.

'substance', filling according to: 235-238.

sum due: 2, 96, 105, 107, 120, 163-165, 171-173, 179, 185 *seqq.*

symmetrical magic squares: *see* magic squares.

T

talismans: vi, 11, 151, 215, 262, 273.

P. Tannery: 8*n*, 37*n*, 303-305.

T̲h̲ābit ibn Qurra: v-vi, 8-10, 13, 13*n*, 84, 100, 101, 104, 105, 107, 111, 120-122, 127, 128, 131, 142, 177, 179, 181*n*, 182, 185, 186, 189, 200, 202, 206, 208, 277, 304.

V

N. Vinel: 9*n*, 305.

W

wafq al-aʿdād: vi, 12.

E. Wiedemann: 8*n*, 305.

Z

al-Zanjānī: 15, 148*n*, 159, 167, 172, 227, 229*n*, 234, 236-237.

al-Zarqālī: 30, 71.

Printed in the United States
By Bookmasters